CAMBRIDGE MONOGRAPHS
ON MATHEMATICAL PHYSICS

General Editors: W. H. McCrea, D. W. Sciama, J. C. Polkinghorne

THE WAVE EQUATION ON A
CURVED SPACE–TIME

THE WAVE EQUATION ON A CURVED SPACE–TIME

F. G. FRIEDLANDER

Department of Applied Mathematics and Theoretical Physics
University of Cambridge

CAMBRIDGE UNIVERSITY PRESS

CAMBRIDGE

LONDON · NEW YORK · MELBOURNE

CAMBRIDGE UNIVERSITY PRESS
Cambridge, New York, Melbourne, Madrid, Cape Town, Singapore,
São Paulo, Delhi, Dubai, Tokyo

Cambridge University Press
The Edinburgh Building, Cambridge CB2 8RU, UK

Published in the United States of America by Cambridge University Press, New York

www.cambridge.org
Information on this title: www.cambridge.org/9780521136365

First published 1975
This digitally printed version 2010

A catalogue record for this publication is available from the British Library

Library of Congress Catalogue Card Number: 74–14435

ISBN 978-0-521-20567-2 Hardback
ISBN 978-0-521-13636-5 Paperback

In memoriam Y.M.F.

Contents

Preface   *page* ix

1   Differential geometry   1
1.1   Differentiable manifolds   2
1.2   Geodesics   12
1.3   Exterior forms and integration on manifolds   20

2   Distribution theory   28
2.1   Distributions in $\mathbf{R}^n$ 29
2.2   Operations on distributions   34
2.3   Distributions with compact support   38
2.4   Tensor products of distributions   40
2.5   Convolution and regularization   43
2.6   Linear differential operators   49
2.7   Examples   53
2.8   Distributions on a manifold   59
2.9   A special type of distribution   65

**3   Characteristics and the propagation of
       discontinuities**   72
3.1   Lorentzian geometry   73
3.2   Characteristics and bicharacteristics   76
3.3   The initial value problem for characteristics and
      null fields   81
3.4   Caustics   87
3.5   The propagation of discontinuities   91
3.6   Progressing waves   101
3.7   Simple progressing wave solutions of the ordinary
      wave equation   110

4 Fundamental solutions *page* 116

4.1 The ordinary wave equation 117

4.2 The singular part of the fundamental solutions 129

4.3 The parametrix 138

4.4 Causal domains 146

4.5 The fundamental solutions 154

4.6 Trivial transformations 159

5 Representation theorems 166

5.1 Existence and uniqueness theorems 167

5.2 The reciprocity theorem 178

5.3 The Cauchy problem 181

5.4 Characteristic initial value problems 193

5.5 Tensor wave equations 203

5.6 The field of a time-like line source 212

5.7 Huygens' principle 221

6 Wave equations on n-dimensional space–times 228

6.1 The ordinary wave equation in n dimensions 229

6.2 Fundamental solutions 235

6.3 Existence, uniqueness and representation theorems 242

6.4 The method of descent 251

Appendix 262

Bibliography 273

Notation 277

Index 281

Preface

In this book, a wave equation is a linear hyperbolic equation on a Lorentzian manifold, that is to say on a given space–time. The classical theory of linear hyperbolic second-order equations is largely the creation of Hadamard and M. Riesz. This book is an attempt to present it in modern language, by using the theory of distributions, and a little differential geometry. There is virtually no physics in the book, but it is hoped that an account of a basic aspect of the mathematics of wave propagation will be of interest to physicists.

The classical theory is local, and constructive. These two features are linked, as the construction of fundamental solutions (Green's functions), on which it is based, requires certain geometrical elements that can only be defined locally. No attempt has therefore been made to discuss global questions.

The first three chapters are preparatory. Chapter 1 deals with differential geometry, and chapter 2 with the theory of distributions, as far as either of these is needed in the sequel. Neither chapter can, of course, replace a detailed exposition, for which one must go to the literature, but the inclusion of this material does make the book more self-contained. Chapter 3 is concerned with characteristics and bicharacteristics; for a wave equation, these are, respectively, the null hypersurfaces and the null geodesics of the underlying space–time. It also introduces, in a rather elementary way, the connection between characteristics and the singularities of solutions of the equation, a topic that has received much attention recently.

The main matter of the book is in the last three chapters. This is the general construction of fundamental solutions, and their application to the Cauchy problem. (There are, I am afraid, very few examples. The best chance of getting solutions in closed form is to utilize, when possible, groups of motions of the space–time. This is a subject that would require another, and quite different, book.) Because of its physical interest, the theory is set out (in chapters 4 and 5) for a four-dimensional space–time. This makes for some simplifications, although, as M. Riesz' method of analytic continuation is reserved for the last chapter, there are also one or two tedious technical arguments.

In chapter 4, I have in all essentials followed Hadamard and M. Riesz. There are a few departures; in particular, the emphasis is on the C^∞ case, where an integral equation method must be used, rather than on the analytic case. It would have been possible to include space–times of finite differentiability class. But it seems to me that these are best treated in terms of Sobolev spaces of functions and distributions, and once these are introduced, it seems altogether more logical to adopt the abstract approach. This is based on integral estimates and the powerful methods of functional analysis. To do justice to it would have made a book that is already too long, quite unwieldy.

There are two points that should perhaps be mentioned. By a method that is now very familiar (for example, in the theory of pseudo-differential operators), one can construct a C^∞ parametrix. The advantage of this is that one gets precise information on the behaviour of the tail term of the fundamental solution near the surface of the null cone. The second point is that I have tried to bring out the causality hypotheses that are implicit in the classical theory, by introducing a class of neighbourhoods that are here called causal domains.

The fundamental solutions derived in chapter 4 are used in chapter 5 to prove the basic local existence and uniqueness theorems. Here, I have used a local version of Leray's past-compact and future-compact sets. There then follow several detailed representation theorems, which may be of interest in applications, in spite of the somewhat heavy computations that are needed to derive them.

It must be pointed out that the theory is mainly developed for a scalar equation. The extension to vector and tensor equations then presents no difficulty, once transport bitensors have been introduced.

Although the corresponding theory for space–times of arbitrary dimension is of less physical interest, it is set out briefly in the last chapter. Apart from the clarification which this brings to the mathematics, it also makes Hadamard's method of descent available in the four-dimensional case.

I want to thank all those who have sat through graduate lectures, have read earlier versions of the book, and have generally helped with discussion and advice; particularly, I want to thank Dr D. W. Sciama, who invited me to contribute the book to this series, and encouraged me during its writing.

Cambridge 1974   F. G. FRIEDLANDER

1
Differential geometry

A space–time can be described naively as a four-dimensional space in which a metric is defined by means of a line element ds, whose square is a non-degenerate indefinite quadratic differential form,

$$ds^2 = g_{ij} dx^i dx^j \quad (g_{ij} = g_{ji}). \tag{1.0.1}$$

The summation convention is used here, and the g_{ij} are functions of $x = (x^1, x^2, x^3, x^4)$. The signature of the form is -2, so that

$$ds^2 = (l_{1j} dx^i)^2 - (l_{2j} dx^i)^2 - (l_{3j} dx^j)^2 - (l_{4j} dx^j)^2,$$

where the linear differential forms $l_{ij} dx^j$ are linearly independent. (Many authors take the signature of ds^2 to be 2.) All coordinate systems that are related to each other by sufficiently differentiable coordinate transformations are considered as equivalent frames of reference, and as ds^2 is invariant, the g_{ij} transform as the components of a symmetric covariant tensor.

Denote the determinant of the g_{ij} by g, and the inverse of the matrix (g_{ij}) by (g^{ij}). It is well known that the linear second-order differential operator

$$\Box u = |g|^{-\frac{1}{2}} \frac{\partial}{\partial x^i} \left(|g|^{\frac{1}{2}} g^{ij} \frac{\partial u}{\partial x^j} \right) \tag{1.0.2}$$

is invariant under coordinate transformations. For the metric of special relativity,

$$ds^2 = (dx^1)^2 - (dx^2)^2 - (dx^3)^2 - (dx^4)^2,$$

the equation $\Box u = 0$ is the ordinary wave equation,

$$\left(\left(\frac{\partial}{\partial x^1} \right)^2 - \left(\frac{\partial}{\partial x^2} \right)^2 - \left(\frac{\partial}{\partial x^3} \right)^2 - \left(\frac{\partial}{\partial x^4} \right)^2 \right) u = 0.$$

One can therefore regard the differential equation $\Box u = 0$ as a wave equation in the space–time with the metric (1.0.1). One can also consider the more general equation

$$Pu = \Box u + a^i \frac{\partial u}{\partial x^i} + bu = f, \tag{1.0.3}$$

[1]

where the a^i are the components of a vector field, b is a scalar field, and f (the 'source term') is a given function. It is this equation which will here be called a wave equation, or a scalar wave equation. There are, also, analogous vector and tensor wave equations. Such equations occur in general relativity, where they govern the propagation of test fields, small disturbances whose effect on the space–time background can be neglected.

Because of (1.0.2), the equation (1.0.3) is, in a fixed coordinate system, an equation of hyperbolic type. Hyperbolic equations occur in many contexts, and usually govern wave propagation phenomena. Clearly, it is natural to associate a space–time with such an equation. This has the formal advantage that it facilitates coordinate transformations. But, in fact, the connection is much deeper, because the characteristics of the equation (1.0.3) are the null hypersurfaces of the metric (1.0.1).

The description of space–time given above is inadequate, and it is better to define a space–time from the outset as a differentiable manifold with a Lorentzian metric. This chapter is an outline of the relevant concepts and results in differential geometry.

The first section outlines the basic definition and concepts, such as differentiable structure, tangent and cotangent spaces and bundles, tensor fields, and metrics. The second section deals in rather more detail with geodesics and the exponential map. It includes a proof of Whitehead's theorem on the existence of geodesically convex domains, and the derivation of the properties of the square of the geodesic distance between points in a geodesically convex domain. This material will be of particular importance later on. In the last section, exterior forms are introduced, and the integral of an exterior form of maximal degree is defined. For a manifold with a metric, a form which can serve as an invariant volume element is defined; this will be constantly used in the sequel. The section ends with a statement of the divergence theorem.

The reader's attention is also drawn to the appendix to this book, which summarizes some elementary topological definitions and results.

1.1 Differentiable manifolds

An *n-dimensional manifold* M is a topological space, every point of which has a neighbourhood that is homeomorphic to an open set in $\mathbf{R}^n$.

If Ω is such a neighbourhood, and $\pi\colon \Omega \to \mathbf{R}^n$ the homeomorphism in question, then π sends every point $p \in \Omega$ to a point

$$\pi p = x = (x^1, \ldots, x^n) \in \mathbf{R}^n;$$

the x^i are *local coordinates* of p, Ω is a *coordinate neighbourhood*, π is a *coordinate system*, and the pair (Ω, π) will be called a *chart*. (A chart at p will be any chart (Ω, π) such that $p \in \Omega$.) Suppose that (Ω, π) and $(\tilde{\Omega}, \tilde{\pi})$ are charts, and that $\Omega \cap \tilde{\Omega}$ is not empty. Then the respective local coordinates $x = \pi p$ and $\tilde{x} = \tilde{\pi}p$ of a point $p \in \Omega \cap \tilde{\Omega}$ are related by the map

$$\tilde{\pi} \circ \pi^{-1}\colon \pi(\Omega \cap \tilde{\Omega}) \to \tilde{\pi}(\Omega \cap \tilde{\Omega}), \tag{1.1.1}$$

which is called a *coordinate transformation*, and is a homeomorphism between open sets in $\mathbf{R}^n$.

A C^∞ *structure* on a manifold M is an indexed family of charts $\{\Omega_\nu, \pi_\nu\}$ that has the following properties:

(i) $\{\Omega_\nu\}$ is a covering of M, $\bigcup \Omega_\nu = M$.

(ii) The coordinate transformations $\pi_\nu \circ \pi_\mu^{-1}$ are infinitely differentiable.

(iii) If (Ω, π) is a chart such that $\pi_\nu \circ \pi^{-1}$ is infinitely differentiable for all ν, then (Ω, π) is a chart belonging to the structure.

A manifold with a C^∞ structure is called a C^∞ manifold. If one drops condition (iii), one obtains a covering of M by C^∞ charts, which can be extended to a C^∞ structure; this extension is unique. If (ii) is replaced by the requirement that the coordinate transformations are to be C^k, which means that they have continuous derivatives of all orders less than or equal to k, one has a C^k structure. A C^∞ structure is also a covering of M by C^k charts; hence it has a unique extension to a C^k structure. We shall usually work with C^∞ manifolds, but occasionally extend the C^∞ structure to a C^k structure with finite k.

If $f(p)$ is a function on M and (Ω, π) is a chart, then

$$f \circ \pi^{-1}(x) = f(\pi^{-1}x)$$

is a function on $\pi\Omega \subset \mathbf{R}^n$. (In this chapter, functions are assumed to be real-valued; in subsequent chapters, complex-valued functions will also occur.) If $f \circ \pi^{-1}$ is infinitely differentiable for all charts, then we say that $f \in C^\infty$ or that $f \in C^\infty(M)$; it is enough for this to hold in each one of a family of charts covering M. The class $C^k(M)$ is defined similarly. If $D \subset M$ is an open set, and f is a function defined on D, then $f \in C^\infty(D)$ (respectively, $f \in C^k(D)$) means that $f \circ \pi^{-1}\colon \pi(\Omega \cap D) \to \mathbf{R}$ is C^∞ (respectively, C^k) for all charts (Ω, π) such that Ω meets D.

The *support* of a function $f \colon M \to \mathbf{R}$ is the closure of the set $\{p; f(p) \neq 0\}$; it is denoted by supp f. If $D \subset M$ is an open set, then $C_0^\infty(D)$ is the class of all $f \in C^\infty(D)$ whose supports are compact subsets of D; $C_0^k(D)$ is defined in the same way.

An open covering $\{\Omega_\nu\}$ of M is called *locally finite* if every compact set meets only a finite number of the Ω_ν. A topological space is called *paracompact* if every open covering has a refinement that is locally finite; C^∞ manifolds will be assumed, throughout, to be paracompact. A *partition of unity* is an indexed family of functions $\phi_\nu \in C_0^\infty(M)$ whose supports are a locally finite covering of M, and which is such that

$$0 \leqslant \phi_\nu \leqslant 1, \quad \sum_\nu \phi_\nu = 1. \tag{1.1.2}$$

Note that, at each point of M, at least one of the ϕ_ν is non-zero, and all but a finite number of the ϕ_ν are zero. It can be shown that, given a covering $\{\Omega_j\}$ of M by open sets, there is a partition of unity $\{\phi_\nu\}$ subordinated to this covering. This means that, for each j, there is a ν such that supp $\phi_\nu \subset \Omega_j$. If the covering is locally finite, and the Ω_j are relatively compact, then one can choose a partition of unity $\{\phi_j\}$ such that supp $\phi_j \subset \Omega_j$ for each j. This will be the usual situation in the sequel, and the Ω_j will generally be coordinate neighbourhoods.

Let M' and M be C^∞ manifolds, of dimensions m and n respectively, and let g be a map $M' \to M$. Such a map is said to be C^∞ if the map $\pi \circ g \circ \pi'^{-1} \colon \pi'\Omega' \to \mathbf{R}^n$ is C^∞ for all pairs of charts (Ω', π') and (Ω, π), in M' and M respectively, such that $g\Omega' \subset \Omega$. Let p' be a point in M', $p = gp'$ its image under g in M, and (Ω', π'), (Ω, π) be charts at p' and p respectively, such that $\pi\Omega' \subset \Omega$. The rank of g at p' is then, by definition, the rank of the Jacobian matrix $D(\pi \circ g \circ \pi'^{-1})$ at $\pi'^{-1}p'$; it is evidently independent of the choice of the charts (Ω', π') and (Ω, π). The map g is called an *imbedding* if it is one-to-one (injective) and its rank is equal to m, the dimension of M', at all points of M'. If g is an imbedding and onto (surjective), then it is a *diffeomorphism*; one then has, necessarily, $m = n$. With the obvious modifications, one can also define maps, imbeddings, and diffeomorphisms, of class C^k.

An m-dimensional *sub-manifold* of M, where $m \leqslant n$, is a set $M' \subset M$ which can be covered by coordinate charts (Ω, π) such that

$$\Omega \cap M' = \{p; p \in \Omega, \quad x = \pi p, \quad x^{m+1} = \ldots = x^n = 0\}; \tag{1.1.3}$$

if $m = n$, M' is just an open set in M. One can consider M' as an m-dimensional manifold, with the differentiable structure inherited from that of M; the inclusion map $M' \to M$ is then obviously an imbedding. Conversely, the image of an imbedding is always a sub-manifold. The non-negative integer $n - m$ is the *codimension* of M'. If $m = 1$, M' will be called a *curve*; if $m = n - 1$, a *hypersurface*; and if $1 < m < n - 1$, it will be sometimes be called an *m-surface*. A sub-manifold M' of codimension k can be specified by giving k C^∞ functions $S_1(p), ..., S_k(p)$ and setting $M' = \{p; S_1(p) = 0, ..., S_k(p) = 0\}$, provided that the rank of the map $M \to \mathbf{R}^k$ which sends $p \in M$ to

$$(S_1(p), ..., S_k(p)) \in \mathbf{R}^k$$

is equal to k at all points of M'; for a hypersurface ($k = 1$) this means that the gradient of $S(p)$ must be non-zero at all points of $\{p; S(p) = 0\}$. (The gradient of a function is defined below.) Locally, any sub-manifold can be given in this form, by (1.1.3).

So far, we have only considered unbounded manifolds. A *manifold-with-boundary* is defined in the same way as an unbounded manifold, except that the sets $\pi\Omega$ are only required to be open subsets of the closed half-space $\overline{\mathbf{R}_+^n} = \{x; x \in \mathbf{R}^n, x^n \geqslant 0\}$. If $\pi_\nu(\Omega_\nu \cap \Omega_\mu)$ contains points in $\partial\overline{\mathbf{R}_+^n} = \{x; x \in \mathbf{R}^n, x^n = 0\}$, condition (ii) above must be understood to mean that the coordinate transformation $\pi_\nu \circ \pi_\mu^{-1}$ can be extended to a diffeomorphism between open sets in $\mathbf{R}^n$. It is obvious that, for any point $p \in M$, πp is either in $\mathbf{R}_+^n = \{x; x \in \mathbf{R}^n, x^n > 0\}$ for all charts at p, or in $\partial\overline{\mathbf{R}_+^n}$. In the former case, p is an interior point, in the latter, it is a boundary point. The set of boundary points is the *boundary* ∂M of M; it is an $(n - 1)$-dimensional C^∞ manifold, with the differentiable structure inherited from that of M.

An n-dimensional manifold M can always be imbedded in a Euclidean space $\mathbf{R}^N$, where N is sufficiently large; by Whitney's embedding theorem, one can take $N = 2n + 1$. Considering $\mathbf{R}^N$ as a vector space, one can see intuitively that the tangent vectors to curves that go through a fixed point $p \in M$ form an n-dimensional vector space, the tangent space at p. For an intrinsic definition, we consider parametrized curves through p. By this we mean a C^1 map $t \to f(t)$ of an interval $I_\delta = (-\delta, \delta) \in \mathbf{R}$ into M, such that $f(0) = p$. If (Ω, π) is a chart at p, and δ is sufficiently small, then $f(I_\delta) \in \Omega$, and $t \to \pi \circ f(t)$ is a curve in $\mathbf{R}^n$ that has a tangent vector

$$\xi = \frac{d}{dt}(\pi \circ f(t))|_{t=0} \qquad (1.1.4)$$

at the point $\pi \circ f(0) \in \mathbf{R}^n$. Let $(\tilde{\Omega}, \tilde{\pi})$ be another chart at p and suppose that $f(I_\delta) \in \tilde{\Omega}$; we obtain another Euclidean tangent vector $\tilde{\xi}$, to the curve $t \to \tilde{\pi} \circ f(t)$ at $\tilde{\pi} \circ f(0)$. If one reads ξ and $\tilde{\xi}$ as column vectors, and the Jacobian $D(\tilde{\pi} \circ \pi^{-1})$ of the coordinate transformation as an $n \times n$ matrix, then it follows from the chain rule for partial derivatives that ξ and $\tilde{\xi}$ are related by the law of contravariance,

$$\tilde{\xi} = D(\tilde{\pi} \circ \pi^{-1})|_{\pi p} \xi. \tag{1.1.5}$$

Guided by this relation, we now consider the set of triples (Ω, π, ξ), where (Ω, π) is a coordinate chart at p and ξ is a vector in $\mathbf{R}^n$. It is easily verified that the relation

$$(\Omega, \pi, \xi) \cong (\tilde{\Omega}, \tilde{\pi}, \tilde{\xi}) \quad \text{if} \quad \tilde{\xi} = D(\tilde{\pi} \circ \pi^{-1})|_{\pi p} \xi$$

is an equivalence relation; a *tangent vector* v at p is then, by definition, an equivalence class with respect to this relation. One usually writes $\xi = \pi_* v$ for the components of ξ. The tangent vectors at p obviously form an n-dimensional vector space over $\mathbf{R}$; addition, and multiplication by a number, can be defined, componentwise, in any coordinate system. This is the *tangent space* at p, and is denoted by TM_p. It follows from (1.1.4) that a parametrized C^1 curve that goes through p determines a tangent vector at p, which will sometimes be denoted by $f'(0)$ or by $(d/dt)f(0)$. Conversely, it is evident that any tangent vector at p can be considered to be the tangent vector to some parameterized curve through p.

The set of all tangent vectors to M can be made into a C^∞ manifold, which is called the *tangent bundle*, and denoted by TM. If $v \in TM$, then it is in one and only one TM_p; let Π denote the projection, which is the map $v \to p$. Let (Ω, π) be a coordinate chart, and suppose that $v \in \Pi^{-1}\Omega$. Assign to v the coordinates $\kappa v = (\pi \circ \Pi v, \pi_* v) = (x, \xi)$, say. One can easily show that there is a unique topology for TM such that, for all charts in the C^∞ structure of M, κ is a homeomorphism on an open set in $\mathbf{R}^n \times \mathbf{R}^n$. If $(\tilde{\Omega}, \tilde{\pi})$ is another chart, and $\Omega \cap \tilde{\Omega}$ is not empty, then it follows from (1.1.5) that the coordinate transformation in TM is

$$\tilde{x} = \tilde{\pi} \circ \pi^{-1} x, \quad \tilde{\xi} = D(\tilde{\pi} \circ \pi^{-1}) \xi \tag{1.1.6}$$

so that the tangent bundle has a C^∞ structure.

A *vector field* can now be defined as a *cross-section* of TM; this is a map $V: M \to TM$ such that $\pi \circ V$ is the identity. In local coordinates, this just means that $V \circ \pi^{-1} x = (x, \xi(x))$. In view of what has already been said about maps of one C^∞ manifold into another one, it is evident how C^k and C^∞ vector fields are defined. In local coordinates, the components $V^i(x)$ of a C^∞ vector field are, of course, C^∞ functions of x, and transform by contravariance.

The *cotangent space* T^*M_p at a point $p \in M$ is the dual of TM_p; it consists of all linear maps $w: TM_p \to \mathbf{R}$, made into an n-dimensional vector space over $\mathbf{R}$ in the natural way. Its elements are called *covectors*. The value of the covector w at $v \in TM_p$ will be denoted by $\langle w, v \rangle$, and called the *scalar product* of w and v. An important example of a covector is the *gradient* of a C^1 function at p. Let u be a C^1 function, defined on a neighbourhood of p, and let $v \in TM_p$. One can always find a parameterized curve $t \to f(t)$ such that $f(0) = p$, and $f'(0) = v$. The composite function $t \to u \circ f(t)$ is a C^1 function, on some interval $(-\delta, \delta) \in \mathbf{R}$, and

$$f'(0) = v \to du(p) = \frac{d}{dt}(u \circ f(t))|_{t=0}$$

is a linear form on TM_p, and so a covector. We denote this map by $\operatorname{grad} u(p)$ or $\nabla u(p)$, and write

$$du(p) = \langle \operatorname{grad} u(p), v \rangle \equiv \langle \nabla u(p), v \rangle. \tag{1.1.7}$$

Let (Ω, π) be a coordinate chart at p. The coordinate curves are the images, under π^{-1}, of the coordinate lines in $\mathbf{R}^n$ at πp,

$$x^i = (\pi p)^i + \delta^i_j t \quad (i,j = 1, ..., n),$$

where the δ^i_j are the Kronecker deltas. The tangent vectors $e_{(i)}$ to these curves are a basis of TM_p, associated in a natural way with (Ω, π); we have already used this basis, for if $\xi = \pi_* v$ then the ξ^i are just the components of v with respect to the basis $\{e_{(i)}\}_{1 \leqslant i \leqslant n}$. The dual basis of $\{e_{(i)}\}_{1 \leqslant i \leqslant n}$ consists of the covectors $e^{*(j)}$ such that $\langle e^{*(j)}, e_{(i)} \rangle = \delta^j_i$. So, if one writes

$$v = \xi^i e_{(i)}, \quad w = \omega_i e^{*(i)}, \tag{1.1.8}$$

using the summation convention, then one obtains the usual identity

$$\langle w, v \rangle = \omega_i \xi^i. \tag{1.1.9}$$

In particular, it follows from (1.1.7) and

$$du(p) = \frac{d}{dt}(u \circ \pi^{-1}) \circ (\pi \circ f)|_{t=0}$$

that $$du = \langle \operatorname{grad} u(p), v \rangle = \xi^i \frac{\partial}{\partial x^i} u \circ \pi^{-1}(x), \tag{1.1.10}$$

which is the classical form of the gradient.

The components ω_i of a covector transform by covariance, as can immediately be inferred from (1.1.9) and (1.1.5); if one reads $\{\omega_i\}_{1 \leqslant i \leqslant n}$ as a row matrix, this is

$$\tilde{\omega} = \omega D\pi \circ \tilde{\pi}^{-1}|_{\tilde{\pi}p}. \tag{1.1.11}$$

We shall frequently use classical notation, which, although ambiguous, is more convenient in computations. By an abuse of language, $u \circ \pi^{-1}(x), u \circ \tilde{\pi}^{-1}(\tilde{x}), \ldots$ are denoted by $u(x), u(\tilde{x}), \ldots$; a tangent vector is written as a differential $dx = (dx^1, \ldots, dx^n)$, so that (1.1.10) assumes the familiar form $du = (\partial u/\partial x^i)\, dx^i$. Of course, dx^i has two distinct meanings; it may be a component of a tangent vector, or a covector, which is the gradient of a coordinate function $(\pi p)^i = x^i$. But no confusion is likely to arise in practice if this is borne in mind.

The *cotangent bundle* is constructed like the tangent bundle; it is a C^∞ manifold T^*M of dimension $2n$. Its coordinate transformations are obtained by combining the coordinate transformations of M and the law of covariance (1.1.11). A covector field is then a cross-section of T^*M. (A covector is of course the same as a 'classical' covariant vector.) A gradient field is an example of a covector field.

Suppose that S is a sub-manifold of M, of dimension $m < n$. The tangent space to S at a point $p \in S$, TS_p, is obviously the subspace of TM_p that is spanned by the tangent vectors to parametrized curves through p that are in S. The annihilator of TS_p is the subspace of T^*M_p defined by

$$N_p = \{w;\ w \in T^*M_p, \langle w, v \rangle = 0 \quad \text{for all} \quad v \in TS_p\},$$

and is, clearly, the *normal* to S. If S is a hypersurface ($m = n-1$), then N_p is one-dimensional; if this hypersurface is given as $\{p;\ u(p) = 0\}$ then N_p is the one-dimensional subspace of T^*M_p which contains $\mathrm{grad}\, u(p)$. (Recall that, by definition, $\mathrm{grad}\, u(p) \neq 0$ at all points of a hypersurface $S = \{p;\ u(p) = 0\}$.)

Covectors have been introduced as linear forms on TM_p. By duality, vectors can also be considered as linear forms on T^*M_p. A *tensor of type* (r, s) is a multilinear form

$$S(w_{(1)}, \ldots, w_{(r)}, v_{(1)}, \ldots, v_{(s)}),$$

that is to say, a map

$$S: \underbrace{T^*M_p \times \ldots \times T^*M_p}_{r \text{ factors}} \times \underbrace{TM_p \times \ldots \times TM_p}_{s \text{ factors}} \to \mathbf{R}$$

that is linear when restricted to any one of the factors TM_p or T^*M_p. Tensors of fixed type are again a vector space over $\mathbf{R}$, addition and multiplication by a number being carried out with the multilinear forms. To define the *tensor product* of a tensor S of type (r, s) and a tensor T of type (k, l), one simply puts

$$S \otimes T(\omega_{(1)}, \ldots, \omega_{(r+k)}, v_{(1)}, \ldots, v_{(s+l)})$$
$$= S(\omega_{(1)}, \ldots, w_{(r)}, v, \ldots, v_{(s)})\, T(\omega_{(r+1)}, \ldots, w_{(r+k)}, v_{(s+1)}, \ldots, v_{(s+l)}).$$

As tensors of type $(1, 0)$ are vectors, and tensors of type $(0, 1)$ are covectors, tensors of all types can be built up by forming tensor products of vectors and covectors. So, if (Ω, π) is a chart at p, and $\{e_{(i)}\}_{1 \leqslant i \leqslant n}$, $\{e^{*(i)}\}_{1 \leqslant i \leqslant n}$ are the associated dual bases of TM_p and T^*M_p respectively, then a basis of the vector space of tensors of type (r, s) is provided by the tensors $e_{(i_1)} \otimes \ldots \otimes e_{(i_r)} \otimes e^{*(j_1)} \otimes \ldots \otimes e^{*(j_s)}$, and a tensor T of type (r, s) can be written as

$$T = T^{i_1 \ldots i_r}{}_{j_1 \ldots j_s} e_{(i_1)} \otimes \ldots \otimes e_{(i_r)} \otimes e^{*(j_1)} \otimes \ldots \otimes e^{*(j_s)},$$

with the usual transformation law for the tensor components,

$$\tilde{T}^{i_1 \ldots i_r}{}_{j_1 \ldots j_s} = \frac{\partial \tilde{x}^{i_1}}{\partial x^{k_1}} \ldots \frac{\partial \tilde{x}^{i_r}}{\partial x^{k_s}} T^{k_1 \ldots k_r}{}_{l_1 \ldots l_s} \frac{\partial x^{l_1}}{\partial \tilde{x}^{j_1}} \ldots \frac{\partial x^{l_s}}{\partial \tilde{x}^{j_s}}.$$

Finally, tensor fields are defined as cross-sections of the appropriate tensor bundle over M. The scalar product is a tensor of type $(1, 1)$, whose components, at all points, and in all coordinate systems, are the Kronecker deltas.

A tensor of type $(0, 2)$ at a point $p \in M$, $G(v, v')$ is called symmetric if $G(v, v') = G(v', v)$ for all tangent vectors v, v' at p; it is called non-degenerate if $G(v, v') = 0$ for all $v' \in TM_p$ implies that $v = 0$. In local coordinates, this means that the components are a symmetric non-singular matrix (g_{ij}). A *metric*, on a C^∞ manifold M, is a non-degenerate symmetric C^∞ tensor field $G(p; v, v')$ of type $(0, 2)$. We define the *signature* of G at a point p to be the sum of the signs of the eigenvalues of the matrix $(g_{ij}(x))$, where the x^i are the local coordinates of p. By the law of inertia for quadratic forms, the signature is independent of the choice of the local coordinate system, and it follows from continuity that it is the same at all points of M, if M is connected. If the signature is equal to n, the dimension of the manifold, so that $G(p; v, v')$ is positive definite, then the metric is *Riemannian*. Otherwise, it is called *pseudo-Riemannian* (unless it is equal to $-n$, when one can replace G by $-G$). If the modulus of the signature of the metric is $n - 2$, then the metric is called *hyperbolic* or *Lorentzian*. A space–time is a C^∞ manifold with a Lorentzian metric; we shall always take the signature to be $2 - n$.

Any C^∞ manifold admits a Riemannian metric, and any non-compact C^∞ manifold admits a Lorentzian metric.

At a point $p \in M$, the metric furnishes an *inner product* for TM_p, which is just $G(v, v')$. By equating this to the scalar product, one obtains an isomorphism of TM_p and T^*M_p which is the classical operation of raising and lowering tensor sub- and superscripts. To see this, note that, for fixed v, $v' \to G(v, v')$ is a linear form on TM_p, and so a covector w, such that

$$G(v, v') = \langle w, v' \rangle \quad \text{for all} \quad v' \in TM_p. \tag{1.1.12}$$

As G is non-degenerate, the linear map $TM_p \ni v \to w \in T^*M_p$ is surjective and hence, as both spaces have the same dimension, it is also bijective. Defining w' in the same way, one sees that $(w, w') \to G(v, v')$ is a tensor $G^*(w, w')$ of type $(2, 0)$, also symmetric and non-degenerate, and it is obvious that the inverse of the map defined by (1.1.12) is given by

$$G^*(w, w') = \langle w', v \rangle \quad \text{for all} \quad w' \in T^*M_p. \tag{1.1.13}$$

It is easy to see that, in local coordinates, the components of G^* are the matrix (g^{ij}) which is the inverse of (g_{ij}). Also, (1.1.12) and (1.1.13) are, in components,

$$\xi_i = g_{ij}\xi^j, \quad \xi^i = g^{ij}\xi_j \quad (i = 1, \dots, h), \tag{1.1.14}$$

where the ξ^i are the components of v and the ξ_i are the components of w. These operations extend to tensors of all types.

From now on, we shall generally only consider manifolds with metrics, and treat (1.1.12) and (1.1.13) as an identification of TM_p and T^*M_p, which extends to the corresponding bundles, and fields. One thus has the classical notation, in which the ξ^i are 'contravariant components' and the ξ_i are the 'covariant components' of the same vector (or covector) v. This is convenient in calculations. Note that, with $\pi_* v = \xi$ and $\pi_* w = \eta$,

$$\langle v, w \rangle = \xi_i \eta^i = \xi^i \eta_i = g_{ij}\xi^i\eta^j = g^{ij}\xi_i\eta_j.$$

Also, in classical notation, $\langle v, v \rangle$ becomes the line element,

$$ds^2 = \langle dx, dx \rangle = g_{ij}(x)\, dx^i dx^j. \tag{1.1.15}$$

At this point, the affine connection, covariant differentiation, curvature tensor and other basic concepts of differential geometry, can be introduced in a coordinate-free way. However, as they will only be needed marginally in the sequel, we merely give a brief summary in terms of local coordinates and components.

The components of the affine connection determined by the metric are

$$\Gamma^i_{jk} = \tfrac{1}{2}g^{il}\left(\frac{\partial g_{lj}}{\partial x^k} + \frac{\partial g_{kl}}{\partial x^j} - \frac{\partial g_{jk}}{\partial x^l}\right). \tag{1.1.16}$$

They are not the components of a tensor, but transform according to the law

$$\tilde{\Gamma}^i_{jk} = \Gamma^a_{bc}\frac{\partial \tilde{x}^i}{\partial x^a}\frac{\partial x^b}{\partial \tilde{x}^i}\frac{\partial x^c}{\partial \tilde{x}^k} + \frac{\partial^2 x^a}{\partial \tilde{x}^j \partial \tilde{x}^k}\frac{\partial \tilde{x}^i}{\partial x^a}. \qquad (1.1.17)$$

The covariant derivative of a covector field w is a tensor field of type $(0, 2)$, denoted by ∇w, with the components

$$(\nabla w)_{ij} = \nabla_i \omega_j = \frac{\partial \omega_j}{\partial x^i} - \Gamma^k_{ij}w_k. \qquad (1.1.18)$$

The covariant derivative of a vector field v is a tensor field ∇v of type $(1, 1)$, with the components

$$(\nabla v)^j_i = \nabla_i v^j = \frac{\partial v^i}{\partial x^j} + \Gamma^j_{ik}v^k. \qquad (1.1.19)$$

The covariant derivative of a scalar field is its gradient. For a tensor field T of type $(r, s) \neq (0, 0)$, the covariant derivative ∇T is a tensor field of type $(t, s + 1)$, whose components can be written down by combining (1.1.18) and (1.1.19).

We shall also write

$$\nabla^i = g^{ij}\nabla_j \quad (i = 1, ..., n). \qquad (1.1.20)$$

Covariant differentiation is not commutative, except in the case of scalar fields. For a covector field w, one has

$$(\nabla_j \nabla_k - \nabla_k \nabla_j)w_i = R^l_{ijk}\omega_l, \quad R^l_{ijk} = g^{la}R_{aijk}, \qquad (1.1.21)$$

where the R_{aijk} are the components of the curvature tensor. The Ricci tensor is obtained by contraction,

$$R_{ij} = R^l_{ilj}, \qquad (1.1.22)$$

and the curvature scalar is

$$R = g^{il}g^{jk}R_{ijkl}. \qquad (1.1.23)$$

We conclude the section by introducing the two differential operators that will be particularly important later on. The *divergence* of a vector field v is a scalar field (a function on M), which is, in local coordinates,

$$\text{div } v = \nabla_i v^i = |g|^{-\frac{1}{2}}\frac{\partial}{\partial x^i}(|g|^{\frac{1}{2}}v^i), \qquad (1.1.24)$$

where $g = \det g_{ij}$. The divergence of the gradient of a scalar field is an invariant second order differential operator,

$$\nabla_i \nabla^i u = |g|^{-\frac{1}{2}}\frac{\partial}{\partial x^i}\left(|g|^{\frac{1}{2}}g^{ij}\frac{\partial u}{\partial x^j}\right). \qquad (1.1.25)$$

For a Riemannian metric, this is the *Laplacian* Δ; for a Lorentzian metric – the case with which we shall be concerned – it is the *d'Alembertian*, and will be denoted by $\square$. An equation of the form

$$Pu = \square u + \langle a, \nabla u \rangle + bu = f, \qquad (1.1.26)$$

on a space–time (a Lorentzian C^∞ manifold) will be called a scalar wave equation; here a is a C^∞ vector field and b and f are C^∞ scalar fields, and all of these may be complex valued.

1.2 Geodesics

Let M be a C^∞ manifold with a metric. A parametrized C^2 curve $\mathbf{R} \ni r \to f(r) \in M$ is a *geodesic* if, in any local coordinate system, the coordinates $x^i(r)$ of f are solutions of the n ordinary differential equations

$$\frac{d^2 x^i}{dr^2} + \Gamma^i_{jk}(x)\frac{dx^i}{dr}\frac{dx^k}{dr} = 0 \quad (i = 1, \ldots, n). \qquad (1.2.1)$$

These equations, which express the geometric property that the tangent vector to a geodesic remains parallel to itself along the curve, are invariant under coordinate transformations; this follows from the transformation laws for the components of the affine connection, (1.1.17). They are also invariant under affine parameter transformations $r \to kr + k'$, where k and k' are constants. The parameter r is therefore called an *affine parameter*. Note that (1.2.1) implies

$$\frac{d}{dr}\left(g_{ij}(x)\frac{dx^i}{dr}\frac{dx^j}{dr} \right) = 0. \qquad (1.2.2)$$

One can write the equations (1.2.1) as a first order system for $2n$ dependent variables,

$$\frac{dx^i}{dr} = \xi^i, \quad \frac{d\xi^i}{dr} = -\Gamma^i_{jk}(x)\,\xi^j\xi^k \quad (i = 1, \ldots, n). \qquad (1.2.3)$$

The ξ^i are the components, in local coordinates, of a vector tangent to an integral curve of (1.2.1). They are invariant under the coordinate transformations of the tangent bundle, (1.1.6), and so a solution of (1.2.3) is a parametrized curve $r \to (x(r), \xi(r)) \in TM$ whose projection on M is a geodesic.

A solution of (1.2.3) is determined by its initial values, say

$$x^i(0) = y^i, \quad \xi^i(0) = \eta^i \quad (i = 1, \ldots, n). \qquad (1.2.4)$$

Working, for the time being, in a fixed coordinate chart (Ω, π), we set

$$B_\alpha = \{x; |x| < \alpha\}, \quad \Sigma_\beta = \{\xi; |\xi| < \beta\}, \tag{1.2.5}$$

where $|x|$, $|\xi|$ are Euclidean (or equivalent) norms in $\mathbf{R}^n$. The following proposition is an immediate consequence of the basic existence theorem for ordinary differential equations.

Lemma 1.2.1. *Let α and β be positive numbers, and suppose that $B_{2\alpha} \subset \Omega$. Then there is a $\rho = \rho(\alpha, \beta)$ such that, if $(y, \eta) \in B_\alpha \times \Sigma_\beta$ and $|r| < \rho(\alpha, \beta)$, the equations (1.2.3) have a unique solution*

$$x = F(r, y, \eta), \xi = G(r, y, \eta),$$

where $F(0, y, \eta) = y$, $G(0, y, \eta) = \eta$; moreover, $(F, G) \in B_{2\alpha} \times \Sigma_{2\beta}$, and both F and G are C^∞ functions of r, y and η. $\square$

Now the equations (1.2.3) are unchanged if r is replaced by kr and ξ by ξ/k, where k is a constant. Hence

$$x^*(r) = F(kr, y, \eta/k), \quad \xi^* = kG(kr, y, \eta/k)$$

is also a solution of (1.2.3), and clearly

$$x^*(0) = F(0, y, \eta/k) = y, \quad \xi^*(0) = kG(0, y, \eta/k) = \eta.$$

By uniqueness, it follows that

$$x^*(r) = F(kr, y, \eta/k) = F(r, y, \eta) = x(r).$$

One can now put $k = 1/r$, and deduce that F is a function of y and of ηr only; as $G = (\partial/\partial r) F$, the solution of the initial value problem (1.2.3), (1.2.4) is therefore of the form

$$x = h(y, \eta r), \quad \xi = D_2 h(y, \eta r) \eta, \tag{1.2.6}$$

where $D_2 h(y, \eta)$ denotes the Jacobian matrix of $h(y, \eta)$ with respect to the second argument η, and η is read as a column vector. As (1.2.6) satisfies (1.2.4), one also has

$$h(y, 0) = y, \quad D_2 h(y, 0) = I, \tag{1.2.7}$$

where I is the identity matrix. By Lemma 1.2.1, the map $(y, \eta) \to h(y, \eta)$ is defined for $|y| < \alpha$, $|\eta| < \beta\rho(\alpha, \beta)$. (A detailed examination of the proof of the existence theorem shows that $\beta\rho(\alpha, \beta)$ can be replaced by a function of α alone.) Writing β for $\beta\rho$, we have the following lemma.

Lemma 1.2.2. *Suppose that* $B_{2\alpha} \subset \Omega$. *Then there is a* β *such that, for* $(y, \eta) \in B_\alpha \times \Sigma_\beta$, *there is a unique geodesic* $r \to x(r)$, *defined on* $|r| \leq 1$, *with* $x(0) = y$ *and* $x'(0) = \eta$. *Also*, $x(r) = h(y, \eta r)$ *where*

$$h(y, \eta) \colon B_\alpha \times \Sigma_\beta \to B_{2\alpha}$$

is a C^∞ *function of both arguments. Finally,* $\xi(r) = x'(r) \neq 0$ *for* $(r) \leq 1$. $\square$

The last assertion follows from the fact that, for $\eta = 0$, the initial value problem (1.2.3), (1.2.4) has the unique solution $x = y = \text{const.}$, $\xi = 0$.

Now let q be any point of M, and let (Ω, π) be a coordinate chart at q, such that $\pi q = 0$. Let $v \in TM_q$ and set $\pi_\alpha v = \eta$. From Lemma 1.2.2, and the invariance of the differential equations (1.2.3) under tangent bundle coordinate transformations, it follows that there is a neighbourhood $\Sigma_q' = \pi^{-1}\{\eta; |\eta| < \beta\}$ on which one can define a function $f(v) = \pi^{-1} \circ h(0, \eta)$ such that $p = f(vr)$, $(r) \leq 1$, is the unique geodesic γ_1 through q with $p'(0) = v$. As the argument can be repeated at any point of γ_1, it is clear that there is a unique maximal geodesic γ through q that contains γ_1. Thus $f(v)$ can be extended to a map $\Sigma_q \to M$, where $\Sigma_q \supset \Sigma_q'$, and the maximal geodesics through q are given by $p = f(vr)$, $vr \in \Sigma_q$. The map $f \colon \Sigma_q \to M$ is called the *exponential map* at q, and written as $\exp_q v$. To state this result, we introduce the projection $\Pi \colon TM \to M$ which assigns the point $q = \Pi v$ to a vector $v \in TM$ that is in the tangent space TM_q.

Theorem 1.2.1. *There exists a* C^∞ *map* $\exp_q v \colon TM \to M$, *where* $q = \Pi v$, *defined on a neighbourhood* Σ *of the zero section of* TM, *such that the geodesics through a point* $q \in M$ *are given by* $p = \exp_q rv$, *where* $v \in \Pi^{-1} q$, $r \in \mathbf{R}$, *and* $rv \in \Sigma \cap \Pi^{-1} q$. $\square$

The zero section of TM maps $p \in M$ on the zero vector at p, and can be identified with M. If $\Sigma = TM$, then the geodesics $p = \exp_q rv$ are defined for all $r \in \mathbf{R}$, and M is called geodesically complete.

In local coordinates, we write, again, $h(y, \eta)$ for $\exp_q v$. By (1.2.7), h has the important property that $D_2 h(0, \eta) = I$, whence

$$\det D_2 h(0, \eta) = 1.$$

By the inverse function theorem, $\eta \to x = h(y, \eta)$ therefore has, for fixed y, a unique inverse when $|\eta|$ is sufficiently small, and as $h \in C^\infty$, this is a diffeomorphism. Let us take $y = 0$ and set $h(0, y) = h_0(\eta)$. Then, for sufficiently small δ, $\eta \to x = h_0(\eta)$ is a diffeomorphism

$$\Sigma_\delta \to h_0(\Sigma_\delta) = N_\delta,$$

say. Also, if $\eta \in \Sigma_\delta$ then $r\eta \in \Sigma_\delta$ for $0 \leqslant r \leqslant 1$, and so if $h_0(\eta) = z$, then the geodesic $x = h_0(r\eta)$, $0 \leqslant r \leqslant 1$, that joins y to z, is in N_δ, and is unique. A neighbourhood N of a point $q \in M$ such that, for each $p \in N$, there is just one geodesic joining q to p in N, is called a *normal neighbourhood* of q. So we can conclude that every point of M has a normal neighbourhood. We shall now prove a stronger form of this result.

A connected open set $D \subset M$ is called a *geodesically convex domain* if any two points q and p in D are joined by a unique geodesic in D; this will be denoted by $\widehat{qp}$. There may of course be other geodesics that join q and p and contain points of M that are not in D.

Theorem 1.2.2. (Whitehead, 1932). *Every point of M has a neighbourhood that is a geodesically convex domain.*

PROOF. Consider, in the notation of Lemma 1.2.2, the map

$$\left.\begin{array}{c} B_\alpha \times \Sigma_\beta \to B_\alpha \times B_{2\alpha}, \\ (y, \eta) \to (y, h(y, \eta)). \end{array}\right\} \qquad (1.2.8)$$

This maps $(0, 0)$ to $(0, 0)$, and its Jacobian matrix is

$$\left\| \begin{array}{cc} I & 0 \\ D_1 h(y, \eta) & D_2 h(y, \eta) \end{array} \right\|.$$

By (1.2.7), the determinant of this matrix at $y = 0$, $\eta = 0$ is unity. Hence, by the inverse function theorem, there is a connected neighbourhood N of $(0, 0) \in B_\alpha \times \Sigma_\beta$ on which (1.2.8) is a diffeomorphism, with a unique inverse $(y, \theta(y, x)) \in B_\alpha \times B_{2\alpha}$, where $\theta(y, x) \in C^\infty$ and $|\theta| < \beta$. For sufficiently small δ, $B_\delta \times B_\delta \subset B_\alpha \times B_{2\alpha}$ will be contained in the image of N under (1.2.8). As $|\theta| < \beta$ gives $|\theta r| < \beta$ for $|r| \leqslant 1$, we then have the following: if y and x are in B_δ, then there is a geodesic

$$[0, 1] \ni r \to z(r) = h(y, \theta(y, x) r) \qquad (1.2.9)$$

joining y to x, and z is a continuous function (in fact, C^∞) of y, x and r. We now prove that, provided δ is sufficiently small, this geodesic is in B_δ.

Let
$$\rho = \tfrac{1}{2}|z(r)|^2,$$

then
$$\frac{d^2\rho}{dr^2} = \left|\frac{dz}{dr}\right|^2 + \sum_{i=1}^{n} z^i \frac{d^2 z^i}{dr^2}.$$

As $r \to z(r)$ satisfies the differential equations of the geodesics (1.2.1), this is

$$\frac{d^2\rho}{dr^2} = \left|\frac{dz}{dr}\right|^2 - \sum_{i=1}^{n} \Gamma^i_{jk}(z)\, z^i \frac{dz^j}{dr} \frac{dz^k}{dr}.$$

We can obviously choose δ so small that

$$\left| \sum_{i=1}^{n} \Gamma^i_{jk}(z)\, z^i \xi^j \xi^k \right| \leqslant \tfrac{1}{2}|\xi|^2$$

for $|z| < \delta$. Suppose that this is so; then it follows that, if the geodesic $r \to z(r)$ is in B_δ, for $0 \leqslant r \leqslant 1$, then

$$\frac{d^2\rho}{dr^2} \geqslant \frac{1}{2}\left|\frac{dz}{dr}\right|^2 > 0. \qquad (1.2.10)$$

(Here we are using the last part of Lemma 1.2.2.) Hence $2\rho = |z(r)|^2$ cannot have a maximum at an interior point, that is to say for $0 < r < 1$. Let $W \subset B_\delta \times B_\delta$ denote the set of all $(y, x) \in B_\delta \times B_\delta$ for which the geodesic (1.2.9) is in B_δ. We must then have

$$(y, x) \in W \Rightarrow |z(r)|^2 \leqslant \max\left(|y|^2, |z|^2\right) \quad \text{for} \quad 0 \leqslant r \leqslant 1,$$

and this implies that W is open in $B_\delta \times B_\delta$. On the other hand, (1.2.9) and the continuity of the functions h and θ imply that if $\{y_\nu\}$ and $\{z_\nu\}$ are sequences of points in B_δ that converge to $y \in B_\delta$ and $z \in B_\delta$ respectively, and $(y_\nu, z_\nu) \in W$, then $(y, z) \in W$; hence W is also closed in $B_\delta \times B_\delta$. Now $B_\delta \times B_\delta$ is connected, and so W is either empty, or equal to $B_\delta \times B_\delta$. It is obviously not empty, as the geodesic from $y = 0$ to z is clearly in B_δ if $|z|$ is small enough. So $W = B_\delta \times B_\delta$, that is to say the geodesic (1.2.9) joining two points of B_δ is in B_δ. But $B_\delta \subset N$, and so this geodesic is unique. Hence B_δ is a geodesically convex neighbourhood of the point $q = \pi^{-1}(0)$, and the theorem is proved. □

Geodesically convex domains will play an important part in the construction of fundamental solutions of wave equations, and we therefore examine them in more detail. We note first that a geodesically convex domain Ω is a normal neighbourhood of each of its points. Hence the exponential map $TM_q \ni v \to \exp_q v$ is, for every $q \in \Omega$, a diffeomorphism between an open set $\Sigma \subset TM_q$ and Ω. Let π be a coordinate system that is admissible in a neighbourhood of q, and put $\pi_* v = \eta$. Then $\pi_* \Sigma$ is a star-shaped neighbourhood of $\eta = 0$ in $\mathbf{R}^n$, and the η^i are local coordinates valid throughout Ω, with the property that the geodesics through q are the straight lines $r \to \eta r$ (where r is such that $r\eta \in \pi_* \Sigma$). Such a coordinate system is called *normal at q*. The argument also shows that a geodesically convex domain can be made into a coordinate neighbourhood, by taking a coordinate system that is normal at a point of Ω.

Let Ω be a geodesically convex domain. Let (Ω, π) be a coordinate chart, and denote the local coordinates of two points p and q of Ω by x and y respectively. If $\gamma: \tau \to z(\tau)$, $0 \leqslant \tau \leqslant 1$, is a parametrized C^2 curve joining q and p in Ω, we define the arc length of γ to be

$$s = \int_0^1 (g_{jk}(z(\tau))\,\dot{z}^j(\tau)\,\dot{z}^k(\tau))^{\frac{1}{2}}\,d\tau, \qquad (1.2.11)$$

where $\dot{z}^j(\tau) = dz^j/d\tau$. Note that s may be complex; it will be seen presently that this is immaterial. When γ is the unique geodesic $\widehat{qp} \subset \Omega$, then s is, by definition, the *geodesic distance* of q and p. We shall denote the square of the geodesic distance by $\Gamma(p,q)$, and write it as $\Gamma(x,y)$ in local coordinates. This function on $\Omega \times \Omega$ will be particularly important later on, and we therefore summarize those of its properties that are independent of the signature of the metric, in the following theorem.

Theorem 1.2.3. *Let Ω be a geodesically convex domain, and let $\Gamma(p,q)$ denote the square of the geodesic distance of two points q and p of Ω, measured along the geodesic from q to p in Ω. Then*

(i) *Γ is a real-valued C^∞ function on $\Omega \times \Omega$;*

(ii) $$\Gamma(p,q) = \Gamma(q,p); \qquad (1.2.12)$$

(iii) *as a function of either argument, Γ satisfies the equation*

$$\langle \nabla\Gamma, \nabla\Gamma \rangle = 4\Gamma; \qquad (1.2.13)$$

(iv) *again, let q be fixed, and let $r \to p(r)$ be a geodesic such that $p(0) = q$, r being an affine parameter, then*

$$\nabla\Gamma = 2r\frac{dp}{dr}, \qquad (1.2.14)$$

(dp/dr denotes the velocity vector of $r \to p(r)$ at $p(r)$);

(v) *let $x = h(y, \eta)$ denote the exponential map in local coordinates, and put*

$$\xi = D_2 h(y, \eta)\,\eta = \left.\frac{\partial h(y, \eta r)}{\partial r}\right|_{r=1},$$

Then

$$g^{ij}(x)\frac{\partial\Gamma(x,y)}{\partial x^j} = 2\xi^i, \quad g^{ij}(y)\frac{\partial\Gamma(x,y)}{\partial y^j} = -2\eta^i \quad (i = 1,...,n); \ (1.2.15)$$

(vi) *if the local coordinates in (v) are normal at q, so that $y = 0$, then*

$$g_{ij}(x)\,x^j = g_{ij}(0)\,x^j \quad (i = 1,...,n), \qquad (1.2.16)$$

and $$\Gamma(p,q) = g_{ij}(0)\,x^i x^j = g_{ij}(x)\,x^i x^j. \qquad (1.2.17)$$

PROOF. As the geodesic distance is the integral (1.2.11), where γ is the geodesic $\widehat{qp}$, one can compute it by taking γ to be given by (1.2.9). Then $z(r)$ satisfies the basic differential equations (1.2.1), and it follows from (1.2.2) that, on γ,

$$g_{jk}(z(r)\,\dot{z}^j(r)\,\dot{z}^k(r) = g_{ik}(y)\,\eta^j\eta^k = A, \qquad (1.2.18)$$

say. Hence $s = A^{\frac{1}{2}}$, whence $\Gamma = A$, which is independent of r, and as also $\eta = \theta(y,x)$, the unique inverse of $\eta \to x = h(y,\eta)$ in Ω, it follows that
$$\Gamma = g_{jk}(y)\,\theta^j(y,x)\,\theta^k(y,x).$$
This proves (i).

The equations (1.2.1) are invariant under the parameter transformation $r \to 1-r$, and so, by the uniqueness of the geodesic join of p and q, $r \to z(1-r)$ is the equation of the geodesic $\widehat{pq}$, which is just $\widehat{qp}$ with reversed orientation. Hence (1.2.12) holds.

We next prove (iv), which will imply all the other assertions. If $\Gamma(p,q) \neq 0$, then the geodesic $\widehat{qp}$ is an extremal of the variational problem $\delta s = 0$: this is, of course, the classical definition of the geodesics. By a well-known argument in the calculus of variations, one can relate the gradient of Γ at $x = \pi p$ to the velocity vector of the geodesic at x. Consider geodesics from $y = \pi q$, and, with a change of notation, write such geodesics as

$$t \to x = x(t) = h(y, t\eta).$$

By (1.2.11) and (1.2.18) one then has, at the point $x(r)$,

$$\Gamma(y,x(r)) = (g_{ij}(y)\,\eta^i\eta^j)\,r^2 = Ar^2. \qquad (1.2.19)$$

Now let $t \to X(t,\lambda)$ be a smooth one-parameter family of parametrized curves such that
$$X(0,\lambda) = y, \quad X(t,0) = x(t).$$

Put
$$s(\lambda) = \int_0^r \left(g_{jk}(X)\frac{\partial X^j}{\partial t}\frac{\partial X^k}{\partial t} \right)^{\frac{1}{2}} dt,$$

r being kept fixed for the time being, and define the variations in the usual way by

$$\delta s = \frac{ds}{d\lambda}\bigg|_{\lambda=0}, \quad \delta x = \frac{\partial X}{\partial \lambda}\bigg|_{\lambda=0}.$$

Then
$$\delta s = \int_0^r \left((g_{ij}(x)\,\delta\dot{x}^i\dot{x}^j + \frac{1}{2}\frac{\partial g_{jk}(x)}{\partial x^i}\,\delta x^i\dot{x}^j\dot{x}^k \right) (g_{jk}(x)\,\dot{x}^j\dot{x}^k)^{-\frac{1}{2}} dt.$$

Now by (1.2.18), $g_{jk}(x)\,\dot{x}^j\dot{x}^k = A$ is independent of t; the first variation

δs is, however, only defined if $A \neq 0$. Assuming this to be the case, integration by parts gives

$$\delta s = \frac{1}{A} g_{ij}(x) \, \delta x^i \delta x^j |_{t=0} + \frac{1}{A} \int_0^r \left(\frac{1}{2} \frac{\partial g_{jk}}{\partial x^i} \dot{x}^j \dot{x}^k - \frac{\partial}{\partial t} (g_{ij} \dot{x}^j) \right) \delta x^i \, dt,$$

as $\delta x = 0$ when $t = 0$. The integrand vanishes, for all δx, because $t \to x(t)$ is a geodesic. One is therefore left with

$$A \delta s = \langle \dot{x}(r), \, \delta x \rangle. \qquad (1.2.20)$$

Now one can suppose that the curves $t \to X(t, \lambda)$ are also geodesics; then $s(\lambda) = \Gamma(y, X(t, \lambda))$, and so

$$\delta s = \frac{ds}{d\lambda}\bigg|_{\lambda=0} = \tfrac{1}{2} \Gamma^{-\frac{1}{2}} \langle \nabla \Gamma, \delta x \rangle.$$

But, by (1.2.19), $\Gamma^{\frac{1}{2}} = A^{\frac{1}{2}} r$, and so comparison with (1.2.20) shows that

$$\langle \nabla \Gamma, \delta x \rangle = 2r \langle \dot{x}(r), \delta x \rangle.$$

As δx is an arbitrary smooth vector field on the geodesic $r \to x(r)$, this implies (1.2.14), provided that $\Gamma(y, x(r)) \neq 0$. But as $\Gamma \in C^\infty(\Omega \times \Omega)$, (1.2.14) also holds on null geodesics, by continuity.

In local coordinates, (1.2.14) is

$$g^{ij}(x) \frac{\partial r(x, y)}{\partial x^j} = 2r \frac{dx^i}{dr}. \qquad (1.2.21)$$

Hence $\qquad \langle \nabla \Gamma, \nabla \Gamma \rangle = 4r^2 \langle \dot{x}(r), \dot{x}(r) \rangle = 4Ar^2 = 4\Gamma,$

by (1.2.19), and the constancy of $\langle \dot{x}, \dot{x} \rangle$ along a geodesic, and so (1.2.13) is proved; it also holds when p and q are interchanged, by (1.2.12).

Next, we prove (v). If one puts $r = 1$ in (1.2.21), one at once obtains the first equation (1.2.15). The second equation then follows from the symmetry of Γ, (1.2.12), and the fact that the geodesic $\widehat{xy}$ is just $\widehat{yx}$, with the orientation reversed.

It remains to discuss (vi). It has already been noted that, in coordinates that are normal at q, one has $\pi q = y = 0$, and $x = r\eta$ for the geodesics through q. But then (1.2.19) becomes

$$\Gamma = g_{ij}(0) \, x^i x^j. \qquad (1.2.22)$$

Now (1.2.21) must hold with this substituted for Γ, and $x(r) = r\eta$. So one obtains (1.2.16). Finally, (1.2.16) and (1.2.22) imply that (1.2.17) holds in normal coordinates. This completes the proof of the theorem. $\square$

Remark. It can be shown that normal coordinates are also characterized by the identities (1.2.16).

It has been assumed throughout that the manifold M has a C^∞ structure. If M has a $C^{(k+2)}$, structure, where $k \geqslant 1$, and the metric tensor is of class $C^{(k+1)}$, then the affine connection is C^k. It then follows from the standard existence theorem for ordinary differential equations that Lemma 1.2.1 still holds, but the functions $F(r,y,\eta)$ and $G(r,y,\eta)$ defined in the lemma are of class C^k. An examination of the proof of Theorem 1.2.3 shows that one also has $\Gamma(p,q) \in C^k(\Omega \times \Omega)$.

1.3 Exterior forms and integration on manifolds

Let M be a C^∞ manifold, and let p be a point of M. A tensor of type $(0, k)$ at p, that is to say, a multilinear form $\alpha(v_{(1)}, ..., v_{(k)})$ on TM_p, is called alternating if it changes sign when two of its arguments $v_{(i)}, v_{(j)}$ $(i \neq j)$, are interchanged. It follows that $\alpha = 0$ when the arguments are linearly dependent, and so the degree k of an alternating form cannot exceed n, the dimension of M. Forms of degree zero are scalars, and forms of degree one are covectors. The forms of degree k are a subspace of the vector space of tensors of type $(0, k)$ at p. There is a corresponding sub-bundle of the tensors of this type over M, and a cross-section of this sub-bundle is an *exterior form* of degree k or k-form, for short, on M.

The components $\alpha_{i_1 ... i_k}(x)$ of a k-form in any local coordinate system are antisymmetric,

$$\alpha_{\sigma(i_1), ..., \sigma(i_k)} = \epsilon(\sigma)\,\alpha_{i_1 ... i_k},$$

where $i \to \sigma(i)$ is a permutation of the i_j and $\epsilon(\sigma)$ is the parity of σ. (Components for which a pair of subscripts are equal vanish identically.) The support of a k-form is the closure of the set of points of M where it is different from zero.

If T is any tensor field of type $(0, k)$, then its antisymmetric, or alternating, part

$$\text{Alt}\, T = \frac{1}{k!}\Sigma\epsilon(\sigma)\,T(v_{\sigma(1)}, ..., v_{\sigma(k)}),$$

where σ runs over all permutations of $1, 2, ..., k$, is a k-form. The *exterior product* (wedge product) of a k-form α and an l-form ρ is defined by

$$\alpha \wedge \beta = \frac{(k+l)!}{k!\,l!}\,\text{Alt}\,\alpha \otimes \beta.$$

It is easily proved that exterior multiplication is distributive, associative, and anticommutative. So

$$\begin{aligned} \alpha \wedge (\beta + \gamma) &= \alpha \wedge \beta + \alpha \wedge \gamma, \\ \alpha \wedge (\beta \wedge \gamma) &= (\alpha \wedge \beta) \wedge \gamma = \alpha \wedge \beta \wedge \gamma. \end{aligned} \right\} \tag{1.3.1}$$

If f is a scalar and α is a k-form, then one defines

$$f \wedge \alpha = \alpha \wedge f = f\alpha. \tag{1.3.2}$$

If α is a k-form and β is an l-form, then

$$\alpha \wedge \beta = (-1)^{kl} \beta \wedge \alpha. \tag{1.3.3}$$

Exterior multiplication can also be defined axiomatically by (1.3.1) and (1.3.2).

In local coordinates, k-forms will be written in classical notation. The dual basis of T^*M_p consists of the differentials of the coordinate functions, $dx^1, \ldots, dx^n$. It follows from (1.3.1) that

$$\begin{aligned} dx^i \wedge dx^j &= -dx^j \wedge dx^i \quad (i \neq j); \quad dx^i \wedge dx^i = 0 \quad \text{(no summation),} \\ dx^i \wedge (dx^j \wedge dx^k) &= (dx^i \wedge dx^j) \wedge dx^k = dx^i \wedge dx^j \wedge dx^k. \end{aligned} \right\} \tag{1.3.4}$$

A basis of the vector space of k-forms at a point $p \in M$ is therefore given by the $dx^{i_1} \wedge \ldots \wedge dx^{i_k}$ with $i_1 < \ldots < i_k$, and any k-form α has a unique ('reduced') expansion

$$\alpha = k! \sum_{i_1 < \ldots < i_k} \alpha_{i_1 \ldots i_k} dx^{i_1} \wedge \ldots \wedge dx^{i_k}. \tag{1.3.5}$$

The dimension of the vector space of k-forms is thus the binomial coefficient $n! / k! (n-k)!$; in particular, the dimension of the space of forms of maximal degree n is unity. (There is only one n-form, up to multiplication by a scalar field.) The law of covariance implies that, in order to compute the components of a form α in a coordinate system $\tilde{x}$ in terms of those in another coordinate system x, one simply replaces the dx^i by their differentials,

$$\begin{aligned} \alpha &= \tilde{\alpha}_{i_1 \ldots i_k} d\tilde{x}^{i_1} \wedge \ldots \wedge d\tilde{x}^{i_k} \\ &= \alpha_{j_1 \ldots j_k} \left(\frac{\partial x^{j_1}}{\partial \tilde{x}^{i_1}} d\tilde{x}^{i_1} \right) \wedge \ldots \wedge \left(\frac{\partial x^{j_k}}{\partial \tilde{x}^{i_k}} d\tilde{x}^{i_k} \right). \end{aligned} \tag{1.3.6}$$

The *exterior differential* of a k-form α is a $(k+1)-$form $d\alpha$, defined as follows. If $\alpha = f$, a scalar field, then $d\alpha = df$, the gradient of f. In general, one has, in local coordinates,

$$d\alpha = d\alpha_{i_1 \dots i_k} dx^{i_1} \wedge \dots \wedge dx^{i_k}$$

$$= \frac{\partial \alpha_{i_1 \dots i_k}}{\partial x^i} dx^i \wedge dx^{i_1} \wedge \dots \wedge dx^{i_k}. \tag{1.3.7}$$

It follows easily that exterior differentiation has the properties

$$\left. \begin{aligned} d(\alpha + \beta) &= d\alpha + d\beta, \\ d(\alpha \wedge \beta) &= d\alpha \wedge \beta + (-1)^k \alpha \wedge d\beta \quad (k = \text{degree of } \alpha,) \\ d^2\alpha &= 0. \end{aligned} \right\} \tag{1.3.8}$$

The last of these is a consequence of the antisymmetry of the $\alpha_{i_1 \dots i_k}$ and the symmetry of the second derivatives. (It is assumed that $\alpha \in C^2(M)$.) One can show that these three rules determine the operator d uniquely (in a coordinate-free manner), and that $d\alpha$ is then a $(k+1)$-form.

Let S and M be two manifolds, and let $\sigma : S \to M$ be a map of class $C^j, j \geqslant 1$. (We shall usually take $j = \infty$.) If $f(p)$ is a function on M, then the pull-back $\sigma^* f$ is the function on S defined by $f \circ \sigma : S \to \mathbf{R}$. One can operate in the same way with k-forms. Let, in particular, Σ and Ω be coordinate neighbourhoods in S and M respectively, such that $\Sigma \subset \Omega$, and let λ and x be the respective local coordinates. Then

$$\sigma^*\alpha = \alpha_{i_1 \dots i_k}(x(\lambda)) \left(\frac{\partial x^{i_1}}{\partial \lambda^{j_1}} d\lambda^{j_1} \right) \wedge \dots \wedge \left(\frac{\partial x^{i_k}}{\partial \lambda^{j_k}} d\lambda^{j_k} \right). \tag{1.3.9}$$

It follows easily that σ^* preserves the algebraic structure and the exterior differential

$$\sigma^*(\alpha + \beta) = \sigma^*\alpha + \sigma^*\beta, \quad \sigma^*(\alpha \wedge \beta) = \sigma^*\alpha \wedge \sigma^*\beta, \quad \sigma^*(d\alpha) = d(\sigma^*\alpha). \tag{1.3.10}$$

Let p be a point of M, and let μ be an n-form that does not vanish at p. The bases of TM_p, that is to say, the sets $\{e_{(j)}\}_{1 \leqslant j \leqslant n}$ of linearly independent tangent vectors at p, can be divided into two classes, those for which $\mu(e) > 0$, and those for which $\mu(e) < 0$. It has already been pointed out that the forms of maximal degree are a one-dimensional vector space; there is essentially only one n-form at p, up to a scalar factor. Hence any other non-vanishing n-form at p determines the same pair of classes of bases of TM_p. Each of these classes is an *orientation* of TM_p. If $\{e_{(j)}\}_{1 \leqslant j \leqslant n}$ and $\{e'_{(j)}\}_{1 \leqslant j \leqslant n}$ are two bases of TM_p, then $e'_j = c^i_j e_j$, $i = 1, ..., n$, where the transition matrix (c^j_i) is non-singular. Hence

$$\mu(e') = \mu(e) \det c^j_i.$$

Therefore bases with the same orientation are related by transition matrices with positive determinants. One can thus also determine an orientation by giving one of the bases; the corresponding class of bases consists of all those that have a transition matrix with positive determinant relative to the given one.

Euclidean space $\mathbf{R}^n$ will always be given its 'usual' or canonical orientation, which is defined by $dx = dx^1 \wedge ... \wedge dx^n > 0$, or by the ordered n-tuple of unit vectors $E_1 = (1, 0, ..., 0), ..., E_n = (0, ..., 0, 1)$. Let M be a manifold, and let (Ω, π) be a coordinate chart in M. The orientation of (Ω, π), or of the coordinate system π, will be, by definition, the orientation of TM_p $(p \in \Omega)$ that is determined by the vectors $\pi_* E_1, ..., \pi_* E_n$ at p. Two charts (Ω, π) and $(\tilde{\Omega}, \tilde{\pi})$ at p then have the same orientation if the Jacobian determinant of the co-ordinate transformation $\det D(\tilde{\pi} \circ \pi^{-1})$ is positive, and opposite orientations if this determinant is negative.

A manifold M is called *orientable* if there is a covering by coordinate charts that all have the same orientation; such a covering is called an oriented covering of M. There are obviously just two classes of oriented coverings of M, and each one of these defines an orientation of M. It can be shown that there exist forms of maximal degree that do not vanish anywhere, on an orientable manifold. If μ is such a form, then one can orient M by the rule that $\mu > 0$. Similarly, an orientable sub-manifold S of M, of dimension k, can be oriented by means of a k-form that does not vanish on S.

Let M be an orientable manifold, and suppose that it has been given an orientation. A coordinate chart (Ω, π) is then called *orientation preserving* if it has the given orientation. We shall now define the integral of a form α of maximal degree n, with compact support. If (Ω, π) is a coordinate chart, then

$$\pi^* \alpha = n!(\pi^* \alpha)_{1...n} dx^1 \wedge ... \wedge dx^n.$$

The form is said to be *locally integrable* if the function $x \to (\pi^* \alpha)_{1...n}$ is integrable on any compact subset of Ω, for all charts. (Integrals will always be understood to be Lebesgue integrals.) Suppose that (Ω, π) is an orientation-preserving chart, and that $\text{supp}\,\alpha \subset \Omega$. Then the *integral of α* is defined as

$$\int \alpha = n! \int (\pi^* \alpha)_{1\,...\,n}\, dx, \qquad (1.3.11)$$

where dx denotes the Lebesgue measure (the 'volume element') in $\mathbf{R}^n$. Let $(\tilde{\Omega}, \tilde{\pi})$ be another orientation-preserving chart, such that $\operatorname{supp}\alpha \subset \Omega$. Then $D\tilde{x}/Dx = \det D(\tilde{\pi} \circ \pi^{-1})$ is positive, and so, by the rule for changing variables in a multiple integral,

$$\int (\tilde{\pi}^*\alpha)_{1\ldots n}\,d\tilde{x} = \int (\tilde{\pi}^*\alpha)_{1\ldots n}\frac{D\tilde{x}}{Dx}\,dx.$$

But
$$(\tilde{\pi}^*\alpha)_{1\ldots n}\frac{D\tilde{x}}{Dx} = (\pi^*\alpha)_{1\ldots n},$$

by the law of covariance. Hence

$$n!\int (\tilde{\pi}^*\alpha)_{1\ldots n}\,d\tilde{x} = n!\int (\pi^*\alpha)_{1\ldots n}\,dx = \int\alpha. \qquad (1.3.12)$$

So the definition of the integral of α, (1.3.12), is independent of the choice of the chart (Ω, π).

In general, the integral of a locally integrable n-form α with compact support can be defined by means of a partition of unity. Let $\{\Omega_\nu, \pi_\nu\}$ be a covering of M by orientation-preserving charts, and let $\{\phi_\nu\}$ be a partition of unity subordinated to this covering. Then we set

$$\int\alpha = \sum_\nu \int \alpha\phi_\nu, \qquad (1.3.13)$$

where each term on the right-hand side is evaluated by means of (1.3.11), in the appropriate coordinate chart of the covering. (Note that the sum is finite, as the supports of the ϕ_ν are a locally finite covering of M.) The integral is independent of the choice of the covering and of the partition of unity. For, let $\{\tilde{\Omega}_\mu, \tilde{\pi}_\mu\}$ be another covering of M, and $\{\tilde{\phi}_\mu\}$ a partition of unity subordinated to it. Then it follows from (1.3.12) that

$$n!\int \tilde{\pi}_\mu^*(\alpha\phi_\nu\tilde{\phi}_\mu)_{1\ldots n}\,d\tilde{x}_\mu = n!\int \pi_\nu^*(\alpha\phi_\nu\tilde{\phi}_\mu)_{1\ldots n}\,dx_\nu.$$

Summing over μ and ν, and taking into account that

$$\sum_\nu \phi_\nu = 1, \quad \sum_\mu \tilde{\phi}_\mu = 1,$$

one obtains the identity

$$\sum_\mu \int \alpha\tilde{\phi}_\mu = \sum_\nu \int \alpha\phi_\nu = \int\alpha.$$

The integral of an n-form α whose support is not compact can also be defined by (1.3.13), provided that the series, which is then infinite, converges absolutely.

One can also use (1.3.13) to define the integral of α over an open set $D \subset M$, which is supposed to have the orientation induced by that of M. One merely has to replace α by $\chi_D \alpha$, where $\chi_D(p)$ is the characteristic function of D, $\chi_D(p) = 1$ if $p \in D$ and $\chi_D(p) = 0$ if $p \notin D$, and set

$$\int_D \alpha = \int \chi_D \alpha. \tag{1.3.14}$$

The integral of an n-form has been defined with respect to an orientation of M. If M is given the opposite orientation, then the sign of the integral of α is reversed. This remark also applies to the integral of α over an open set $D \subset M$. However, if M is an orientable manifold with a metric, then an *invariant volume element* μ can be defined, as follows.

A pseudo-scalar is a quantity that transforms like a scalar, except that it changes sign under an orientation-reversing coordinate transformation. So, if (Ω, π) and $(\tilde{\Omega}, \tilde{\pi})$ are overlapping coordinate charts, and $x \to \tilde{x}$ is the coordinate transformation $\tilde{\pi} \circ \pi^{-1}$, one has, on $\Omega \cap \tilde{\Omega}$,

$$\epsilon \circ \tilde{\pi}^{-1}(\tilde{x}) = \frac{J}{|J|} \epsilon \circ \pi^{-1}(x), \quad J = \frac{Dx}{D\tilde{x}}. \tag{1.3.15}$$

In particular, there are just two pseudo-scalars such that $\epsilon^2 = 1$; if ϵ denotes one of them, then the other one is $-\epsilon$. One can now define an n-form μ by setting, in each coordinate chart (Ω, π),

$$\pi^* \mu = \epsilon \circ \pi^{-1}(x) \, |g(x)|^{\frac{1}{2}} \, dx^1 \wedge \ldots \wedge dx^n, \tag{1.3.16}$$

where $g(x)$ denotes the determinant of the components of the metric tensor in (Ω, π). For it is well known that

$$|\tilde{g}(\tilde{x})|^{\frac{1}{2}} = |g(x)|^{\frac{1}{2}} |J|, \tag{1.3.17}$$

and so the transformation law (1.3.15) ensures that (1.3.15) defines an n-form. If one defines the pseudo-scalar ϵ so that $\epsilon = 1$ when M has its canonical orientation, then this orientation also corresponds to $\mu > 0$. It is clear that $-\mu$ is another invariant volume element, associated with the other orientation of M. The essential point is that two such forms exist, that one can select one of them, and then use it to orient M. This has the effect of making μ into an invariant measure on M. We shall usually suppose that M is oriented in this way, and suppress the factor ϵ.

The integral of a locally integrable function f on M, with compact support, can now also be defined, as $\int f \mu$; it one multiplies f by the characteristic function of an open set $D \subset M$, one obtains the integral of f over D. If D is relatively compact, then the restriction that f is to have compact support can of course be dropped.

We must finally consider the divergence theorem. Let v be a C^1 vector field, and put, in a coordinate chart (Ω, π),

$$\pi^*(*v) = \epsilon \circ \pi^{-1}(x) \sum_{i=1}^{n} (-1)^{i+1} |g|^{\frac{1}{2}} (\pi_* v)^i \, dx^1 \wedge \ldots \wedge \widehat{dx^i} \wedge \ldots \wedge dx^n,$$

$$(1.3.18)$$

where $\widehat{dx^i}$ means that this differential is to be omitted. It follows from (1.3.15) and (1.3.17) that this defines an $(n-1)$-form $*v$ on M, which may be called the flux of v. The exterior differential of $*v$ is

$$d(*v) = \epsilon \circ \pi^{-1}(x) \sum_{i=1}^{n} (-1)^{i+1} d(|g|^{\frac{1}{2}} (\pi_* v)^i) \, dx' \wedge \ldots \wedge \widehat{dx^i} \wedge \ldots \wedge dx^n$$

$$= \epsilon \circ \pi^{-1}(x) \frac{\partial}{\partial x^i} (|g|^{\frac{1}{2}} (\pi_* v)^i) \, dx' \wedge \ldots \wedge dx^n,$$

or, by (1.3.16) and (1.1.24),

$$d(*v) = (\operatorname{div} v) \mu. \qquad (1.3.19)$$

The divergence theorem is obtained by applying Stokes' theorem to this identity. Let D be a relatively compact open set in M (so that $\bar{D}$ is compact), and suppose that its boundary ∂D is smooth. It is actually enough to suppose that ∂D is C^1, but we shall take it to be C^∞. Then ∂D is an $(n-1)$-dimensional sub-manifold of M. As M is orientable, D is orientable, and so ∂D is also orientable. Let D have the canonical orientation, determined by $\mu > 0$. Then ∂D can be given an orientation, called the *induced orientation of the boundary*, as follows. At every point of ∂D there is a chart (Ω, π) such that

$$\pi(\Omega \cap D) \subset \mathbf{R}^n_+ = \{x; x^1 > 0\}$$

and $$\pi(\Omega \cap \partial D) \subset \partial \mathbf{R}^n_+ = \{x; x^1 = 0\}.$$

(If ∂D is C^j, with $j < \infty$, one must first extend the differentiable structure of M to a C^j structure.) The induced orientation of ∂D is then the opposite of the orientation of $\partial \mathbf{R}^n_+$ that is determined by the coordinate vectors $e_2 = (0, 1, 0, \ldots, 0), \ldots, e_n = (0, \ldots, 0, 1)$. One can also say that the 'positive normal' $-e_1$ points away from D.

The restriction of $*v$ to ∂D is a form of maximal degree, and ∂D is compact. Hence the integral of $*v$ over ∂D is well defined. By Stokes' theorem,

$$\int_{\partial D} *v = \int_D d(*v)$$

provided that ∂D is given the induced orientation. Substituting for $d(*v)$ from (1.3.19), we obtain the *divergence theorem*,

$$\int_{\partial D} *v = \int_D (\operatorname{div} v)\,\mu. \qquad (1.3.20)$$

Finally, Stokes' theorem, and hence also the divergence theorem, can be extended to compact manifolds whose boundaries have certain singularities. In particular, ∂D may be allowed to be piece-wise smooth, which means that it consists of a finite number of smooth hypersurfaces that intersect transversally. Another admissible type of singularity, of special importance in the context of this book, is a point of ∂D that has a neighbourhood in which ∂D is diffeomorphic to a cone in $\mathbf{R}^n$.

NOTES

Most of the material in this chapter will be found in any book on differential geometry or differential manifolds, for example Sternberg (1964) or Lang (1972). A summary, covering much of the same ground, is given in Hawking and Ellis (1973).

Section 1.2. For the basic theorems on differential equations, and the inverse function theorem, see Dieudonné (1960) pp. 279–300 and 265–267. In Theorem 1.2.3 the properties of the square of the geodesic distance are classical, and will be found in Hadamard (1923, 1932).

Section 1.3. For exterior forms and integration, and the operator $*$, see also de Rham (1960). A very comprehensive form of Stokes' theorem, with singularities allowed on the boundary, is given by Whitney (1957) p. 100, Theorem 14A; see also Chapter IX, Section 3 of Lang (1972).

2

Distribution Theory

In the classical theory, a function that is a solution of a partial, differential equation must possess derivatives of all the orders that appear in the equation. This requirement, although patently logical, is frequently ignored in mathematical physics. It has also proved too tight for the development of the mathematical theory. It is now usual to extend differential operators on functions to operators on the members of a suitable function space. There are various ways of doing this, each suited to a particular class of problems. In this book, wave equations on space–times will be considered in the context of distribution theory. One immediate, and important, advantage of this approach is that, if a sequence of solutions of a wave equation converges in the distribution topology, then its limit is again a solution of the equation.

Generally, it may be said that distribution theory enlarges the scope of the theory of partial differential equations in a useful way. For instance, it makes it possible to treat systematically such objects as discontinuous solutions, point sources (monopoles) and layers of monopoles and multipoles, which otherwise have to be introduced by means of ad hoc definitions, or treated heuristically. Furthermore, distribution theory often clarifies and simplifies the classical theory; the simple way in which a fundamental solution is defined is an example of this. Finally, it may be noted that the employment of distributions say sometimes also have some technical advantages.

This chapter contains a short account of the elements of the theory of distributions, as far as they will be needed. The first eight sections deal with distributions in $\mathbf{R}^n$. The exposition is a little more detailed than that in chapter 1, as a number of proofs, most of them quite simple ones, have been supplied. Distributions are defined in section 2.1. In section 2.2 the fundamental operations of differentiation, multiplication by a function and translation are introduced. Distributions with compact support are discussed in section 2.3. The next two sections deal with tensor products, convolution, and regularization, and include the proof of the important fact that distributions are 'weak' limits of sequences of functions. Some examples are treated in section 2.7, both to illustrate the general theory, and because they will be useful later on. An extension of distribution theory to pseudo-Riemannian manifolds is set out in section 2.8. The last section deals with a special type of distribution, analogous to a composite function $f(S(p))$, where $f(t)$ is a function on $\mathbf{R}$ and $S(p)$ is a function defined on an open subset of a manifold. Such distributions are of particular importance in the theory of wave equations on a space–time.

2.1 Distributions in $\mathbf{R}^n$

Except in the last two sections of this chapter, $x, y, \ldots$ will denote points in $\mathbf{R}^n$, with the coordinates $(x^1, \ldots, x^n), (y^1, \ldots, y^n), \ldots$; the Euclidean norm of x will be denoted by $|x|$. Functions on $\mathbf{R}^n$ will generally be complex valued. The derivatives $\partial f/\partial x^i$ of a function f will be denoted by $\partial_i f, i = 1, \ldots, n$. For derivatives of higher order, the multi-index notation will be used. A *multi-index* α is an n-tuplet of non-negative integers, $\alpha = (\alpha_1, \ldots, \alpha_n)$ whose sum $\alpha_1 + \ldots + \alpha_n$ is denoted by $|\alpha|$. If α and β are multi-indices, then $\alpha + \beta = (\alpha_1 + \beta_1, \ldots, \alpha_n + \beta_n)$, and one writes $\alpha \leqslant \beta$ to indicate that $\alpha_i \leqslant \beta_i$ for $i = 1, \ldots, n$. For the derivatives of f one then writes

$$\partial^\alpha f = \partial_1^{\alpha_1} \ldots \partial_n^{\alpha_n} f \equiv (\partial/\partial x^1)^{\alpha_1} \ldots (\partial/\partial x^n)^{\alpha_n} f.$$

Note that, formally, $\partial^\alpha \partial^\beta f = \partial^{\alpha+\beta} f$. Conventionally, the derivative $\partial^0 f$ is f itself. It is also convenient to put

$$x^\alpha = (x^1)^{\alpha_1} \ldots (x^n)^{\alpha_n}, \quad \alpha! = \alpha_1! \ldots \alpha_n!.$$

The formal Taylor series of $f(x)$ then becomes

$$f(x+h) = \sum_{\alpha \geqslant 0} \frac{h^\alpha}{\alpha!} \partial^\alpha f(x). \tag{2.1.1}$$

Furthermore, it follows from Leibniz' theorem, by induction, that

$$\partial^\alpha (fg) = \sum_{\beta+\gamma=\alpha} \frac{\alpha!}{\beta!\gamma!} \partial^\beta f \partial^\gamma g. \tag{2.1.2}$$

The class C^k consists of all complex valued functions $f(x)$ defined on $\mathbf{R}^n$ for which the $\partial^\alpha f$ are continuous if $|\alpha| \leqslant k$; C^∞ consists of all infinitely differentiable functions. The *support* of a function $f(x)$, denoted by $\mathrm{supp} f$, is again the closure of $\{x; f(x) \neq 0\}$. The class C_0^k consists of all functions in C^k with compact support; here k may be infinite.

The class L_1 consists of all integrable functions. (Integrals are Lebesgue integrals; the Lebesgue measure is denoted by dx.) When no domain of integration is specified, an integral is taken over $\mathbf{R}^n$. A function is called locally integrable if it is integrable on any compact set; the class of locally integrable functions is denoted by L_1^{loc}.

It is obvious that C_0^∞ is a vector space over the field of complex numbers $\mathbf{C}$. Distributions, which are sometimes also called generalized functions, will be defined as continuous linear forms on C_0^∞; the members of C_0^∞ are called *test functions*. An example of a test function is

$$\psi(x) = \exp\left(1/(|x|^2 - 1)\right) \text{ if } |x| < 1, \quad \psi(x) = 0 \text{ if } |x| \geq 1. \quad (2.1.3)$$

The following lemma shows that any function in C^k can be approximated by test functions.

Lemma 2.1.1. *Let* $\rho(x) \in C_0^\infty$ *be such that*

$$\rho(x) \geq 0, \quad \operatorname{supp}\rho \subset \{x; |x| \leq 1\}, \quad \int \rho\, dx = 1. \quad (2.1.4)$$

Suppose that $f(x) \in C^k$, *and put*

$$f_\epsilon(x) = \epsilon^{-n} \int f(y)\rho\left(\frac{x-y}{\epsilon}\right) dy, \quad (2.1.5)$$

where ϵ *is a positive number. Then* $f_\epsilon \in C^\infty$, *and its support is contained in an* ϵ-*neighbourhood of* $\operatorname{supp} f$. *Moreover, if* $\epsilon \downarrow 0$, *then* $\partial^\alpha f_\epsilon \to \partial^\alpha f$ *for all* α *with* $|\alpha| \leq k$, *uniformly in any compact set.*

PROOF. Functions that satisfy (2.1.4) exist, for instance the function defined by (2.1.3), multiplied by a suitable constant. The integral is over the ball $\{y; |x-y| < \epsilon\}$, and as $\rho \in C_0^\infty$, it follows from a standard theorem that it can be differentiated repeatedly under the integral sign with respect to x; hence $f_\epsilon \in C^\infty$. Also, $f_\epsilon = 0$ if the distance of x from the support of f exceeds ϵ.

By a simple change of the variable of integration, (2.1.5) can be replaced by

$$f_\epsilon(x) = \int f(x - \epsilon y)\rho(y)\, dy. \quad (2.1.6)$$

It follows from this and (2.1.4) that

$$|f_\epsilon(x) - f(x)| = \left|\int (f(x-\epsilon y) - f(x))\rho(y)\, dy\right|$$

$$\leq \int |f(x-\epsilon y) - f(x)|\,\rho(y)\, dy,$$

whence $|f_\epsilon(x)-f(x)| \leqslant \sup_{|z|\leqslant\epsilon} |f(x+z)-f(x)|,$

by the third condition (2.1.4). The uniform continuity of f now implies that $f_\epsilon \to f$ as $\epsilon \downarrow 0$, uniformly on any compact set. Finally, if $k > 0$ and $|\alpha| \leqslant k$, then (2.1.6) gives

$$\partial^\alpha f_\epsilon(x) = \int \partial^\alpha f(x-\epsilon y)\,\rho(y)\,dy,$$

and a repetition of the argument shows that $\partial^\alpha f_\epsilon \to \partial^\alpha f$ as $\epsilon \downarrow 0$, uniformly on any compact set. $\square$

Remark. The function f_ϵ also exists if $f \in L_1^{\mathrm{loc}}$, and is then again in C^∞. One can prove that $f_\epsilon \to f$ in L_1 norm in any finite ball,

$$\lim_{\epsilon\downarrow 0} \int_{|x|<R} |f_\epsilon(x)-f(x)|\,dx = 0,$$

where $R < \infty$.

Suppose that $f(x)$ is continuous, and that

$$\int f(x)\,\phi(x)\,dx = 0, \tag{2.1.7}$$

for all $\phi \in C_0^\infty$. As $y \to \phi((x-y)/\epsilon)$ is in C_0^∞, it follows that $f_\epsilon(x) = 0$ for all $\epsilon > 0$, and so, by Lemma 2.1.1, that $f(x) = 0$. By a more elaborate argument one can establish the following generalization of this result, which is stated without proof:

Lemma 2.1.2. *Suppose that $f \in L_1^{\mathrm{loc}}$, and that (2.1.7) holds. Then $f = 0$, almost everywhere.* $\square$

Now $(f,\phi) = \int f(x)\,\phi(x)\,dx \tag{2.1.8}$

is a linear form $C_0^\infty \ni \phi \to (f,\phi) \in \mathbf{C}$. So Lemma 2.1.2 asserts that a locally integrable function f is determined, up to a null function, by the linear form (f,ϕ). The essential step in distribution theory is to replace a function f, which is by definition a map $f\colon \mathbf{R}^n \to \mathbf{C}$, by the linear form (2.1.8). Other 'generalized functions' are then introduced as linear forms on C_0^∞, with an appropriate continuity property.

It may be remarked that (f,ϕ) is a weighted average of f. If, for example, $f(x)$ is a component of a gravitational field, and $\phi(x)$ is the density distribution of a test particle, then (f,ϕ) is the corresponding component of the force exerted by the field on the particle. This is exactly how the field is measured, and so the step of replacing a function f by the linear form (f, ϕ) also makes sense in a physical context. If f is continuous, then its value at x can be recovered from (f, ϕ) by means of Lemma 2.1.1. However, for many purposes, and particularly in the theory of partial differential equations, it is sufficient, and easier, to work with the linear form (f, ϕ) rather than with the function f.

We now need a precise statement of the continuity requirement.

Definition 2.1.1. *A distribution is a linear form (u, ϕ) on C_0^∞ such that, for every compact set K, there are constants $C = C(K)$ and $N = N(K)$ such that*

$$|(u, \phi)| \leqslant C \sum_{|\alpha| \leqslant N} \sup |\partial^\alpha \phi| \qquad (2.1.9)$$

for all $\phi \in C_0^\infty$ with $\operatorname{supp} \phi \subset K$. The set of all distributions is denoted by $\mathscr{D}'$. A locally integrable function $f(x)$ is identified with the distribution (2.1.8).

It is obvious that (2.1.8) defines a distribution, as (2.1.9) holds with $N = 0$ and

$$C = \int_K |f| \, dx.$$

Note that functions which differ only by a null function are identified with the same distribution.

An inequality such as (2.1.9) is called a *semi-norm estimate*, because the quantities $\sup |\partial^\alpha \phi|$ are semi-norms for the vector space

$$\{\phi; \; \phi \in C_0^\infty, \operatorname{supp} \phi \subset K\}.$$

The relation of the definition to the general concept of continuous linear maps on topological vector spaces is indicated in the appendix.

A simple example of a distribution that is not of the form (2.1.8) is the *Dirac delta function* (Dirac measure). Let y be a fixed point; then $C_0^\infty \ni \phi(x) \to \phi(y) \in \mathbf{C}$ is a distribution which we shall denote by

$$(\delta(x - y), \phi(x)) = \phi(y). \qquad (2.1.10)$$

Other examples will be given in section 2.7.

There is another way of defining distributions.

Theorem 2.1.1. *A linear form $\phi \to (u, \phi)$ is a distribution if and only if $(u, \phi_\nu) \to 0$ for every sequence of test functions $\{\phi_\nu\}$ that has the following properties.*

(i) *The supports of all the ϕ_ν are in a fixed compact set.*

(ii) *$\partial^\alpha \phi_\nu \to 0$, uniformly in x as $\nu \to \infty$, for every α. (Such a sequence is said to tend to zero in C_0^∞.)*

PROOF. The necessity of the condition follows from the semi-norm estimates (2.1.9). To prove sufficiency, suppose that (u, ϕ) satisfies the hypotheses of the theorem, but that (2.1.9) is false for some compact set K. Then for every N, $N = 1, 2, \ldots$, there exists a $\phi_N \in C_0^\infty$ with supp $\phi_N \subset K$ such that

$$|(u, \phi_N)| \geqslant N \sum_{|\alpha| \leqslant N} \sup |\partial^\alpha \phi_N|.$$

Set $\phi_N/(u, \phi_N) = \psi_N$; then supp $\psi_N \subset K$, and

$$\sum_{|\alpha| \leqslant N} \sup |\partial^\alpha \psi_N| \leqslant \frac{1}{N}.$$

For fixed α, this implies that $\partial^\alpha \psi_N \to 0$, uniformly as $N \to \infty$. Hence $\psi_N \to 0$ in C_0^∞. But $|(u, \psi_N)| = 1$ so that (u, ψ_N) does not tend to zero. This contradiction proves the assertion. $\square$

Remark. The space C_0^∞ is sequentially complete with respect to convergence in C_0^∞. For if $\{\phi_\nu\}$ is a sequence of test functions with supports in a fixed compact set, and $\partial^\alpha(\phi_\nu - \phi_\mu) \to 0$ uniformly as $\nu, \mu \to \infty$, for each α, then it follows from the theory of uniform convergence that the ϕ_ν tend to a limit function which is in C_0^∞.

One often speaks of 'a distribution f', meaning the map that sends $\phi \in C_0^\infty$ to the number (f, ϕ). If (f, ϕ) happens to be of the form (2.1.8), one sometimes says that the distribution f is of the type of the function $x \to f(x)$. Distributions will usually be written like functions,

$$f(x), u(x), \ldots;$$

the independent variable x then refers to the space on which the test functions are defined.

A distribution u can be localized, to some extent, by considering the restriction of u to an open set $\Omega \subset \mathbf{R}^n$. This is just the linear form (u, ϕ), for all $\phi \in C_0^\infty$ with supports in Ω. Thus $u = 0$ in Ω if $(u, \phi) = 0$ for all $\phi \in C_0^\infty$ with supp $\phi \subset \Omega$, and $u = v$ in Ω if $u - v = 0$ in Ω. It is easy to prove, by means of a partition of unity, that if $\Omega \subset \mathbf{R}^n$ is open, if $u \in \mathscr{D}'$, and if every point of Ω has an open neighbourhood in Ω in which $u = 0$, then $u = 0$ in Ω. Hence one can define the *support of a distribution* as the complement of the maximal open set in which the distribution vanishes. The support of a distribution that is of the type of a continuous function f is equal to the support of the function; the support of $\delta(x - y)$ is (y), the set consisting of the single point y.

The space $\mathscr{D}'$ of distributions becomes a vector space over $\mathbf{C}$ if addition and multiplication by a number are defined in the natural way:

$$(f+g, \phi) = (f, \phi) + (g, \phi) \quad (f, g \in \mathscr{D}', \phi \in C_0^\infty),$$

$$(\lambda f, \phi) = \lambda (f, \phi) \quad (f \in \mathscr{D}', \phi \in C_0^\infty, \lambda \in \mathbf{C}).$$

This space will always be given the *weak topology*. For sequences $\{u_\nu\}_{1 \leqslant \nu < \infty}$ this means that $u_\nu \to u$ in $\mathscr{D}'$ if $(u_\nu, \phi) \to (u, \phi)$ for all $\phi \in C_0^\infty$. A basis for the neighbourhoods of zero in $\mathscr{D}'$ consists of sets of the form

$$\{u; u \in \mathscr{D}', |(u, \phi_1)| < \epsilon_1, \dots, |(u, \phi_k)| < \epsilon_k\},$$

where $\{\phi_j\}_{1 \leqslant j \leqslant k}$ is some finite set of test functions, and the ϵ_j are positive numbers.

A distribution is said to be of *finite order* if, in the semi-norm estimates (2.1.9), one can choose N independently of the set K. The smallest N is then the order of the distribution. By Lemma 2.1.1, a distribution of order k can be extended, by continuity, to a continuous linear form on C_0^k.

2.2 Operations on distributions

A number of operations on functions can be extended to distributions. The most important of these is differentiation. If $u(x)$ is a C^1 function, then both u and its derivatives $\partial_i u$ can be identified with distributions. Now, for any $\phi \in C_0^\infty$,

$$(\partial_i u, \phi) = \int \phi \, \partial_i u \, dx = \int (\partial_i(u\phi) - u \, \partial_i \phi) \, dx.$$

As $u\phi$ has compact support, the integral of $\partial_i(u\phi)$ vanishes. Hence

$$(\partial_i u, \phi) = -\int u \, \partial_i \phi \, dx = -(u, \partial_i \phi).$$

The derivative of a distribution is defined by generalizing this identity.

Theorem 2.2.1. *If $u \in \mathscr{D}'$, then the linear forms*

$$(\partial_i u, \phi) = -(u, \partial_i \phi) \quad (i = 1, \dots, n), \tag{2.2.1}$$

are distributions; they are, by definition, the derivatives of u. If $u \in C^1$, then its distributional derivatives are identical with its classical derivatives. The maps $u \to \partial_i u, i = 1, \dots, n$, are continuous maps $\mathscr{D}' \to \mathscr{D}'$.

PROOF. It follows from the semi-norm estimates (2.1.9) that the $(\partial_i u, \phi)$, which are linear forms, are distributions. If $u \in C^1$, then

$$(\partial_i u, \phi) = -\int u \, \partial_i \phi \, dx = \int \phi \, \partial_i u \, dx$$

(where, in the integrand, $\partial_i u$ is the classical derivative) holds for all $\phi \in C_0^\infty$. By Lemma 2.1.2, this implies the identity of the classical and the distributional derivatives. As to continuity, one has, if $u_\nu \to u$ in $\mathscr{D}'$, that

$$(\partial_i u_\nu, \phi) = -(u_\nu, \partial_i \phi) \to -(u, \partial^i \phi) = (\partial_i u, \phi),$$

and the general statement also follows easily if one considers neighbourhoods of zero in $\mathscr{D}'$. So the theorem is proved. $\square$

Remark. By iteration, one obtains the derivatives of higher order as

$$(\partial^\alpha u, \phi) = (-1)^{|\alpha|} (u, \partial^\alpha \phi). \tag{2.2.2}$$

Note that a distribution has derivatives of all orders. If $u = f$, a locally integrable function, then (2.2.2) means that

$$(\partial^\alpha f, \phi) = (-1)^{|\alpha|} \int f \partial^\alpha \phi \, dx.$$

Conversely, any distribution can be written locally in this form:

Theorem 2.2.2. *Let $u \in \mathscr{D}'$, and let Ω be an open set with compact closure. Then there exists a continuous function $f(x)$ and a multi-index α such that $u = \partial^\alpha f$ in Ω.* $\square$

This is sometimes called the structure theorem; the proof will be found in the literature. The hypothesis that Ω is compact is essential. For example, if $n = 1$, then

$$u(t) = \sum_{j=1}^\infty \delta^{(j)}(t-j)$$

is a distribution that is not a derivative of a function.

As a simple example, for $n = 1$, let $H(t)$ denote the Heaviside function,

$$H(t) = 1 \text{ if } t \geqslant 0, \quad H(t) = 0 \text{ if } t < 0.$$

Then $$(H'(t), \phi(t)) = -(H(t), \phi'(t)) = -\int_0^\infty \phi'(t) \, dt.$$

As ϕ has compact support, the second member is equal to $\phi(0)$. So the derivative of the distribution $H(t)$ is $\delta(t)$.

Multiplication by a function is also easy to deal with. For a distribution that is of the type of a function, one has

$$(gu, \phi) = \int gu\phi \, dx = (u, g\phi).$$

To carry this over to distributions, one must require g to be sufficiently smooth, so that $g\phi$ is a test function.

Theorem 2.2.3. *If $u \in \mathscr{D}'$ and $g \in C^\infty$ then*

$$(gu, \phi) = (u, g\phi) \tag{2.2.3}$$

is a distribution, which is equal to the product ug when u is a function. The map $u \to gu$ is a continuous linear map $\mathscr{D}' \to \mathscr{D}'$. Leibniz' rule holds for distributions,

$$\partial_i(gu) = u\,\partial_i g + g\,\partial_i u \quad (i = 1, \dots, n),$$

and so does its generalization (2.1.2).

PROOF. All the assertions are immediate consequences of the definitions; the proof that gu is a distribution is a consequence of the semi-norm estimates and the generalized Leibniz' theorem. As to Leibniz' theorem for gu, note first that

$$(\partial_i(ug), \phi) = -(ug, \partial_i \phi) = -(u, g\,\partial_i \phi).$$

But $$-g\,\partial_i \phi = -\partial_i(g\phi) + \phi\,\partial_i g.$$

Hence $$(\partial_i(gu), \phi) = -(u, \partial_i(g\phi)) + (u, \phi\,\partial_i g)$$
$$= (\partial_i u, g\phi) + (u, \phi\,\partial_i g),$$

and by the definition of the product,

$$(\partial_i u, g\phi) + (u, \phi\,\partial_i g) = (g\,\partial_i u + u\,\partial_i g, \phi). \quad \square$$

Remark. It is evident that gu is also well defined by (2.2.3) if u is of finite order k, and $\phi \in C^k$.

EXAMPLE. By (2.1.10) and (2.2.3),

$$(g(x)\,\delta(x), \phi(x)) = (\delta(x), g(x)\,\phi(x)) = g(0)\,\phi(0) = g(0)\,(\delta, \phi).$$

Hence $$g(x)\,\delta(x) = g(0)\,\delta(x),$$

and this holds for $g \in C^0$. Similarly, it follows from (2.1.2) that

$$g(x)\,\partial^\alpha \delta(x) = (-1)^{|\alpha|} \sum_{\beta \leqslant \alpha} \frac{\alpha!}{\beta!\,(\alpha-\beta)!} (-1)^{|\beta|}\,\partial^{\alpha-\beta} g(0)\,\partial^\beta\,\delta(x).$$

We next consider translation. If $f(x)$ is a function and $h \in \mathbf{R}^n$, then

$$\int f(x-h)\,\phi(x)\,dx = \int f(x)\,\phi(x+h)\,dx.$$

In terms of distributions, this is $(f(x-h), \phi(x)) = (f(x), \phi(x+h))$, and one can extend this identity to distributions.

Theorem 2.2.4. *Let $u \in \mathscr{D}'$. The distribution $u(x-h)$, where $h \in \mathbf{R}^n$, is defined by*

$$(u(x-h), \phi(x)) = (u(x), \phi(x+h)). \qquad (2.2.4)$$

The map $u(x) \to u(x-h)$ is a continuous linear map $\mathscr{D}' \to \mathscr{D}'$ that commutes with differentiation.

PROOF. As all the assertions are immediate consequences of the definitions, this is left to the reader. $\square$

The distribution $u(x-h)$ acts on test functions $\phi(x) \in C_0^\infty$ and also depends on h. Generally, one can consider distributions $u_\lambda(x)$ that act on $\phi(x) \in C_0^\infty$ and are functions of an auxiliary variable λ which will be in some other space. One then says that u_λ is continuous (differentiable, analytic, ...) with respect to λ if, for every $\phi \in C_0^\infty$, the function $\lambda \to (u_\lambda(x), \phi(x))$ has the property in question. In the case of translation, the following important result holds.

Theorem 2.2.5. *As a function of $h \in \mathbf{R}^n$, $u(x-h)$ is in C^∞, and*

$$(\partial/\partial h^i)\,(u(x-h), \phi(x)) = -(\partial_i u(x-h), \phi(x)) \quad (i = 1, ..., n). \qquad (2.2.5)$$

PROOF. For $\phi \in C_0^\infty$, set

$$F(h) = (u(x-h),\,\phi(x)) = (u(x),\,\phi(x+h)).$$

Then
$$F(h+k) - F(k) = (u(x),\,\phi(x+h+k) - \phi(x+h)).$$

For fixed h and bounded $|k|$, the supports of the functions

$$x \to \phi(x+h+k) - \phi(x+h)$$

are in a fixed compact set. Hence it follows from a semi-norm estimate of the type (2.1.9) and the mean value theorem that

$$F(h+k) - F(h) = O(|k|)$$

as $k \to 0$. So $F(h)$ is continuous. A similar argument shows that $F(h)$ is differentiable and that its derivatives are given by (2.2.5). As the

$\partial_i u$ are distributions, the argument can now be repeated, and it evidently follows by induction that $F(h) \in C^\infty$. $\square$

Remark. For $h = 0$, (2.2.5) is equivalent to

$$\lim_{\epsilon \to 0} \frac{u(x + \epsilon e_i) - u(x)}{\epsilon} = \partial_i u(x) \quad (i = 1, \ldots, n), \qquad (2.2.6)$$

where the limit is in the distribution topology, and e_i is the ith unit vector in $\mathbf{R}^n$.

Finally, let $A = \{a_j^i\}_{1 \leqslant i, j \leqslant n}$ be a constant non-singular matrix, and set $(Ax)^i = a_j^i x^j$; $A^{-1}x$ is defined similarly.

Theorem 2.2.6. *Let $u \in \mathscr{D}'$ and let A be a constant non-singular $n \times n$ matrix. Then*
$$(u(Ax), \phi(x)) = |\det A|^{-1} (u(x), \phi(A^{-1}x)) \qquad (2.2.7)$$
defines a distribution which, when $u(x)$ is a function, is identical with the function $u(Ax)$.

PROOF. This is straightforward, and left to the reader. $\square$

Remark. With the usual identification of $n \times n$ matrices and points in $\mathbf{R}^N$, where $N = h^2$, the non-singular matrices become an open set in $\mathbf{R}^N$. One can then show that $A \to u(Ax)$ is a C^∞ function on this set.

There are some particular cases of (2.2.7) that are worth noting separately. If $A = \mathrm{diag}\,(-1, \ldots, -1)$, then (2.2.7) defines $u(-x)$, as
$$(u(-x), \phi(x)) = (u(x), \phi(-x)). \qquad (2.2.8)$$

A distribution is called even or odd according as $u(-x) = u(x)$ or $u(-x) = -u(x)$ respectively. If $A = \mathrm{diag}\,(\lambda, \ldots, \lambda)$, where $0 < \lambda \in \mathbf{R}$, then
$$(u(\lambda x), \phi(x)) = \lambda^{-n} (u(x), \phi(x/\lambda)). \qquad (2.2.9)$$

Combining (2.2.8) and (2.2.9) one obtains, for $0 \neq \lambda \in \mathbf{R}$,
$$(u(\lambda x), \phi(x)) = |\lambda|^{-n} (u(x), \phi(x/y)). \qquad (2.2.10)$$

A distribution is homogeneous of degree k if $u(\lambda x) = |\lambda|^k u(x)$. It is easy to verify that the Dirac delta function is even, and homogeneous of degree $-n$.

2.3   Distributions with compact support

A distribution with compact support can be extended to a linear form on C^∞. We need a simple lemma.

Lemma 2.3.1. *Let K be a compact set, ϵ a positive number, and let K_ϵ be the set of all x whose distance from K is less than ϵ. Then there exists a $\sigma \in C_0^\infty$ such that $\sigma = 1$ on K_ϵ.*

PROOF. The function

$$\sigma(x) = \epsilon^{-n} \int \chi(y) \rho\left(\frac{x-y}{\epsilon}\right) dy,$$

where χ is the characteristic function of $K_{2\epsilon}$, and ρ satisfies the conditions (2.1.4) of Lemma 2.1.1, has the required properties.

Now suppose that $u \in \mathscr{D}'$ and that $\operatorname{supp} u = K$, a compact set. Choose any $\sigma \in C_0^\infty$ such that $\sigma = 1$ on a neighbourhood of K, and set, for all $\phi \in C^\infty$

$$(u, \phi) = (u, \sigma\phi). \tag{2.3.1}$$

This defines (u, ϕ), because $\sigma\phi \in C_0^\infty$. Moreover, (u, ϕ) is defined uniquely. For if $\sigma' \in C_0^\infty$ is another function such that $\sigma' = 1$ on a neighbourhood of K, then $(\sigma - \sigma')\phi$ is in C_0^∞ and vanishes on an open neighbourhood of K. Hence

$$(u, \sigma\phi) - (u, \sigma'\phi) = (u, (\sigma - \sigma')\phi) = 0. \quad \square$$

For a fixed cut-off function σ, one has $\operatorname{supp}(\sigma\phi) \subset \operatorname{supp} \sigma$, a fixed compact set. Hence there is a semi-norm estimate

$$|(u, \phi)| = |(u, \sigma\phi)| \leqslant C^1 \sum_{|\epsilon| \leqslant N} \sup |\partial^\alpha(\sigma\phi)|.$$

By Leibniz' theorem, this implies that, for all $\phi \in C^\infty$,

$$|(u, \phi)| \leqslant C \sum_{|\alpha| \leqslant N} \sup_K |\partial^\alpha\phi|, \tag{2.3.2}$$

where C is another constant, and K now denotes the compact set $\operatorname{supp} \sigma$. This suggests the following definition.

Definition 2.3.1. *A linear form $C^\infty \ni \phi \to (u, \phi)$ is said to be continuous if there exist a compact set K and constants C and N such that (2.3.2) holds for all $\phi \in C^\infty$. The vector space consisting of continuous linear forms on C^∞ is denoted by $\mathscr{E}'$.*

Remark. In L. Schwartz' original notation, C_0^∞ was denoted by $\mathscr{D}$ and C^∞ by $\mathscr{E}$. The notations $\mathscr{D}'$ and $\mathscr{E}'$ indicate that these spaces are the duals of $\mathscr{D}$ and $\mathscr{E}$ respectively.

Theorem 2.3.1. *The sub-space of $\mathscr{D}'$ consisting of distributions with compact support can be identified with $\mathscr{E}'$.*

PROOF. It has already been shown that a distribution with compact support can be extended uniquely to a continuous linear form on C^∞. Conversely, suppose that $u \in \mathscr{E}'$. If $\phi_\nu \in C_0^\infty$ and $\phi_\nu \to 0$ in C_0^∞, then it follows from (2.3.2) that $(u, \phi_\nu) \to 0$. By Theorem 2.1.1, this implies that the restriction of (u, ϕ) to C_0^∞ is a distribution. It also follows from (2.3.2) that if $\phi \in C_0^\infty$ and the support of ϕ does not meet K, then $(u, \phi) = 0$. Hence the restriction of $u \in \mathscr{E}'$ to C_0^∞ is a distribution with compact support. $\square$

By an argument similar to the proof of Theorem 2.1.1, one can establish the following alternative characterization of $\mathscr{E}'$.

Theorem 2.3.2. *A linear form $C^\infty \ni \phi \to (u, \phi)$ is in $\mathscr{E}'$ if and only if $(u, \phi_\nu) \to 0$ when $\phi_\nu \to 0$ in C^∞, in the sense that $\partial^\alpha \phi_\nu \to 0$ as $\nu \to \infty$, uniformly on every compact set, for each multi-index α.* $\square$

It is an immediate consequence of (2.3.2) that a distribution with compact support is of finite order. By applying Theorem 2.2.2 to $(u, \sigma\phi)$, where $u \in \mathscr{E}'$ and σ is a cut-off function, one can show that u is a finite sum of derivatives of continuous functions whose supports are contained in a pre-assigned neighbourhood of the support of u.

An obvious example of a distribution with compact support is $\partial^\alpha \delta(x)$; the support of this is a point, the origin. There is a useful converse, which is stated without proof.

Theorem 2.3.3. *Suppose that the support of $u \in \mathscr{D}'$ is the origin. Then u is a finite linear combination of $\delta(x)$ and its derivatives.*

2.4   Tensor products of distributions

If $\mathbf{R}^n \ni x \to f(x)$ and $\mathbf{R}^m \ni y \to g(y)$ are two locally integrable functions, then their product $f(x)g(y)$ is a locally integrable function on

$$\mathbf{R}^n \times \mathbf{R}^m = \mathbf{R}^{n+m}.$$

This kind of multiplication can be extended to distributions. Let the underlying spaces be denoted by $\mathbf{R}_x^n$ and $\mathbf{R}_y^m$ respectively, so that

$x = (x^1, ..., x^n) \in \mathbf{R}_x^n$ and $y = (y^1, ..., y^m) \in \mathbf{R}_y^m$. (One may, of course, have $n = m$.) The distribution determined by $f(x)\,g(y)$ is

$$\iint f(x)\,g(y)\,\phi(x,y)\,dx\,dy.$$

By Fubini's theorem, this can be written in either of the forms

$$\int f(x)\,dx \int g(y)\,\phi(x,y)\,dy = (f(x), (g(y), \phi(x,y))),$$

$$\int g(y)\,dy \int f(x)\,\phi(x,y)\,dx = (g(y), (f(x), \phi(x,y))).$$

For the distribution case, one needs the following lemma.

Lemma 2.4.1. *Suppose that* $v(y) \in \mathscr{D}'(\mathbf{R}_y^m)$ *and that*

$$\phi(x,y) \in C_0^\infty(\mathbf{R}_x^n \times \mathbf{R}_y^m),$$

and put $$\psi(x) = (v(y), \phi(x,y));$$ (2.4.1)

in the second member, ϕ *is treated as a function in* $C_0^\infty(\mathbf{R}_y^m)$ *that depends on a parameter* $x \in \mathbf{R}_x^n$. *Then* $\psi \in C_0^\infty(\mathbf{R}_x^n)$, *and*

$$\partial_i \psi(x) = (v(y), (\partial/\partial x^i)\,\phi(x,y)) \quad (i = 1, ..., n).$$ (2.4.2)

PROOF. For any set $K \subset \mathbf{R}_x^n \times \mathbf{R}_y^m$, let $Pr_1 K$ and $Pr_2 K$ denote the projections on the first and second factor respectively. It is evident that $\psi(x)$ is well defined, and that supp $\psi \subset Pr_1(\text{supp } \phi)$, a fixed compact set. Again,

$$\psi(x+h) - \psi(x) - \left(v(y), \sum_{i=1}^n h^i(\partial/\partial x^i)\,\phi(x,y)\right)$$

$$= \left(v(y), \phi(x+h,y) - \phi(x,y) - \sum_{i=1}^n h^i(\partial/\partial x^i)\,\phi(x,y)\right).$$ (2.4.3)

For fixed x, the support of the function

$$y \to \phi(x+h,y) - \phi(x,y) - \sum_{i=1}^n h^i(\partial/\partial x^i)\,\phi(x,y)$$

is in the fixed compact set $Pr_2(\text{supp } \phi)$. A semi-norm estimate and the mean value theorem therefore imply that the first member of (2.4.3) is $O(|h|^2)$ as $h \to 0$. Hence $\psi(x)$ is continuous and differentiable for all x, and (2.4.2) holds. Repeated application of the argument shows that $\psi \in C_0^\infty(\mathbf{R}_x^n)$, and the proof is complete. $\square$

Theorem 2.4.1. *Suppose that* $u(x) \in \mathcal{D}'(\mathbf{R}_x^n)$ *and* $v(y) \in \mathcal{D}'(\mathbf{R}_y^m)$. *Then the linear form defined by*

$$(u(x) \otimes v(y), \phi(x,y)) = (u(x), \psi(x)), \quad \phi \in C_0^\infty(\mathbf{R}_x^n \times \mathbf{R}_y^m), \quad (2.4.4)$$

where ψ *is given by* (2.4.1), *is a distribution, called the tensor product of* u *and* v.

PROOF. The existence of (u, ψ) is assured by Lemma 2.4.1. If

$$\operatorname{supp} \phi \subset K,$$

a fixed compact set in $\mathbf{R}_x^n \times \mathbf{R}_y^m$, then $\operatorname{supp} \psi \subset Pr_1 K$, which is a fixed compact set in $\mathbf{R}_x^n$. Hence there is a semi-norm estimate for (u, ψ).

$$|(u, \psi)| \leqslant B \sum_{|\beta| \leqslant M} \sup |\partial^\beta \psi|,$$

and by (2.4.2),

$$\partial^\beta \psi = (v(y), \partial_x^\beta \phi(x,y)).$$

But the support of $y \to \phi(x,y)$ is in $Pr_2 K$, and so the $\partial^\beta \psi$ can be estimated by

$$|\partial^\beta \psi| \leqslant A \sum_{|\alpha| \leqslant N} \sup |\partial_y^\alpha \partial_x^\beta \phi(x,y)|.$$

Combining these two estimates, one obtains a semi-norm estimate for $u(x) \otimes v(y)$, which is therefore a distribution. $\square$

In the case of functions, one has commutativity: this also extends to distributions.

Theorem 2.4.2. *The tensor product of two distributions is commutative,*

$$u(x) \otimes v(y) = v(y) \otimes u(x), \quad\quad\quad (2.4.5)$$

or, explicitly,

$$(u(x) \otimes v(y), \phi(x,y)) = (u(x), (v(y), \phi(x,y)))$$
$$= (v(y), (u(x), \phi(x,y))). \quad\quad (2.4.6)$$

PROOF. The theorem is true if $\phi = \psi(x)\chi(y)$, where $\psi \in C_0^\infty(\mathbf{R}_x^n)$ and $\chi \in C_0^\infty(\mathbf{R}_y^m)$. For then $(v, \phi) = \psi(x)(v, \chi)$, whence

$$(u(x) \otimes v(y), \psi(x)\chi(y)) = (u, \psi)(v, \chi),$$

and it is obvious that the same result is obtained if one operates with u first. By linearity, it follows that (2.4.6) holds if ϕ is a finite sum of such products. But it can be shown that any $\phi \in C_0^\infty(\mathbf{R}_x^n \times \mathbf{R}_y^m)$ can be approximated, in the sense of convergence in $C_0^\infty(\mathbf{R}_x^n \times \mathbf{R}_q^m)$, by functions of this form.

It is obviously sufficient to prove this when the support of ϕ is contained in the open unit cube $Q = \{x; 0 < x^i < 1, i = 1, \ldots, n\}$. By a form of Weierstrass' approximation theorem, one can then find a sequence of polynomials $p_\nu(x, y)$ such that $\partial^\alpha p_\nu \to \partial^\alpha \phi$ for each α when $\nu \to \infty$, uniformly on Q. As Q is open, there is a $\delta > 0$ such that

$$\operatorname{supp} \phi \subset \{x; \delta \leqslant x^i \leqslant 1 - \delta, i = 1, \ldots, n\}.$$

Let $\rho(t) \in C_0^\infty(\mathbf{R})$ be a cut-off function such that $\operatorname{supp} \rho \subset [0, 1]$ and $\rho = 1$ on $(\frac{1}{2}\delta, 1 - \frac{1}{2}\delta)$ (Lemma 2.3.1). Then, by Leibniz' theorem, the functions
$$\phi_\nu = \rho(x^1) \ldots \rho(x^n) \rho(y^1) \ldots \rho(y^m) p_\nu(x, y)$$

tend to ϕ in $C_0^\infty(\mathbf{R}_x^n \times \mathbf{R}_y^m)$, and they are of the required form. As both $u(x) \otimes v(y)$ and $v(y) \otimes u(x)$ are distributions, and (2.4.6) holds when ϕ is replaced by ϕ_ν, the theorem follows by continuity, if one makes $\nu \to \infty$. $\square$

The following basic properties of the tensor product are immediate consequences of Theorem 2.4.1

Theorem 2.4.3. (i) *The support of $u(x) \otimes v(y)$ is* $\operatorname{supp} u(x) \times \operatorname{supp} v(y)$.
   (ii) *Tensor products are differentiated as follows:*

$$\partial_x^\alpha \partial_y^\beta u(x) \otimes v(y) = \partial_x^\alpha u(x) \otimes \partial_y^\beta v(y). \qquad (2.4.7) \; \square$$

Tensor products can of course also be defined for any finite number of factors; they are associative, distributive, and commutative. A simple example is the identity

$$\delta(x^1, \ldots, x^n) = \delta(x^1) \otimes \ldots \otimes \delta(x^n). \qquad (2.4.8)$$

2.5 Convolution and regularization

The convolution of two functions f and g, both defined on $\mathbf{R}^n$, is the integral

$$f * g(x) = \int f(y) g(x - y) \, dy, \qquad (2.5.1)$$

provided that it exists. This will for instance be the case if one of the factors has compact support, and both are locally integrable. As a distribution, $f * g$ is

$$(f * g, \phi) = \int \phi(x) \, dx \int f(y) g(x - y) \, dy = \iint f(x) g(y) \, \phi(x + y) \, dx \, dy,$$

by a simple change of variables. Note that this also shows that

$f * g = g * f$. One would therefore expect to be able to define the convolution of two distributions u and v by setting

$$(u * v, \phi) = (u(x) \otimes v(y), \phi(x+y)), \quad \phi \in C_0^\infty. \tag{2.5.2}$$

But the function $(x, y) \to \phi(x+y)$, although in $C^\infty(\mathbf{R}_x^n \times \mathbf{R}_y^n)$, does not have compact support. So the second member of (2.5.2) may not exist. We shall show that this difficulty can be overcome by assuming that one of the factors has compact support.

If A and B are sets in $\mathbf{R}^n$, then their vector sum is by definition the set $A + B = \{x; x = y+z, y \in A, z \in B\}$; if B consists of a single point h, then $h+A$ is just A, translated by the vector h. If one also sets $-B = \{x; -x \in B\}$, then $A - B$ can be defined as $A + (-B)$.

Lemma 2.5.1. *Suppose that $A \subset \mathbf{R}^n$ is compact and that $B \subset \mathbf{R}^n$ is closed; then $A + B$ is closed.*

PROOF. Suppose that $x_\nu \in A + B$ and that the sequence $\{x_\nu\}_{1 \leqslant \nu < \infty}$ converges to a point x. For each ν, there is a $y_\nu \in A$ and a $z_\nu \in B$ such that $x_\nu = y_\nu + z_\nu$. As A is compact, $\{y_\nu\}$ has a convergent subsequence whose limit is a point $y \in A$; deleting the other y_ν, one may suppose that $y_\nu \to y \in A$. Hence $z_\nu = z_\nu - y_\nu$ converges to a point z, and as B is closed, and $z_\nu \in B$, z is in B. So $x = y+z \in A+B$, and $A+B$ is a closed set. $\square$

Theorem 2.5.1. *Suppose that $u \in \mathscr{E}'$ and $v \in \mathscr{D}'$. Let $\rho \in C_0^\infty$ be a cut-off function, such that $\rho = 1$ on a neighbourhood of supp u. Then the identity*

$$(u * v, \phi) = (u(x) \otimes v(y), \rho(x) \phi(x+y)) \tag{2.5.3}$$

defines a unique distribution $u * v$ which is called the convolution of u and v. The support of $u * v$ is contained in supp u + supp v.

PROOF. If $\phi \in C_0^\infty$, then $\rho(x) \phi(x+y) \in C^\infty(\mathbf{R}_x^n \times \mathbf{R}_y^n)$, and

$$\text{supp} ((x, y) \to \rho(x) \phi(x+y)) \subset \text{supp}\, \rho \times (\text{supp}\, \phi - \text{supp}\, \rho). \tag{2.5.4}$$

By Lemma 2.5.1, the set supp ϕ − supp ρ is closed; as it is also bounded, it is compact. It follows that $\rho(x) \phi(x+y) \in C_0^\infty(\mathbf{R}_x^n \times \mathbf{R}_y^n)$, and so the second member of (2.5.3) exists. An argument similar to the proof of the first part of Theorem 2.3.1 shows that it is independent of the choice of ρ. Again, (2.5.4) implies that, if the support of ϕ is in a fixed compact set, then the support of $\rho(x) \phi(x+y)$ is in a fixed compact set in $\mathbf{R}_x^n \times \mathbf{R}_y^n$. So the second member of (2.5.4) can be estimated by semi-norms in $\mathbf{R}_x^n \times \mathbf{R}_y^n$, and it is easily seen that this implies a semi-norm estimate of the usual type (2.1.9) for $(u*v, \phi)$. Hence $u*v$ is a distribution.

To prove the last part of the theorem, put $\operatorname{supp} u = A$ and $\operatorname{supp} v = B$. By Lemma 2.5.1, $A + B$ is a closed set; denote its complement in $\mathbf{R}^n$ by Ω. Suppose that $\operatorname{supp} \phi \in \Omega$. Then

$$\operatorname{supp}((x,y) \to \phi(x+y)) \subset \{(x,y); x+y \in \Omega\}.$$

But, by Theorem 2.4.3, $\operatorname{supp} u(x) \otimes v(y) = A \times B$. Hence the sets $\operatorname{supp} \phi(x+y)$ and $\operatorname{supp} u(x) \otimes v(y)$, which are closed in $\mathbf{R}_x^n \times \mathbf{R}_y^n$, do not meet. One can therefore choose ρ so that the intersection of

$$\operatorname{supp} \rho(x) \phi(x+y)$$

and of $\operatorname{supp} u(x) \otimes v(y)$ is also empty, and (2.5.3) then shows that $(u*v, \phi) = 0$. This completes the proof. $\square$

Remark. It follows from Theorems 2.4.1 and 2.4.2 that

$$(u*v, \phi) = (u(x), \rho(x)(v(y), \phi(x+y))) = (v(y), (u(x), \rho(x)\phi(x+y))). \tag{2.5.5}$$

Now, by Theorem 2.2.5, $\psi(x) = (v(y), \phi(x+y))$ is a C^∞ function, and as $u \in \mathscr{E}'$, the second member of (2.5.5) is just (u, ψ). Similarly, we can write

$$\chi(y) = (u(x), \rho(x)\phi(x+y)) = (u(x), \phi(x+y))$$

if we read the second member as a linear form on C^∞; Theorem 2.2.5 and the compactness of the support of u imply that $\chi \in C_0^\infty$. Hence one can suppress the cut-off function ρ in either case, and compute $u*v$ by means of the identities

$$(u*v, \phi) = (u(x), (v(y), \phi(x+y))) = (v(y), (u(x), \phi(x+y))). \tag{2.5.6}$$

The next theorem lists the principal properties of the convolution. In its statement, we use the translation operator τ_h which maps a distribution $u(x)$ to $\tau_h u = u(x-h)$.

Theorem 2.5.2. *Let u and v be distributions, and suppose that one of them, at least, has compact support. Then*

(i) $$u*v = v*u;$$

(ii) $$\delta*u = u, \quad \tau_h \delta*u = \tau_h u;$$

(iii) $$\partial_i(u*v) = \partial_i u*v = u*\partial_i v \quad (i = 1, \dots, n);$$

(iv) $$\tau_h(u*v) = \tau_h u*v = u*\tau_h v;$$

(v) *if $u_\nu \to u$ in $\mathscr{D}'$ and $v \in \mathscr{E}'$, then $u_\nu * v \to u * v$ in $\mathscr{D}'$;*

(vi) *if $u_\nu \to u$ in $\mathscr{D}'$ and the supports of the u_ν are in a fixed compact set, then $u_\nu * v \to u * v$ in $\mathscr{D}'$.*

PROOF. We only prove (v) and (vi); the other assertions are easy consequencies of (2.5.6), and their verification is left to the reader. To prove (v), note that $\psi(x) = (v(y), \phi(x+y)) \in C_0^\infty$, by Theorem 2.2.5 and the compactness of the support of v. Hence, by (2.5.6), for all $\phi \in C_0^\infty$,

$$(u_\nu * v, \phi) = (u_\nu, \psi) \to (u, \psi) = (u * v, \phi).$$

In case (vi), ψ is only C^∞, but the same cut-off function ρ can be used for all u_ν. So (2.5.5) gives

$$(u_\nu * v, \phi) = (u_\nu, \rho\psi) \to (u, \rho\psi) = (u * v, \phi). \ \square$$

A function $f \in C_0^\infty$ can be identified with a distribution $f \in \mathscr{E}'$. Hence $u * f$ will exist for all $u \in \mathscr{D}'$. It is, in fact, a function.

Theorem 2.5.3. *If $u \in \mathscr{D}'$ and $f \in C_0^\infty$, then*

$$u * f(x) = (u(y), f(x-y)), \qquad (2.5.7)$$

and this is a C^∞ function.

PROOF. Denote $f(-x)$ by $\check{f}(x)$; then

$$(u(y), f(x-y)) = (u(y), \ \check{f}(y-x)) = (u(y+x), \check{f}(y)),$$

and this is a C^∞ function, by Theorem 2.2.5.

By (2.5.6),

$$(u * f, \phi) = (u(y), (f(x), \phi(x+y))) = \left(u(y), \int f(x) \, \phi(x+y) \, dx \right).$$

But as f is a test function, one can also form

$$(\check{u} * \phi, f) = \left(u(-y), \int \phi(x) f(x+y) \, dx \right) = \left(u(y), \int (x-y) \, \phi(x) \, dx \right).$$

Now $\quad\quad\quad\quad \int f(x) \, \phi(x+y) \, dx = \int \phi(x) f(x-y) \, dx,$

by a simple change of the variable of integration. Hence

$$(u * f, \phi) = (\check{u} * \phi, f) = (\phi(x), (u(y), f(x-y))).$$

As both ϕ and $x \to (u(y), f(x-y))$ are functions, this is

$$(u * f, \phi) = \int (u(y), f(x-y)) \, \phi(x) \, dx,$$

which proves (2.5.7). $\square$

Remark. The function $u * f(x)$ is called a *regularization* of u; it is a smooth function whose properties are closely related to those of u.

Suppose that ψ satisfies the conditions (2.1.4), and put

$$\psi_\nu = \nu^n \psi(\nu x), \quad \nu = 1, 2, \dots.$$

By Lemma 2.2.1, the ψ_ν converge to $\delta(x)$ in $\mathscr{D}'$, and their supports are contained in the unit ball. So it follows from (ii) and (vi) of Theorem 2.5.2 that

$$\lim_{\nu \to \infty} u * \psi_\nu = u * \delta = u,$$

in the weak topology, for all $u \in \mathscr{D}'$. So every distribution can be approximated in $\mathscr{D}'$ by C^∞ functions. In fact, one can prove a little more.

Theorem 2.5.4. *Every distribution is the limit, in the distribution topology, of a sequence of C_0^∞ functions.*

PROOF. Let $\sigma \in C_0^\infty$ be such that $\sigma = 1$ for $|x| \leqslant 1$. Then, with the ψ_ν just defined, $\sigma(x/\nu) \psi_\nu(x) \in C_0^\infty$, by Theorem 2.5.3, and if $\phi \in C_0^\infty$ then

$$(\sigma(x/\nu) \psi_\nu * u, \phi) = (u * \psi_\nu, \phi),$$

when ν exceeds the radius of the smallest ball $\{x; |x| < R\}$ that contains the support of ϕ. So it follows that

$$\lim_{\nu \to \infty} \sigma(x/\nu) (u * \psi_\nu) = u$$

in $\mathscr{D}'$, and the theorem is proved. $\square$

Remark. This theorem, which can be expressed by saying that C_0^∞ is dense in $\mathscr{D}'$, given another characterization of distributions.

There are a number of other cases in which the convolution of two distributions can be defined. In particular, there is a simple extension of Theorem 2.5.1 which will be needed later on. If $u \in \mathscr{D}'$ and $\phi \in C^\infty$, then (u, ϕ) can be defined, uniquely, provided that the set

$$\operatorname{supp} u \cap \operatorname{supp} \phi$$

is compact. This is done exactly as in the case of distributions with compact support, by introducing a cut-off function. Hence (2.5.2) has a meaning, and defines a distribution $u * v \in \mathscr{D}'$, if the inverse image of every compact set in $\mathbf{R}^n$ under the map

$$\operatorname{supp} u \times \operatorname{supp} v \ni (x, y) \to x + y \in \mathbf{R}^n \tag{2.5.8}$$

is compact. Such a map is called *proper*. An equivalent condition is as follows: for every finite $c > 0$, there are finite positive numbers a and b such that $x \in \operatorname{supp} u$, $y \in \operatorname{supp} v$ and $|x+y| \leqslant c$ implies $|x| \leqslant a$ and $|y| \leqslant b$.

Theorem 2.5.5. *Let u and v be distributions, and suppose that the map (2.5.8) is proper. Then the convolution $u * v \in \mathscr{D}'$ is defined by (2.5.2), and $\operatorname{supp} u * v \subset \operatorname{supp} u + \operatorname{supp} v$. Also, if $u_\nu \to u$ in $\mathscr{D}'$, and the supports of the u_ν are contained in a set A such that the map*

$$A \times \operatorname{supp} v \ni (x, y) \to x + y \in \mathbf{R}^n$$

is proper, then $u_\nu * v \to u * v$ in $\mathscr{D}'$.

The proof, which is similar to the proofs of Theorem 2.5.1 and parts (v) and (vi) of Theorem 2.5.2, is left to the reader. □

Remark. This theorem can be generalized to the convolution of a finite number of distributions $u_1, \ldots, u_k$, which is defined by

$$(u_1 * \ldots * u_k, \phi) = (u_1(x_1) \otimes \ldots \otimes u_k(x_k), \phi(x_1 + \ldots + x_k)), \quad (2.5.9)$$

provided that the map

$$\operatorname{supp} u_1 \times \ldots \times \operatorname{supp} u_k \ni (x_1, \ldots, x_k) \to x_1 + \ldots + x_k \subset \mathbf{R}^n \quad (2.5.10)$$

is proper. Convolution is then commutative, distributive, and associative. However, the condition that (2.5.10) is to be a proper map is essential for associativity. This is shown by the following simple example. Take $n = 1$, and consider the three distributions $u(t) = 1$, $v(t) = H(t)$, the Heaviside function, and $w(t) = \delta'(t)$. By part (iii) of Theorem 2.5.2, one has $\delta' * f = f'$ for all $f \in \mathscr{D}'$. So $1 * \delta' = 0$ and $H * \delta' = H' = \delta$. Hence

$$(1 * \delta') * H = 0, \quad 1 * (\delta' * H) = 1 * \delta = 1.$$

An important case to which Theorem 2.5.5 can be applied is that of distributions in $\mathbf{R}$ whose supports are bounded on the left; this class is denoted by $\mathscr{D}'^+(\mathbf{R})$. So $u \in \mathscr{D}'^+(\mathbf{R})$ means that $u \in \mathscr{D}'(\mathbf{R})$, and that there is a number $a > -\infty$ such that $u(t) = 0$ for $t < a$. Let $v(t)$ be another distribution in $\mathscr{D}'^+$, such that $v(t) = 0$ for $t < b$. Then $s \in \operatorname{supp} u$ and $t \in \operatorname{supp} v$ implies that $s + t \geqslant a + b$, and if $c \geqslant a + b$, one has

$$a \leqslant s \leqslant c - b, \quad b \leqslant t \leqslant c - a.$$

So $u * v$ exists, and its support is contained in $[a+b, \infty)$. Another important example will be found in section 4.1.

2.6 Linear differential operators

Let $\{a^\alpha(x)\}_{|\alpha|\leqslant m}$ be a set of C^∞ functions, and u a distribution. By Theorems 2.2.1 and 2.2.2,

$$\mathscr{D}' \ni u \to Pu = \sum_{|\alpha|\leqslant m} a^\alpha \partial^\alpha u \qquad (2.6.1)$$

is a continuous linear map $\mathscr{D}' \to \mathscr{D}'$, which is called a *linear differential operator of order m*. If $u \in C^k$, $k \geqslant m$, then the $\partial^\alpha u$ in (2.6.1) are identical with the classical derivatives of u, and $u \to Pu$ is a map $C^k \to C^{k-m}$. Explicitly, the distribution Pu is

$$(Pu, \phi) = (u, {}^tP\phi), \ \phi \in C_0^\infty, \qquad (2.6.2)$$

where
$$ {}^tP\phi = \sum_{|\alpha|\leqslant m} (-1)^{|\alpha|} \partial^\alpha(a^\alpha\phi) \qquad (2.6.3)$$

is another mth order differential operator, called the (formal) *adjoint* of P. The adjoint of ${}^tP$ is P itself. For, if $u \in C_0^\infty$, then (2.6.2) becomes

$$\int \phi Pu \, dx = \int u \, {}^tP\phi \, dx.$$

Here, one can replace P by ${}^tP$, and interchange u and ϕ, to obtain

$$\int u \, {}^tP\phi \, dx = \int \phi \, {}^{tt}Pu \, dx.$$

Hence
$$\int (Pu - {}^{tt}Pu) \phi \, dx = 0$$

for all u, ϕ in C_0^∞. By Lemma 2.1.2, this implies that ${}^{tt}Pu = Pu$ for all $u \in C_0^\infty$, and so it follows by Theorem 2.5.4 that also ${}^{tt}Pu = Pu$ for all $u \in \mathscr{D}'$.

If f is a given distribution, then the equation $Pu = f$ is an mth order differential equation for distributions. If it happens that u and f are locally integrable functions, then $Pu = f$ in $\mathscr{D}'$ means, by (2.6.2), that

$$\int u \, {}^tP\phi \, dx = \int f\phi \, dx, \qquad (2.6.4)$$

for all $\phi \in C_0^\infty$. In this case, u is also called a *weak solution* of $Pu = f$. A function $u \in C^m$ that satisfies the equation $Pu = f$ in the usual sense will be called a *classical solution*. There is a simple relation between classical and weak solutions.

Theorem 2.6.1. *Let Ω be an open set. Suppose that $u \in C^m(\Omega)$ and that $Pu = f \in C^0(\Omega)$ in the usual sense. Then $Pu = f$ also holds in the distribu-*

tion sense in Ω. *Conversely, suppose that* u *and* f *are distributions, that* $Pu = f$, *and that the restrictions of* u *and* f *to* Ω *are respectively in* C^m *and in* C^0. *Then* u *is a classical solution of* $Pu = f$ *in* Ω.

PROOF. Note that $C^k(\Omega)$ denotes the class of C^k functions defined on Ω, and $C_0^k(\Omega)$ the class of functions in $C^k(\Omega)$ whose supports are compact subsets of Ω. If $u \in C^m(\Omega)$ and $\phi \in C_0^\infty(\Omega)$ then it follows, by repeated integration by parts, that

$$\int \phi Pu \, dx = \int u^t P\phi \, dx.$$

Hence $u \in C^m(\Omega)$, $f \in C^0(\Omega)$ and $Pu = f$ imply that $Pu = f$ holds also for the distributions u and f, restricted to Ω. Conversely, if u and f are distributions, and $Pu = f$, then, under the hypotheses of the theorem, (2.6.4) holds for all $\phi \in C_0^\infty(\Omega)$. As u is supposed to be in $C^m(\Omega)$, one can convert it into

$$\int (Pu - f)\, \phi \, dx = 0,$$

by repeated integration by parts. An obvious variant of Lemma 2.1.1 then implies, as $Pu - f$ is continuous, that $Pu = f$ in Ω. $\square$

The *Cauchy problem* in the classical theory of partial differential equations is to find a solution of an equation, say of order m, in a neighbourhood of a given hypersurface, such that the solution and its derivatives of all orders up to and including $m - 1$ assume prescribed values on this hypersurface. (These Cauchy data must of course satisfy the appropriate compatibility relations.) Such a problem can be replaced by a differential equation between distributions. We shall describe this procedure when $m = 2$, as this is essentially the only relevant case in this book. The differential operator can then be written in the following form, using the summation convention:

$$Pu = \partial_i(a^{ij} \partial_j u) + b^i \partial_i u + cu, \; a^{ij} = a^{ji} \quad (i, j = 1, \dots, n). \quad (2.6.5)$$

The adjoint of P is

$$^t P\phi = \partial_i(a^{ij} \partial_j \phi) - \partial_i(b^i \phi) + c\phi. \quad (2.6.6)$$

If all the b^i vanish identically, then $P = {}^t P$, and the differential operator P is called self-adjoint. (The definition of the adjoint will be modified when P acts on distributions in a space–time.) It follows from (2.6.5) and (2.6.6) that

$$\phi Pu - u^t P\phi = \partial_i(a^{ij}(\phi \partial_j u - u \partial_j \phi) + b^i u\phi). \quad (2.6.7)$$

Let us take the Cauchy hypersurface to be the hyperplane $x^n = 0$; the Cauchy problem is then often called an *initial value problem*. Write $x' = (x^1, \ldots, x^{n-1})$, so that $x = (x', x^n)$. Suppose also, for the moment, that $a^{nn} \neq 0$ on $x^n = 0$. Then a suitable pair of Cauchy data are the functions

$$u|_{x^n=0} = u_0(x'), \quad \partial_n u|_{x^n=0} = u_1(x'), \qquad (2.6.8)$$

as one has $\quad \partial_i u|_{x^n=0} = \partial_i u_0(x') \quad (i = 1, \ldots, n-1)$.

To convert the Cauchy problem into a distributional problem, we shall put

$$u^+ = u \text{ if } x^n \geqslant 0, \quad u^+ = 0 \text{ if } x^n < 0, \qquad (2.6.9)$$

and compute the distribution Pu^+. As u^+ is a function, this is

$$(Pu^+, \phi) = (u^+, P\phi) = \int_{x^n \geqslant 0} u^t P\phi \, dx.$$

So it follows from (2.6.7) and the divergence theorem that

$$(Pu^+, \phi) = \int_{x^n \geqslant 0} \phi Pu \, dx + \int_{x^n \geqslant 0} (a^{nj}(\phi \, \partial_j u - u \, \partial_j \phi) + b^n u \phi) \, dx',$$
$$(2.6.10)$$

where dx' is the Lebesgue measure on $x^n = 0$. This identity holds for all $u \in C^2$. Suppose now that u is a C^2 solution of $Pu = f$, and that it assumes the Cauchy data (2.6.8) on $x^n = 0$. Then, defining f^+ in the same way as u^+, one can write (2.6.10) as

$$Pu^+ = f^+ + g, \qquad (2.6.11)$$

where g is a distribution whose support is contained in $x^n = 0$ and which is determined by the Cauchy data,

$$(g, \phi) = \int_{x^n = 0} \left(a^{nn} u_1 \phi + \sum_{j=1}^{n-1} a^{nj} \phi \, \partial_j u_0 + (b^n \phi - a^{nj} \partial_j \phi) u_0 \right) dx'. \quad (2.6.12)$$

So the classical initial value problem has been converted into a partial differential equation for a distribution, in which the second member f of the original equation, and the Cauchy data, appear on the same footing. The initial data have in fact been replaced by layers of monopoles and dipoles on the initial hypersurface. There is a similar distribution form of the backward initial value problem, in $x^n \leqslant 0$. An analogous reduction can be carried out for the general mth order equation.

It has been assumed that $a^{nn}(x', 0) \neq 0$. If $a^{nn}(x' \, 0) = 0$, then the hyperplane $x^n = 0$ is a *characteristic* of P. The equation $Pu = f$ then becomes, on $x^n = 0$, a relation between u, $\partial_n u$, and their derivatives tangential to $x^n = 0$. Cauchy data can therefore not be prescribed on a characteristic. But there is a *characteristic initial value problem* in which only u is given on $x^n = 0$. This is also reduced to distribution form by (2.6.10); for as $a^{nn}(x', 0) = 0$, the integral over $x^n = 0$ on the right-hand side is determined by $u(x', 0)$ and its first order derivatives.

Suppose that P is a differential operator with constant coefficients. One then writes $P \equiv P(\partial)$, where $P(\partial)$ is the formal polynomial which results when, in $\xi \to P(\xi)$, the numbers ξ^α are replaced by the operators ∂^α. A distribution G is then called a *fundamental solution* of P if

$$P(\partial)\, G(x) = \delta(x). \qquad (2.6.13)$$

In the constant coefficient case, one has ${}^t P = P(-\partial)$, by (2.6.3). So (2.6.13) means that

$$(G, P(-\partial)\, \phi) = \phi(0), \qquad (2.6.14)$$

for all $\phi \in C_0^\infty$. If $f \in \mathscr{E}'$, then $G * f$ exists, and it follows from Theorem 2.5.2, (iii), that

$$P(\partial)\,(G * f) = P(\partial)\, G * f = \delta * f = f. \qquad (2.6.15)$$

Hence, if G is a fundamental solution of P, and f is a distribution with compact support, then $u = G * f$ is a solution of $P(\partial)\, u = f$. It must however be noted that the identity $G * P(\partial)\, v = v$, which would ensure uniqueness, need not be valid. Writing its first member as $G * P(\partial)\, \delta * v$ one sees that its validity depends on the associative law for convolutions, which may fail. In fact, the example given at the end of the last section is a simple instance of this. But it is clear that the construction of fundamental solution is an important step in the integration of the differential equation $P(\partial)\, u = f$.

For a linear differential operator $P \equiv P(x, \partial)$ with variable coefficients, a fundamental solution will be a distribution $G_y(x)$ which acts on test functions of x and is also a function of y, and satisfies the equation

$$P(x, \partial)\, G_y(x) = \delta(x - y). \qquad (2.6.16)$$

By (2.6.2), this means that

$$(G_y(x), {}^t P \phi(x)) = \phi(y), \qquad (2.6.17)$$

for all $\phi \in C_0^\infty$. Formally, a solution of $Pu = f \in \mathscr{D}'$ can be constructed, given a fundamental solution G_y, by setting

$$(u, \phi) = (f(y), (G_y(x), \phi(x))), \quad \phi \in C_0^\infty. \qquad (2.6.18)$$

For one then has, by (2.6.17),

$$(Py, \phi) = (u, {}^tP\phi) = (f, \phi).$$

Now $$C_0^\infty \ni \phi \to \psi(y) = (G_y(x), \phi(x)) \qquad (2.6.19)$$

maps test functions linearly to some space of functions of y. Also, $C_0^\infty \ni \psi \to (f, \psi) \in \mathbf{C}$ is a continuous linear map which in many cases can be extended to a larger class of functions. It is clear that (2.6.18) will only be valid if the composition of these two maps is a distribution, that is to say a continuous linear map $C_0^\infty \to \mathbf{C}$. Suppose, for example, that (2.6.19) is a continuous linear map $C_0^\infty \to C^\infty$. (Such maps can be characterized by semi-norm estimates.) Then (2.6.18) will be a solution of $Pu = f$ if $f \in \mathscr{E}'$. This will, in fact, be the case for the wave equations which are the subject of the present book.

2.7   Examples

In this section, some special distributions will be defined, and discussed briefly. These will be needed later, and will also serve to illustrate the theory.

Let us first take $n = 1$; the independent variable will then be denoted by t. Let λ be a complex number, and set

$$t_+^{\lambda-1} = t^{\lambda-1} \text{ if } t > 0, \quad t_+^{\lambda-1} = 0 \text{ if } t \leqslant 0. \qquad (2.7.1)$$

For $\operatorname{Re} \lambda > 0$, $t_+^{\lambda-1}$ is locally integrable, and is to be identified with the distribution

$$(t_+^{\lambda-1}, \phi(t)) = \int_0^\infty t^{\lambda-1}\phi(t)\, dt. \qquad (2.7.2)$$

As differentiation with respect to λ under the integral sign is permissible here, $\lambda \to (t_+^{\lambda-1}, \phi)$ is C^1 and satisfies the Cauchy–Riemann equations. It is therefore an analytic function of λ on $\{\lambda; \operatorname{Re}\lambda > 0\}$. It will now be shown that a distribution $(t_+^{\lambda-1}, \phi)$ can be defined, for all

$$\mathbf{C} \ni \lambda \neq 0, -1, \ldots,$$

by analytic continuation.

The reader is reminded that the principle involved is the following one. Let Ω and Ω_1 be open sets in $\mathbf{C}$, such that $\Omega \subset \Omega_1$, and suppose that Ω_1 is connected. Let $f(\lambda)$ and $f_1(\lambda)$ be analytic functions, defined on Ω and on Ω_1 respectively, and suppose that $f_1 = f$ on Ω. Then f_1 is an analytic continuation of f. The analytic continuation is unique. For the difference of two analytic continuations would vanish on Ω, and as Ω_1 is supposed to be connected, this would imply that they are also equal on Ω_1.

Let k be a positive integer, and suppose that $\operatorname{Re}\lambda > -k$. Then $t_+^{\lambda+k-1}$ is locally integrable, and

$$t_+^{\lambda-1} = \frac{1}{(\lambda+k+1)\dots\lambda}\left(\frac{d}{dt}\right)^k t_+^{\lambda+k-1} \qquad (2.7.3)$$

is a distribution; by (2.2.2), it is

$$(t_+^{\lambda-1}, \phi) = \frac{(-1)^k}{(\lambda+k-1)\dots\lambda}\int_0^\infty t^{\lambda+k-1}\phi^{(k)}(t)\,dt, \ \phi\in C_0^\infty. \qquad (2.7.4)$$

If $\operatorname{Re}\lambda > 0$, then this reduces to (2.7.2), by integration by parts. As the second member of (2.7.4) is analytic for $\operatorname{Re}\lambda > -k, \lambda \neq 0, -1, \dots, 1-k$, it is the required analytic continuation of $(t_+^{\lambda-1}, \phi)$. The integer k can be taken arbitrarily large, and so (2.7.4) defines $t_+^{\lambda-1}$ for all

$$\mathbf{C}\ni\lambda \neq 0, -1, \dots.$$

Another expression for $t_+^{\lambda-1}$ can be derived by writing (2.7.2) in the form

$$\int_0^1 t^{\lambda-1}\left(\phi(t) - \sum_{j=0}^{k-1}\frac{t^j}{j!}\phi^{(j)}(0)\right)dt + \sum_{j=0}^{k-1}\int_0^1\frac{t^{j+\lambda-1}}{j!}\phi^{(j)}(0)\,dt + \int_1^\infty t^{\lambda-1}\phi(t)\,dt,$$

whence

$$(t_+^{\lambda-1}, \phi) = \int_0^1 t^{\lambda-1}\left(\phi(t) - \sum_{j=0}^{k-1}\frac{t^j}{j!}\phi^{(j)}(0)\right)dt + \sum_{j=0}^{k-1}\frac{\phi^{(j)}(0)}{(\lambda+j)j!} + \int_1^\infty t^{\lambda-1}\phi(t)\,dt. \qquad (2.7.5)$$

By the mean value theorem, the integrand in the first term on the right-hand side is $O(t^{\lambda+k})$ when $t\downarrow 0$. Hence (2.7.5) is also valid for $\operatorname{Re}\lambda > -k, \lambda \neq 0, -1, \dots, 1-k$, and, by the uniqueness of the analytic continuation, is therefore the same distribution as (2.7.4)

Finally, $(t_+^{\lambda-1}\phi)$, is also the *finite part* of a divergent integral. Let ϵ be a positive number, and put

$$J_\epsilon = \int_\epsilon^\infty t^{\lambda-1}\phi(t)\,dt, \ \phi\in C_0^\infty.$$

When $\epsilon\downarrow 0$, J_ϵ diverges if $\operatorname{Re}\lambda < 0$. However, for $0 < \epsilon < 1$ one can write

$$J_\epsilon = \int_\epsilon^1 t^{\lambda-1}\left(\phi(t) - \sum_{j=1}^{k-1}\frac{t^j}{j!}\phi^{(j)}(0)\right)dt + \int_1^\infty t^{\lambda-1}\phi(t)\,dt$$

$$+ \sum_{j=0}^{k-1}\frac{\phi^{(j)}(0)}{(\lambda+j)j!}(1-\epsilon^{\lambda+j}).$$

Suppose that k is a positive integer, and that $-k < \mathrm{Re}\,\lambda < 1-k$. Then

$$J_\epsilon = F - \sum_{j=0}^{k-1} \frac{\phi^{(j)}(0)}{(\lambda+j)j!} \epsilon^{\lambda+j} + O(\epsilon^{\mathrm{Re}\,\lambda+k}),$$

where F, which is supposed to be independent of ϵ, is the finite part of the divergent integral. Comparison with (2.7.5) shows that

$$F = (t_+^{\lambda-1}, \phi).$$

So $(t_+^{\lambda-1}, \phi)$ can be computed by subtracting a suitable chosen 'fractional infinity' from J_ϵ, and then making $\epsilon \downarrow 0$:

$$(t_+^{\lambda-1}, \phi) = Pf \int_0^\infty t^{\lambda-1}\phi(t)\,dt$$

$$= \lim_{\epsilon \downarrow 0} \left(\int_0^\infty t^{\lambda-1}\phi(t)\,dt - \sum_{j=0}^{k-1} c_j \epsilon^{\lambda+j} \right). \qquad (2.7.6)$$

Here the c_j, which may be constants or C^∞ functions of ϵ, are to be chosen such that the limit exists; the limit is then unique.

The distribution $t_+^{\lambda-1}$ is not analytic at the points $\lambda = -k, k = 0, 1, \ldots$. It can be seen from (2.7.5) that these points are simple poles, and that the residue of $(t_+^{\lambda-1}, \phi(t))$ at $\lambda = -k$ is $\phi^{(k)}(0)/k!$. Hence

$$\mathop{\mathrm{res}}_{\lambda=-k} t_+^{\lambda-1} = \frac{(-1)^k}{k!} \delta^{(k)}(t). \qquad (2.7.7)$$

The gamma function is defined by Euler's integral

$$\Gamma(\lambda) = (\lambda-1)! = \int_0^\infty t^{\lambda-1}e^{-t}\,dt,$$

when $\mathrm{Re}\,\lambda > 0$. By arguing as we did for $(t_+^{\lambda-1}, \phi)$, one easily recovers the well-known result that $\Gamma(\lambda)$ is analytic except for simple poles at $\lambda = -k$, $k = 0, 1, \ldots$, with the residues $(-1)^k/k!$. It is clear that $t_+^{\lambda-1}/\Gamma(\lambda)$ is an entire function (analytic for all λ). Set

$$\left. \begin{aligned} E_\lambda(t) &= \frac{t_+^{\lambda-1}}{\Gamma(\lambda)} \quad (\lambda \neq 0, -1, \ldots), \\ E_k(t) &= \delta^{(k)}(t) \qquad (k = 0, 1, \ldots). \end{aligned} \right\} \qquad (2.7.8)$$

It then follows from (2.7.3) that

$$\frac{dE_\lambda}{dt} = E_{\lambda-1}, \qquad (2.7.9)$$

for all $\lambda \in \mathbf{C}$.

3

The distributions E_λ are in $\mathscr{D}'^+$, the class of distributions on the real line whose supports are bounded on the left. It was noted at the end of section 2.5 that the convolution of such distributions always exists. For $\operatorname{Re}\lambda > 0$, $\operatorname{Re}\mu > 0$, a well known identity between Euler integrals gives

$$E_\lambda * E_\mu = E_{\lambda+\mu}, \qquad (2.7.10)$$

and this holds for all complex numbers λ and μ, by analytic continuation. So the E_λ are a one-parameter group of convolution operators on $\mathscr{D}'^+$. If $\lambda = m$, a positive integer, and f is a locally integrable function whose support is bounded to the left, then $E_m * f$ is the m times iterated integral of f, from $-\infty$ to t. The $E_\lambda * f$ are therefore integrals of f of fractional order. For $0 < \operatorname{Re}\lambda < 1$ and $\mu = 1-\lambda$, (2.7.10), applied as a convolution operator to a C^1 function supported in the positive real line, gives the well-known solution of Abel's integral equation.

Another distribution can be defined by setting

$$t_-^{\lambda-1} = (-t)_+^{\lambda-1}. \qquad (2.7.11)$$

By (2.2.8), this means that

$$(t_-^{\lambda-1}, \phi(t)) = (t_+^{\lambda-1}, \phi(-t)).$$

So one has, for $\operatorname{Re}\lambda > 0$,

$$t_-^{\lambda-1} = 0 \text{ if } t \geqslant 0, \quad t_-^{\lambda-1} = |t|^{\lambda-1} \text{ if } t < 0. \qquad (2.7.12)$$

Alternatively, $t_-^{\lambda-1}$ can be defined from this by analytic continuation in λ. Again, $\lambda \to t_-^{\lambda-1}$ is analytic except for simple poles at $\lambda = -k$, $k = 0, 1, \ldots$. As $\delta^{(k)}(t)$ is even or odd according as k is even or odd, it follows from (2.7.7) that

$$\operatorname*{res}_{\lambda=-k} t_-^{\lambda-1} = \frac{1}{k!} \delta^{(k)}(t). \qquad (2.7.13)$$

One can now also put

$$\begin{rcases} |t|^{\lambda-1} = t_+^{\lambda-1} + t_-^{\lambda-1}, \\ |t|^{\lambda-1}\operatorname{sign}t = t_+^{\lambda-1} - t_-^{\lambda-1}. \end{rcases} \qquad (2.7.14)$$

By (2.7.7) and (2.7.13), $|t|^{\lambda-1}$ is an analytic function of λ except when $\lambda = 0, -2, \ldots$, as the other singularities cancel out. Similarly, $|t|^{\lambda-1}\operatorname{sign}t$ is analytic except when $\lambda = -1, -3, \ldots$. For the special values $\lambda = 0, -2, \ldots$ and $\lambda = -1, -3, \ldots, |t|^{\lambda-1}$ and $|t|^{\lambda-1}\operatorname{sign}t$ respectively can also be defined directly.

Consider, first, the derivative of $\log|t|$,

$$((\log|t|)', \phi) = -(\log|t|, \phi') = -\int \phi'(t) \log|t|\, dt.$$

One can also write this as

$$((\log|t|)', \phi) = -\lim_{\epsilon \downarrow 0} \left(\int_{-\infty}^{-\epsilon} + \int_{\epsilon}^{\infty} \phi'(t) \log|t| \, dt \right).$$

An integration by parts shows that the second member is

$$\lim_{\epsilon \downarrow 0} \left((\phi(-\epsilon) - \phi(\epsilon)) \log \epsilon + \left(\int_{-\infty}^{-\epsilon} + \int_{\epsilon}^{\infty} \right) \frac{\phi(t)}{t} \, dt \right).$$

Now $\phi(\epsilon) - \phi(-\epsilon) = O(\epsilon)$, and so the limit is equal to the Cauchy principal value (p.v.) of the (divergent) integral of $\phi(t)/t$,

$$\text{p.v.} \int \frac{\phi(t)}{t} \, dt = \lim_{\epsilon \downarrow 0} \left(\int_{-\infty}^{-\epsilon} + \int_{\epsilon}^{\infty} \right) \frac{\phi(t)}{t} \, dt = \int_{0}^{\infty} \frac{\phi(t) - \phi(-t)}{t} \, dt. \quad (2.7.15)$$

We shall adopt (2.7.15) as the definition of the distribution $1/t$, and therefore have the identity

$$(d/dt) \log|t| = 1/t.$$

It can be shown that $|t|^{\lambda-1} \operatorname{sign} t \to 1/t$ in $\mathscr{D}'$ when $\lambda \to 0$. Hence

$$t^{-m-1} = \frac{(-1)^m}{m!} \left(\frac{d}{dt} \right)^m t^{-1} = \frac{(-1)^m}{m!} \left(\frac{d}{dt} \right)^{m+1} \log|t|. \quad (2.7.16)$$

The second example is $r^{\lambda-n}$, where $r = |x|$, $x \in \mathbf{R}^n$, and λ is again a complex parameter. Some preliminary remarks on spherical means are needed. Let $\mathbf{S}^{n-1}$ denote the unit sphere in $\mathbf{R}^n$. If $\theta \in \mathbf{S}^{n-1}$, so that $|\theta| = 1$, then the $(n-1)$-form

$$d\omega_\theta = \sum_{i=1}^{n} (-1)^{i+1} \theta^i \, d\theta^1 \wedge \ldots \wedge \widehat{d\theta^i} \wedge \ldots \wedge d\theta^n \quad (2.7.17)$$

is the surface element on $\mathbf{S}^{n-1}$; the orientation of $\mathbf{S}^{n-1}$ determined by $d\omega_\theta > 0$ is the usual one, with the positive normal pointing outwards. The spherical mean of a function $\phi(x) \in C_0^\infty(\mathbf{R}^n)$ is

$$\check{\phi}(r) = \frac{1}{\omega_n} \int \phi(r\theta) \, d\omega_\theta \quad (r > 0), \quad (2.7.18)$$

where ω_n is the surface area of $\mathbf{S}^{n-1}$,

$$\omega_n = \int d\omega_\theta = 2\pi^{\frac{1}{2}n}/\Gamma(\tfrac{1}{2}n). \quad (2.7.19)$$

It is clear that $\check{\phi}(r) = 0$ for $r > A$ if $\phi(x) = 0$ for $|x| > A$. Also, for $k = 1, 2, \ldots,$

$$\check{\phi}^{(k)}(r) \equiv (d/dr)^k \check{\phi}(r) = \frac{1}{\omega_n} \int \left(\sum_{i=1}^{n} \theta^i \partial_i \right)^k \phi(x) \Big|_{x=r\theta} d\omega_\theta.$$

One can obviously define $\check{\phi}(0)$ and the $\check{\phi}^{(k)}(0)$ by continuity. It is immediate from (2.7.18) that $\check{\phi}(0) = \phi(0)$, and a simple symmetry argument shows that $\check{\phi}^{(k)}(0) = 0$ when k is odd. Hence $\check{\phi}(r)$, extended as an even function to $r < 0$, is in $C_0^\infty(\mathbf{R})$.

For $\operatorname{Re}\lambda > 0$, $r^{\lambda-n}$ is locally integrable, and can be identified with the distribution

$$(r^{\lambda-n}, \phi(x)) = \int r^{\lambda-n}\phi(x)\,dx.$$

As $dx = r^{n-1}\,dr\,d\omega_\theta$, one has, in view of (2.7.18),

$$(r^{\lambda-n}, \phi(x)) = \omega_n \int_0^\infty r^{\lambda-1}\check{\phi}(r)\,dr,$$

and this can be written as

$$(r^{\lambda-n}, \phi(x)) = \omega_n(t_+^{\lambda-1}, \check{\phi}(t)), \qquad (2.7.20)$$

with the understanding that, say, $\check{\phi}(t) = \check{\phi}(-t)$ for $t < 0$. (In the first member of (2.7.20), (,) refers to the duality between $\mathscr{D}'(\mathbf{R}^n)$ and $C_0^\infty(\mathbf{R}^n)$; in the second member, it is the duality between $\mathscr{D}'(\mathbf{R})$ and $C_0^\infty(\mathbf{R})$.) A semi-norm estimate shows that (2.7.20) defines a distribution in $\mathscr{D}'(\mathbf{R}^n)$ for all $\lambda \in \mathbf{C}$ except $\lambda = 0, -1, \dots$. Moreover, as the derivatives of odd order of $\check{\phi}$ vanish for $r = 0$, $r^{\lambda-n}$ has no poles at $\lambda = -1, -3, \dots$.

The residue of $r^{\lambda-n}$ at $\lambda = 0$ can be computed from (2.7.20), which gives

$$\lim_{\lambda\to 0}\lambda(r^{\lambda-n}, \phi(x)) = \omega_n\lim_{\lambda\to 0}\lambda(t_+^{\lambda-1}, \check{\phi}(t))$$

$$= \omega_n\check{\phi}(0),$$

by (2.7.7), with $k = 0$. As also $\check{\phi}(0) = \phi(0)$, it follows that

$$\operatorname*{res}_{\lambda=0} r^{\lambda-n} = \omega_n\delta(x). \qquad (2.7.21)$$

For $\operatorname{Re}\lambda > 0$, an elementary computation gives

$$\Delta r^{\lambda+2-n} \equiv (\partial_1^2 + \dots + \partial_n^2)r^{\lambda+2-n} = (\lambda+2-n)\lambda r^{\lambda-n}. \qquad (2.7.22)$$

This can be extended to the distribution case by analytic continuation. In particular, one can let $\lambda \to 0$, as the second member contains a factor λ. By (2.7.21), one then obtains

$$\Delta r^{2-n} = -(n-2)\omega_n\delta(x). \qquad (2.7.23)$$

(Recall that differentiation is a continuous operation in distribution theory.) For $n > 2$, (2.7.23) shows that r^{2-n} is, up to a constant factor, a fundamental solution of the Laplacian; for $n = 3$, one has $\omega_3 = 4\pi$, and so (2.7.23) reduces to the familiar identity

$$\Delta(1/r) = -4\pi\,\delta(x).$$

For $n = 2$, one can write (2.7.22) as

$$\Delta\frac{r^\lambda - 1}{\lambda} = \lambda r^{\lambda-2}.$$

It is easy to verify that $(r^\lambda - 1)/\lambda \to \log r$ in $\mathscr{D}'(\mathbf{R}^2)$. So one obtains, in the limit, the logarithmic potential, in the distributional form

$$\Delta(\log r) = 2\pi\,\delta(x) \quad x \in \mathbf{R}^2.$$

2.8   Distributions on a manifold

To prepare for the extension of the theory of distributions to manifolds, we must first consider distributions on an open set $\Omega \subset \mathbf{R}^n$ briefly. The function classes $C^k(\Omega)$, $C^\infty(\Omega)$, $C_0^k(\Omega)$ and $C_0^\infty(\Omega)$ have already been introduced, in connection with Theorem 2.6.1. A distribution on Ω is a linear form $C_0^\infty(\Omega) \ni \phi \to (u, \phi) \in \mathbf{C}$ such that, for every compact set $K \subset \Omega$, there are constants C and N with which a seminorm estimate of type (2.1.9),

$$|(u, \phi)| \leqslant C \sum_{|\alpha| \leqslant N} \sup |\partial^\alpha \phi|,$$

holds for all $\phi \in C_0^\infty(\Omega)$ with $\operatorname{supp} \phi \subset K$. The vector space of distributions on Ω is denoted by $\mathscr{D}'(\Omega)$; the weak topology will be used for this. A locally integrable function $f(x)$, defined on Ω, is identified with the distribution

$$(f, \phi) = \int f\phi\, dx, \tag{2.8.1}$$

where the integral is extended over Ω. The restriction of a distribution in $\mathscr{D}'(\mathbf{R}^n)$ to Ω is in $\mathscr{D}'(\Omega)$, but in general a distribution in $\mathscr{D}'(\Omega)$ cannot necessarily be extended to a distribution in $\mathscr{D}'(\mathbf{R}^n)$.

A considerable part of the theory developed so far carries over to $\mathscr{D}'(\Omega)$ when $C_0^\infty(\mathbf{R}^n)$ is replaced by $C_0^\infty(\Omega)$. In particular, differentiation, and multiplication by a function in $C^\infty(\Omega)$, are defined as before, and so is the tensor product of two distributions on open sets $\Omega \subset \mathbf{R}^n$ and $\Sigma \subset \mathbf{R}^m$. The identification of the linear sub-space of $\mathscr{D}'(\Omega)$ consisting of distributions with compact supports, and of the dual $\mathscr{E}'(\Omega)$ of $C^\infty(\Omega)$, also goes through as before. But the translate of a distribution, and the convolution of two distributions, cannot be defined. Nevertheless, Theorem 2.5.4, which asserts that every distribution is the limit, in the distribution topology, of a sequence of test functions, remains valid.

In fact, let $\{K_\nu\}_{1\leqslant\nu<\infty}$ be an expanding sequence of compact subsets of Ω whose union covers Ω, and let $\{\sigma_\nu\}_{1\leqslant\nu<\infty}$ be a sequence of functions in $C_0^\infty(\Omega)$ such that $\sigma_\nu = 1$ on a neighbourhood of K_ν. If $u\in\mathscr{D}'(\Omega)$, then $(u_\nu,\phi) = (u,\sigma_\nu\phi)$, $\phi\in C_0^\infty(\mathbf{R}^n)$, defines a distribution in $\mathscr{E}'(\mathbf{R}^n)$ whose restriction to Ω is equal to $u\sigma_\nu\in\mathscr{E}'(\Omega)$. Now let $\{\epsilon_\nu\}_{1\leqslant\nu<\infty}$ be a sequence of positive numbers, tending to zero, and such that ϵ_ν is less than the distance of supp σ_ν from $\mathbf{R}^n\setminus\Omega$. Suppose that $\psi\in C_0^\infty(\mathbf{R}^n)$ satisfies the conditions (2.1.4). Then the regularizations

$$\epsilon_\nu^{-n}u_\nu(x) * \psi(x/\epsilon_\nu)$$

are in $C_0^\infty(\Omega)$, and obviously converge to u in $\mathscr{D}'(\Omega)$, when $\nu\to\infty$.

We now turn to distributions on an n-dimensional manifold M. As the principal application envisaged here is to distributions on a space–time, it will be assumed from the outset that M is a paracompact orientable C^∞ manifold which is pseudo–Riemannian, or Riemannian. As explained in section 1.3, one can then introduce an invariant volume element μ. Let M be given an orientation; then μ is the exterior form of maximal degree which is equal to $|g(x)|^{\frac12}dx$ in every orientation-preserving coordinate chart. If M is given the opposite orientation, then μ must be replaced by $-\mu$. The integral of a locally integrable function f, defined on M, and with compact support, can therefore be defined to be $\int f\mu$, where μ is an invariant volume element, and M is to have the orientation determined by $\mu > 0$. The natural analogue of (2.8.1) is then

$$(f,\phi) = \int f\phi\mu, \qquad (2.8.2)$$

where f is locally integrable, and $\phi\in C_0^\infty(M)$. If (Ω,π) is a coordinate chart, and supp $\phi\subset\Omega$, then this becomes

$$(f,\phi) = \int f\circ\pi^{-1}(x)\,\phi\circ\pi^{-1}(x)\,|g(x)|^{\frac12}dx,$$

and this suggests the following definition.

Definition 2.8.1. *A linear form $C_0^\infty(M)\ni\phi\to(u,\phi)\in\mathbf{C}$ is a distribution if, for every coordinate chart (Ω,π), there is a distribution*

$$u\circ\pi^{-1}(x)\in\mathscr{D}'(\pi\Omega)$$

such that

$$(u, \phi) = (u \circ \pi^{-1}(x), \phi \circ \pi^{-1}(x) |g(x)|^{\frac{1}{2}}) \qquad (2.8.3)$$

for all $\phi \in C_0^\infty(\Omega)$. *A locally integrable function is to be identified with the distribution* (2.8.2). *The vector space of distributions on* M *is denoted by* $\mathscr{D}'(M)$.

Some comments on this definition are required. The first concerns coordinate transformations. Let (Ω, π) and $(\tilde{\Omega}, \tilde{\pi})$ be overlapping coordinate charts, and suppose that $\phi \in C_0^\infty(\Omega \cap \tilde{\Omega})$. Then it follows from (2.8.3) that

$$(u \circ \pi^{-1}(x), \phi \circ \pi^{-1}(x) |g(x)|^{\frac{1}{2}}) = (u \circ \tilde{\pi}^{-1}(\tilde{x}), \phi \circ \tilde{\pi}^{-1}(\tilde{x}) |\tilde{g}(\tilde{x})|^{\frac{1}{2}}), \quad (2.8.4)$$

and this is the transformation law for distributions on M. One can also express it in terms of the coordinate transformation, as follows. Set $\pi(\Omega \cap \tilde{\Omega}) = A$ and $\tilde{\pi}(\Omega \cap \tilde{\Omega}) = \tilde{A}$, so that A and $\tilde{A}$ are open sets in $\mathbf{R}^n$, and denote the inverse coordinate transformation $\pi \circ \tilde{\pi}^{-1}$: $\tilde{x} \to x$ by κ. Put.

$$u \circ \pi^{-1}(x) = v(x) \in \mathscr{D}'(A), \quad \phi \circ \pi^{-1}(x) |g(x)|^{\frac{1}{2}} = \psi(x) \in C_0^\infty(A).$$

Then

$$\phi \circ \tilde{\pi}^{-1}(\tilde{x}) |\tilde{g}(\tilde{x})|^{\frac{1}{2}} = \psi \circ \kappa(\tilde{x}) |Dx/D\tilde{x}|.$$

So (2.8.4) becomes

$$(v(x), \psi(x)) = (v \circ \kappa(\tilde{x}), \psi \circ \kappa(\tilde{x}) |Dx/D\tilde{x}|).$$

When v is a function, this is

$$\int_A v(x) \psi(x) \, dx = \int_{\tilde{A}} v \circ \kappa(\tilde{x}) \, \psi \circ \kappa(\tilde{x}) |Dx/D\tilde{x}| \, d\tilde{x},$$

which is, in effect, the rule for changing variables in a multiple integral.

In general, (u, ϕ) can be evaluated by means of a partition of unity $\{\phi_\nu\}$ subordinated to a covering $\{\Omega_\nu, \pi_\nu\}$ of M by coordinate charts. Then, if $\phi \in C_0^\infty(M)$, one has

$$\phi = \sum_\nu \phi\phi_\nu,$$

where only a finite number of terms are non-zero, as the supports of the ϕ_ν are, by definition, a locally finite covering of M. One can then put

$$(u, \phi) = \sum_\nu (u \circ \pi_\nu^{-1}, (\phi\phi_\nu) \circ \pi_\nu^{-1} |g_\nu|^{\frac{1}{2}}), \qquad (2.8.5)$$

where g_ν denotes the determinant of the components of the metric tensor in the chart (Ω_ν, π_ν). It follows from (2.8.4) that the right-hand side of (2.8.5) has the same value for all partitions of unity subordinated to coverings of M by coordinate charts. The proof will be omitted, as it is virtually identical with the proof of the corresponding result for integrals, which was given in section 1.3.

The second comment on Definition 2.8.1. concerns semi-norm estimates. Let (Ω_ν, π_ν) be a fixed covering of M by coordinate charts, and let $\{\phi_\nu\}$ be a partition of unity subordinated to this covering. For $\phi \in C_0^\infty(M)$, set

$$\|\phi\|_k = \sum_\nu \sum_{|\alpha| \leqslant k} \sup |\partial^\alpha(\phi\phi_\nu)|,$$

where k is a non-negative integer, and the $\partial^\alpha(\phi\phi_\nu)$ are computed in the appropriate local coordinate system. Then it follows from Definition 2.8.1 and (2.8.5) that, if $u \in \mathscr{D}'(M)$ and K is a compact set, then there are constants C and N such that

$$|(u, \phi)| \leqslant C \|\phi\|_N \qquad (2.8.6)$$

for all $\phi \in C_0^\infty(M)$ whose supports are contained in K. For a different partition of unity, a similar estimate is obtained, with a different C but the same N. Conversely, a linear form on (u, ϕ) which satisfies the transformation law (2.8.4), and for which there is a semi-norm estimate of the type (2.8.6) for every compact set, is a distribution.

Finally, we note that convergence in $C^\infty(M)$ and in $C_0^\infty(M)$ can also be defined. Choose a fixed partition of unity, and define semi-norms for $C^\infty(M)$ by

$$\|\phi\|_{k, K} = \sum_\nu \sum_{|\alpha| \leqslant k} \sup_{\Omega_\nu \cap K} |\partial^\alpha(\phi\phi_\nu)|,$$

where k is a non-negative integer, and K is a compact set. Then a sequence $\{\phi_\nu\}_{1 \leqslant \nu < \infty}$ of C^∞ functions is said to tend to zero in $C^\infty(M)$ if $\|\phi_\nu\|_{k, K} \to 0$, when $\nu \to \infty$, for all integers $k \geqslant 0$ and all compact sets K. Similarly, if the ϕ_ν are in $C_0^\infty(M)$, then they are said to tend to zero in $C_0^\infty(M)$ if they do so in $C^\infty(M)$, and their supports are contained in a fixed compact set. One can then deduce from Theorem 2.1.1 that a linear form (u, ϕ) on $C_0^\infty(M)$ is a distribution if and only if $(u, \phi_\nu) \to 0$ when $\phi_\nu \to 0$ in $C_0^\infty(M)$. There is a similar characterization of $\mathscr{E}'(\Omega)$. An equivalent criterion for convergence to zero in $C^\infty(M)$ is that, for every relatively compact coordinate chart (Ω, π), the $\partial_\alpha(\phi_\nu \circ \pi^{-1})$ are to tend to zero uniformly on Ω when $\nu \to \infty$, for each multi-index α.

The distributions in $\mathscr{D}'(M)$ are scalar distributions; tensor distributions can be defined in the same way. First, one can introduce the scalar product of two tensors T and ϕ, of types (r, s) and (s, r), respectively, by setting

$$\langle T, \phi \rangle = T^{i_1 \cdots i_r}{}_{j_1 \cdots j_s} \phi^{j_1 \cdots j_s}{}_{i_1 \cdots i_r} \qquad (2.8.7)$$

in local coordinates. At a point $p \in M$, the second member is an invariant; the extension of (2.8.7) to fields follows in the usual way by means of a partition of unity. Now, let $\mathscr{D}^{(s,r)}(M)$ denote the class of C^∞ tensor fields with compact supports. A locally integrable tensor field T of type (r, s) can be identified with the linear form

$$(T, \phi) = \int \langle T, \phi \rangle \mu, \quad \phi \in \mathscr{D}^{(s,r)}(M). \tag{2.8.8}$$

This immediately suggests the following definition.

Definition 2.8.2. *A tensor distribution of type (r, s) on M is a linear form on $\mathscr{D}^{(s,r)}(M)$ such that, for every coordinate chart (Ω, π), there is a collection of distributions*

$$(T \circ \pi^{-1})^{i_1 \cdots i_r}{}_{j_1 \cdots j_s} \in \mathscr{D}'(\pi \Omega)$$

such that

$$(T, \phi) = ((T \circ \pi^{-1})^{i_1 \cdots i_r}{}_{j_1 \cdots j_s}, (\phi \circ \pi^{-1})^{j_1 \cdots j_s}{}_{i_1 \cdots i_r} |g(x)|^{\frac{1}{2}})$$

for all $\phi \in \mathscr{D}^{(s,r)}(\Omega)$; a locally integrable tensor field T is to be identified with the distribution (2.8.8). The components of T transform according to the usual tensor transformation laws. The vector space of tensor distributions of type (r, s) is denoted by $\mathscr{D}'^{(r,s)}(M)$.

It may be added that the tensor sub- and superscripts of the components of tensor distributions may be raised and lowered in the usual way.

In the sequel, classical notation will be used frequently, as for functions. The definitions which have been given reduce distribution theory on a manifold, in effect, to the case of open sets in $\mathbf{R}^n$. The only additional comment needed concerns differentiation, which must be replaced by covariant differentiation. It is obviously sufficient to work with test fields whose supports are contained in a coordinate neighbourhood. Then (2.8.8) becomes, in classical notation,

$$(T, \phi) = \int T^{i_1 \cdots i_r}{}_{j_1 \cdots j_s} \phi^{j_1 \cdots j_s}{}_{i_1 \cdots i_r} |g|^{\frac{1}{2}} dx.$$

Suppose that $T \in C^1(M)$. Then the covariant derivative ∇T of T is the distribution

$$(\nabla T, \phi) = \int (\nabla_i T^{i_1 \cdots i_r}{}_{j_1 \cdots j_s}) \phi^{i j_1 \cdots j_s}{}_{i_1 \cdots i_r} |g|^{\frac{1}{2}} dx,$$

where $\phi \in \mathscr{D}^{(s+1,r)}$ as ∇T is of type $(r, s+1)$. Now

$$(\nabla_i T^{i_1 \cdots i_r}{}_{j_1 \cdots j_s}) \phi^{i j_1 \cdots j_s}{}_{i_1 \cdots i_r} + T^{i_1 \cdots i_r}{}_{j_1 \cdots j_s} \nabla_i \phi^{i j_1 \cdots j_s}{}_{i_1 \cdots i_r}$$

$$= \nabla_i (T^{i_1 \cdots i_r}{}_{j_1 \cdots j_s} \phi^{i j_1 \cdots j_s}{}_{i_1 \cdots i_r}) = |g|^{-\frac{1}{2}} \partial_i (|g|^{\frac{1}{2}} T^{i_1 \cdots i_r}{}_{j_1 \cdots j_s} \phi^{i j_1 \cdots j_s}{}_{i_1 \cdots i_r}).$$

As ϕ has compact support, the integral of this expression, multiplied by $\mu = |g|^{\frac{1}{2}}\,dx$, vanishes, and one obtains

$$(\nabla T, \phi) = \int \langle \nabla T, \phi \rangle \mu = \int \langle T, \delta\phi \rangle \mu, \qquad (2.8.9)$$

where $\delta\phi$, which is called the *coderivative* of ϕ, is given in local coordinates by

$$(\delta\phi)^{j_1 \cdots j_s}{}_{i_1 \cdots i_r} = -\nabla_i \phi^{ij_1 \cdots j_s}{}_{i_1 \cdots i_r}. \qquad (2.8.10)$$

Note that the operator δ converts a tensor field of type $(r+1, s)$ into one of type (r, s). It is evident that δ is defined invariantly by (2.8.9); in fact, it is the adjoint of the covariant differentiation operator ∇ with respect to the scalar product (T, ϕ).

The identity (2.8.9) can now be used to define the covariant derivative of a tensor distribution,

$$(\nabla T, \phi) = (T, \delta\phi), \quad T \in \mathcal{D}'^{(r,s)}, \quad \phi \in \mathcal{D}^{(s+1,r)}. \qquad (2.8.11)$$

It is also easy to show that the adjoint of δ is ∇,

$$(\delta T, \phi) = (T, \nabla\phi). \qquad (2.8.12)$$

In the sequel, we shall be mainly concerned with scalar and vector distributions. If $u \in \mathcal{D}'(M)$ then one has, exactly as for functions,

$$\nabla u = \operatorname{grad} u, \qquad (2.8.13)$$

which means that the components of ∇u in local coordinates are (distribution) derivatives $\partial_i(u \circ \pi^{-1})$. For it follows from (2.8.10) and (2.8.11) that

$$(\nabla u, \phi) = -(u, \operatorname{div}\phi). \qquad (2.8.14)$$

The divergence of a vector distribution $\xi \in \mathcal{D}'^{(1,0)}$ is

$$(\operatorname{div}\xi, \phi) = -(\xi, \operatorname{grad}\phi), \quad \phi \in C_0^\infty(M). \qquad (2.8.15)$$

By combining (2.8.13) and (2.8.15), one obtains the distributional form of the second-order operator (1.1.25). This will again be called the Laplacian of u if M is a Riemannian manifold, and then denoted by Δ. If M is Lorentzian, so that it is a space–time, then the operator is the d'Alembertian, and is denoted by $\square$, so that, in local coordinates,

$$\square u = \nabla_i \nabla^i u = |g|^{-\frac{1}{2}} \partial_i(|g|^{\frac{1}{2}} g^{ij} \partial_j u). \qquad (2.8.16)$$

It follows from (2.8.14) and (2.8.15) that this is a self-adjoint operator,

$$(\square u, \phi) = (u, \square\phi). \qquad (2.8.17)$$

One can form a more general second-order differential operator on M,

$$Pu = \square u + \langle a, \nabla u \rangle + bu, \qquad (2.8.18)$$

where a is a C^∞ vector field and b is a C^∞ scalar field. The adjoint of P must now be defined with respect to the pairing of $\mathscr{D}'(M)$ and $C_0^\infty(M)$ and will again be denoted by ^tP, so that $(Pu, \phi) = (u, {}^tP\phi)$. A simple calculation in local coordinates shows that

$$^tP\phi = \square \phi - \operatorname{div}(a\phi) + b\phi. \qquad (2.8.19)$$

Finally, a fundamental solution of P is a distribution G_q such that

$$PG_q = \delta_q. \qquad (2.8.20)$$

Here, δ_q denotes the Dirac delta function on M. Let $q \in M$ be a fixed point; then δ_q is, by definition,

$$(\delta_q(p), \phi(p)) = \phi(q), \quad \phi \in C_0^\infty(M). \qquad (2.8.21)$$

So (2.8.20) means that

$$(G_q(p), {}^tP\,\phi(p)) = \phi(q), \qquad (2.8.22)$$

for all $\phi \in C_0^\infty(M)$.

2.9   A special type of distribution

In the applications of distribution theory which will be made later on, distributions will not usually be defined globally on a manifold M, but only locally, on an open set $\Omega \subset M$. As such a set can be considered as a manifold, with the differentiable structure, metric, and invariant volume element inherited from M, no further discussion of this situation is required.

In this section, we shall consider a special type of distribution, which will play an important part in the theory of wave equations on a space–time. Let $\Omega \subset M$ be an open set, let $S(p)$ be a real-valued function defined on Ω, and let $f(t)$ be a function on $\mathbf{R}$. Then the composite function $f(S(p)) = f \circ S(p)$ is a function on Ω. The distributions on Ω which we propose to discuss are obtained when f is replaced by a distribution in $\mathscr{D}'(\mathbf{R})$. We first derive some auxiliary results.

Lemma 2.9.1. *Suppose that $S(p) \in C^\infty(\Omega)$, and that $\operatorname{grad} S \neq 0$ on Ω. Then every point of Ω has a neighbourhood in which there exist local coordinates x^i such that $x^1 = S$.*

PROOF. Let q be a point of Ω, and let $\tilde{x}$ be a local coordinate system defined in some neighbourhood of q. By hypothesis, at least one of the derivatives $\partial S/\partial \tilde{x}^i, i = 1, ..., n$, does not vanish at q; without loss of generality, one can assume that it is $\partial S/\partial \tilde{x}^1$. Then it follows from the inverse function theorem that there is a connected neighbourhood of q in which the map

$$\tilde{x} \to x = (S(\tilde{x}), \tilde{x}^2, ..., \tilde{x}^n)$$

is a diffeomorphism, and so the x^i are local coordinates at q which have the required property. $\square$

It follows from the hypothesis that $\operatorname{grad} S \neq 0$ on Ω, that the sets

$$S_t = \{p; \, p \in \Omega, S(p) = t \in \mathbf{R}\} \qquad (2.9.1)$$

are hypersurfaces (sub-manifolds of Ω of codimension $n-1$) unless they are empty. The next lemma introduces an $(n-1)$-form which can be used as an invariant hypersurface element on these hypersurfaces.

Lemma 2.9.2. *Suppose that* $S \in C^{\infty}(\Omega)$ *and that* $\operatorname{grad} S \neq 0$ *on* Ω. *Then there exists an* $(n-1)$-*form* μ_S *such that*

$$dS \wedge \mu_S = \mu. \qquad (2.9.2)$$

The restriction of μ_S *to an* S_t *is unique.*

PROOF. We first show that there exist C^{∞} vector fields v on Ω such that $\langle v, \operatorname{grad} S \rangle \neq 0$. A C^{∞} manifold always admits a Riemannian metric. Let h be such a metric, and set, in local coordinates, $v^i = h^{ij} \partial_j S$. Then

$$\langle v, \operatorname{grad} S \rangle = h^{ij} \partial_i S \partial_j S > 0,$$

as the matrix (h^{ij}) is positive definite and, by hypothesis, $\operatorname{grad} S \neq 0$ on Ω.

The $(n-1)$-form $*v$ was defined, in local coordinates, by (1.3.18); assuming that Ω has been given a canonical orientation, and that the coordinate system is orientation preserving, it is

$$*v = \sum_{i=1}^{n} (-1)^{i+1} |g|^{\frac{1}{2}} v^i \, dx^1 \wedge ... \wedge \widehat{dx^i} \wedge ... \wedge dx^n. \qquad (2.9.3)$$

Hence $\qquad\qquad\qquad dS \wedge *v = \langle v, \operatorname{grad} S \rangle \mu, \qquad (2.9.4)$

and as $\langle v, \operatorname{grad} S \rangle \neq 0$, the form $\mu_S = *v/\langle v, \operatorname{grad} S \rangle$ satisfies (2.9.2).

To show that the restriction $\mu_S|_{S_t}$ is unique, we note that this is a local property, so that it is sufficient to prove it in a coordinate system in which $x^1 = S$, by Lemma 2.9.1. Let μ_S' be another form which satisfies (2.9.2), and put

$$\mu_S' - \mu_S = \sum_{i=1}^{n} \alpha^i \, dx^1 \wedge \dots \wedge \widehat{dx^i} \wedge \dots \wedge dx^n.$$

As $S = x^1$, it follows that

$$0 = dS \wedge (\mu_S' - \mu_S) = \alpha^1 \, dx^1 \wedge \dots \wedge dx^n,$$

whence $\alpha^1 = 0$. But on $S_t = \{x; x^1 = t\}$,

$$(\mu_S' - \mu_S)|_{S_t} = \alpha^1 \, dx^2 \wedge \dots \wedge dx^n = 0,$$

and so the lemma is proved. $\square$

Note that it follows from (2.9.3) that

$$*v|_{S_t} = \langle v, \operatorname{grad} S \rangle \mu_S|_S \tag{2.9.5}$$

holds for all vector fields v on Ω.

Lemma 2.9.3. *Let ψ be a function on Ω which is locally integrable and has compact support. Then*

$$\int \psi\mu = \int dt \int_{S_t} \psi\mu_S, \tag{2.9.6}$$

and, for almost all t,

$$\frac{d}{dt} \int_{S<t} \psi\mu = \int_{S_t} \psi\mu_S, \tag{2.9.7}$$

where, in each case, the orientation of S_t is determined by $\mu_S > 0$.

PROOF. By Lemma 2.9.1, there is a covering of Ω by open coordinate neighbourhoods in each of which $S = x^1$. As the integral of ψ can be evaluated by means of a partition of unity subordinated to this covering, it is sufficient to prove the lemma when the support of ψ is contained in such a neighbourhood. Then $\mu = |g|^{\frac{1}{2}} dx^1 \wedge \dots \wedge dx^n$, and so one can take $\mu_S = |g|^{\frac{1}{2}} dx^2 \wedge \dots \wedge dx^n$. Hence (2.9.6) follows from Fubini's theorem, which also shows that

$$t \to F(t) = \int_{S_t} \psi\mu_S$$

exists for almost all t, and is integrable. Now

$$\int_{S<t} \psi\mu = \int \psi\chi\mu = \int^t F(t') \, dt',$$

where χ is the characteristic function of $\{p; p \in \Omega, S(p) < t\}$. So (2.9.7) follows from the well-known fact that the derivative of the indefinite integral of an integrable function $F(t)$ is almost everywhere equal to $F(t)$. $\square$

An exterior form μ_S that satisfies (2.9.2) will be called a *Leray form*. It can be seen from (2.9.6) that the restriction $\mu_S|_{S_t}$ has a simple geometric meaning. For example, in $\mathbf{R}^n$ it is the area of the base of a volume element that is bounded by two hypersurfaces S_t and $S_{t+\delta t}$, and by a hypercylinder normal to S_t. In this case, one also has

$$\mu_S|_{S_t} = d\sigma/|\text{grad } S|,$$

where $d\sigma$ is the Euclidean hypersurface element.

Now suppose that $f(t)$ is a locally integrable function on $\mathbf{R}$. Then $\Omega \ni p \to f(S(p)) = f \circ S(p)$ is a locally integrable function on Ω. It can be identified with a distribution which is, by (2.9.6),

$$(f(S), \phi) = \int f(S(p))\, \phi(p)\, \mu(p) = \int f(t)\, dt \int_{S_t} \phi(p)\, \mu_S(p), \quad (2.9.8)$$

where $\phi \in C_0^\infty(\Omega)$. The extension of this identity to distributions $f(t) \in \mathscr{D}'(\mathbf{R})$ is immediate.

Theorem 2.9.1. *Let $\Omega \subset M$ be an open set, and suppose that*

$$S(p) \in C_0^\infty(\Omega)$$

is real valued, and that $\text{grad } S \neq 0$ *on* Ω. *Let* $f(t) \in \mathscr{D}'(\mathbf{R})$. *Then*

$$(f(S), \phi) = \left(f(t), \int_{S_t} \phi(p)\, \mu_S(p) \right), \quad \phi \in C_0^\infty(\Omega), \quad (2.9.9)$$

is a distribution, which is equal to the composite function $f(S)$ *when* $f(t)$ *is a function. Also, if* $f_\nu(t) \to f(t)$ *in* $\mathscr{D}'(\mathbf{R})$ *when* $\nu \to \infty$, *then* $f_\nu(S) \to f(S)$ *in* $\mathscr{D}'(\Omega)$.

PROOF. Suppose first that the support of ϕ is contained in a coordinate neighbourhood $\Omega' \subset \Omega$ in which $S = x^1$. Then

$$\tilde{\phi}(t) = \int_{S_t} \phi(p)\, \mu_S(p) = \int \phi(t, x^2, \ldots, x^n) |\, g(t, x^2, \ldots, x^n) |^{\frac{1}{2}} dx^2 \ldots dx^n.$$

It is evident that $\tilde{\phi}(t) \in C_0^\infty(\mathbf{R})$, and also that, if $\phi_\nu \in C_0^\infty(\Omega')$ and $\sigma_\nu \to 0$ in $C_0^\infty(\Omega')$, then $\tilde{\phi}_\nu(t) \to 0$ in $C_0^\infty(\mathbf{R})$. Hence the second member of (2.9.8) is a distribution. The general case now follows from Lemma 2.9.1, as every compact sub-set of Ω can be covered by a finite number of neighbourhoods of this type. The identification of the distribution

$f(S)$ and the composite function $f \circ S$, when $f(t) \in L_1^{\mathrm{loc}}(\mathbf{R})$, is an immediate consequence of (2.9.8). Finally, as $\tilde{\phi}(t) \in C_0^\infty(\mathbf{R})$, it is clear that $f_\nu(t) \to f(t)$ in $\mathscr{D}'(\mathbf{R})$ implies that $f_\nu(S) \to f(S)$ in $\mathscr{D}'(\Omega)$. $\square$

Distributions of this type can be differentiated by the chain rule.

Theorem 2.9.2. *For a distribution $f(S)$, as defined in Theorem 2.9.1, one has*

$$\nabla f(S) = f'(S)\,\nabla S. \tag{2.9.10}$$

PROOF. This can be established by regularization, or directly, as follows. Let $\xi \in \mathscr{D}^{(1,0)}(\Omega)$ be a test vector field. Then

$$(\nabla f(S), \xi) = -(f(S), \operatorname{div} \xi) = -\left(f(t), \int_{S_t} (\operatorname{div} \xi)\,\mu_S\right).$$

Now it follows from (2.9.7) that

$$\int_{S_t} (\operatorname{div} \xi)\,\mu_S = \frac{d}{dt} \int_{S \leqslant t} (\operatorname{div} \xi)\,\mu.$$

This can be evaluated by the divergence theorem; using (2.9.5), one finds

$$\int_{S \leqslant t} (\operatorname{div} \xi)\,\mu = \int_{S_t} * \xi = \int_{S_t} \langle \xi, \operatorname{grad} S \rangle\,\mu_S,$$

as the orientation determined by $\mu_S > 0$ is also the induced orientation of the boundary of $\{p;\ p \in \Omega,\ S(p) \leqslant t\}$. Hence

$$(\nabla f(S), \xi) = -\left(f(t), \frac{d}{dt} \int_{S_t} \langle \xi, \operatorname{grad} S \rangle\,\mu_S\right)$$

$$= \left(f'(t), \int_{S_t} \langle \xi, \operatorname{grad} S \rangle\,\mu_S\right),$$

which is $(f'(S)\,\nabla S, \xi)$, by (2.9.10) extended to covector distributions. So the theorem is proved. $\square$

Leray forms and their integrals can also be defined for m functions on Ω, where $m < n$. It will be sufficient to treat the case $m = 2$. Let S and T be C^∞ functions on Ω, and suppose that, for all real numbers s and t, the sets $\{p;\ p \in \Omega,\ S(p) = s,\ T(p) = t\}$ are either sub-manifolds of Ω, or empty. The necessary and sufficient condition for this is that the forms dS, dT and $dS \wedge dT$ must be non-zero at every point of Ω. Then there exist $(n-2)$-forms $\mu_{S,T}$ such that

$$dS \wedge dT \wedge \mu_{S,T} = \mu. \qquad (2.9.11)$$

Note that $dT \wedge \mu_{S,T}$ is a Leray form for S.

As in the proof of Lemma 2.9.2, one can show that the restriction of such a $\mu_{S,T}$ to a sub-manifold

$$\Sigma_{s,t} = \{p; \, p \in \Omega, S(p) = s \in \mathbf{R}, T(p) = t \in \mathbf{R}\} \qquad (2.9.12)$$

is unique. The integral of $\psi \mu_{S,T}$, where ψ is (say) continuous and has compact support, over $\Sigma_{s,t}$, is then defined by the rule that $\Sigma_{s,t}$ is to be oriented by $\mu_{S,T} > 0$.

It is obvious that if $\mu_{T,S}$ is defined in the same way, then one has

$$\mu_{S,T} = -\mu_{T,S}, \qquad (2.9.13)$$

when $S(p)$ and $T(p)$ are constant. However, the orientation rule shows that $\Sigma_{T,S} = -\Sigma_{S,T}$, which is $\Sigma_{S,T}$ with the orientation reversed. Hence

$$\int_{\Sigma_{s,t}} \psi \mu_{S,T} = \int_{\Sigma_{t,s}} \psi \mu_{T,S} = \frac{\partial}{\partial s} \frac{\partial}{\partial t} \int_{S<s,\,T<t} \psi \mu. \qquad (2.9.14)$$

Given a distribution $f(s,t) \in \mathscr{D}'(\mathbf{R}^2)$, one can now define a distribution $f(S,T) \in \mathscr{D}'(\Omega)$ by setting

$$(f(S,T), \phi) = \left(f(s,t), \int_{\Sigma_{s,t}} \phi(p)\mu_{S,T}(p)\right), \quad \phi \in C_0^\infty(\Omega). \qquad (2.9.15)$$

The properties of $f(S,T)$ are similar to those of $f(S)$ (sequential continuity, and differentiation by the chain rule); the verification is straightforward, and is left to the reader.

NOTES

The basic reference for distribution theory is Schwartz (1966), who also gives a brief history of the antecedents of the theory. See also Chapter I of Hörmander (1963) and Yosida (1971).

Section 2.7. The method of analytic continuation is due to M. Riesz (1949); a detailed exposition, in terms of distributions, will be found in Gelfand and Shilov (1964). The introduction of the finite part of a divergent integral was a key step in Hadamard's theory. In his essay 'The psychology of invention in the mathematical field' (Hadamard (1954) p. 100n), he writes 'I could not avoid it any more than the prisoner in Poe's tale *The Pit and the Pendulum* could avoid the hole at the center of his cell.'

Section 2.8. The reader's attention is drawn to the remarks on pp. 338–9 of Schwartz (1966) on the different ways of defining distributions on manifolds. The treatment of vector and tensor distributions follows Lichnérowicz (1961).

Section 2.9. Leray forms were introduced by Leray (1952); they are used extensively in Chapter 3 of Gelfand and Shilov (1964). See also pp. 61 and 143 of Hadamard (1923).

3

Characteristics and the propagation of discontinuities

We now turn to the theory of wave equations on a space–time. A *space–time* will be, by definition, an n-dimensional orientable C^∞ manifold with a Lorentzian metric. The signature of the metric will always be taken to be $2 - n$. It is obvious that a pseudo–Riemannian manifold M will be a space–time if and only if there is a coordinate chart at every point $p \in M$ such that

$$g_{ij}|_p = \eta_{ij} = \begin{cases} 1, & \text{if } i = j = 1, \\ -1, & \text{if } i = j = 2, \dots, n, \\ 0, & \text{if } i \neq j. \end{cases}$$

Such a chart will be called *Minkowskian* at p. In a Minkowskian chart, the scalar product of two tangent vectors ξ and η at p has the form familiar from special relativity,

$$\langle \xi, \eta \rangle = \xi^1 \eta^1 - \xi^2 \eta^2 - \dots - \xi^n \eta^n.$$

In physics, the important case is, of course, $n = 4$. For the present, no material simplification results from this choice, and it will only be assumed that $n > 2$.

In a flat space–time, with the usual metric, which is Minkowskian at all points, one has

$$\Box = \partial_1^2 - \partial_2^2 - \dots - \partial_n^2,$$

so that $\Box u = 0$ is the *ordinary wave equation*. It is therefore natural to call an equation of the form

$$Pu = \Box u + \langle a, \nabla u \rangle + bu = f,$$

on a general space–time, where a is a C^∞ vector field and b is a C^∞ scalar field, a (scalar) *wave equation*; tensor wave equations will be defined in section 3.2. This chapter deals with the theory of the characteristics of wave equations, which is developed in the first four sections. The remainder of the chapter contains some simple results on the propagation of discontinuities, and an account of the progressing wave formalism which links the wave equation with geometrical optics.

3.1 Lorentzian geometry

Let M be a space–time, and let p be a point of M. The tangent vectors ξ at p can be divided into three types, according to the sign of $\langle \xi, \xi \rangle$. A vector $\xi \in TM_p$ is called *time-like* if $\langle \xi, \xi \rangle > 0$, *null* if $\xi \neq 0$ and $\langle \xi, \xi \rangle = 0$, and *space-like* if $\langle \xi, \xi \rangle < 0$. (No type will be assigned to the zero vector.) The set of all null vectors will be called the *tangent null cone* at p; the time-like vectors are its interior, and the space-like vectors are its exterior.

In a coordinate chart that is Minkowskian at p, $\langle \xi, \xi \rangle > 0$ splits into the two inequalities

$$\xi^1 > ((\xi^2)^2 + \ldots + (\xi^n)^2)^{\frac{1}{2}}, \quad \xi^1 < -((\xi^2)^2 + \ldots + (\xi^n)^2)^{\frac{1}{2}}.$$

Hence the interior of the tangent null cone consists of two components. One can introduce a *time orientation* at p by choosing a time-like vector θ (the 'arrow of time') and calling all vectors that are time-like and in the same component of the interior of the tangent null cone as θ *future-directed*; the other time-like vectors are *past-directed*. This time orientation extends in the obvious way to null vectors. It is easily proved that each of the components of the interior of the tangent null cone is convex. In fact, one has more: if ξ and η are, say, future-directed and time-like, and if $A \geqslant 0$, $B \geqslant 0$, $A + B > 0$, then $A\xi + B\eta$ is also future-directed and time-like.

A continuous vector field $v(p)$, defined on a connected open set $\Omega \subset M$, will be called time-like on Ω if it is time-like at every point $p \in \Omega$; note that this implies that $v(p) \neq 0$ on Ω. When such a vector field exists, one can give Ω a time-orientation, by taking $v(p)$ to be the arrow of time at each point of Ω. The time orientation is reversed when v is replaced by $-v$. It will always be assumed, when a space–time M or a connected open set $\Omega \subset M$ is considered, that M or, respectively, Ω, is time-orientable. It is useful to note that, physical considerations apart, every result that refers to the time-orientation has a time-reversed counterpart.

The classification of vectors into types can be extended to linear sub-space of TM_p. A k-dimensional sub-space Π of TM_p, where $1 \leqslant k \leqslant n - 1$, is space-like if it contains only space-like vectors, time-like if it contains both space-like and time-like vectors (so that it meets the interior of the tangent null cone) and null if it touches the tangent null cone.

Let $\xi_{(\alpha)}, \alpha = 1, \ldots, k$ be a basis of Π. Then every $\xi \in \Pi$ can be written in one and only one way as

$$\xi = \sum_{\alpha=1}^{k} c^\alpha \xi_{(\alpha)},$$

where the c^α are real numbers. The restriction of the scalar product to Π is therefore

$$\langle \xi, \xi \rangle|_\Pi = \sum_{\alpha,\beta=1}^{k} c^\alpha c^\beta \langle \xi_{(\alpha)}, \xi_{(\beta)} \rangle.$$

Now one can also obtain $\langle \xi, \xi \rangle|_\Pi$ from $\langle \zeta, \zeta \rangle$, where $\zeta \in TM_p$, by imposing $n - k$ homogeneous linear relations on ζ. Hence the quadratic form $c \to \langle \xi, \xi \rangle|_\Pi$ has at most one positive eigenvalue. It follows that the signature of $c \to \langle \xi, \xi \rangle|_\Pi$ is $2 - k$ if Π is time-like, $1 - k$ if Π is null, and $-k$ if Π is space-like. Also, $c \to \langle \xi, \xi \rangle|_\Pi$ is negative semi-definite when Π is null, and negative definite when Π is space-like. Note that a null sub-space Π contains just one null one-dimensional sub-space (null line), which is the generator along which Π touches the tangent null cone.

Two vectors ξ and η are called *orthogonal* if $\langle \xi, \eta \rangle = 0$. A null vector is orthogonal to itself, and to all the vectors in the one-dimensional sub-space which it generates. If Π is a k-dimensional sub-space of TM_p, then the sub-space

$$\Pi' = \{\eta; \eta \in TM_p, \langle \xi, \eta \rangle = 0 \quad \text{for all} \quad \xi \in \Pi\}$$

will be called the normal of Π. There is a simple relation between the type of Π and that of its normal.

Lemma 3.1.1. *A sub-space of TM_p is space-like, null, or time-like respectively, if and only if its normal is time-like, null, or space-like respectively.*

PROOF. Let k be the dimension of Π, $1 \leqslant k \leqslant n - 1$. It will be convenient to let the sub- and superscripts α, β run from 1 to k, and the sub- or superscripts A, B from $k + 1$ to n. Choose a basis of TM_p such that Π is spanned by the first k coordinate vectors. Then

$$\Pi = \{\xi; \xi \in TM_p, \xi^A = 0\}$$

and

$$\langle \xi, \xi \rangle|_\Pi = g_{\alpha\beta} \xi^\alpha \xi^\beta. \tag{3.1.1}$$

If $\eta \in \Pi'$, the normal of Π, then $\langle \xi, \eta \rangle = \xi^i \eta_i = 0$ for all $\xi \in \Pi$, and hence the covariant components η_α must vanish. Hence

$$\langle \eta, \eta \rangle|_{\Pi'} = g^{AB} \eta_A \eta_B. \tag{3.1.2}$$

Now one can show that

$$g \det g^{AB} = \det g_{\alpha\beta}. \tag{3.1.3}$$

Postponing the proof of this identity, we show that it implies the lemma. If Π is null, then (3.1.1) is negative semi-definite, and so $\det g_{\alpha\beta} = 0$. By (3.1.3) one then has $\det g^{AB} = 0$, so that (3.1.2) is also negative semi-definite, and so Π' is null. If Π is space-like, then (3.1.1) is negative definite, and so $(-1)^k \det g_{\alpha\beta} > 0$. Hence $\det g^{AB} \neq 0$, and as the dimension of Π' is $n-k$, there are only two possibilities. If Π' is time-like, then (3.1.2) has the signature

$$1 - (n-k-1) \quad \text{and} \quad (-1)^{n-k-1} \det g^{AB} > 0;$$

if it is space-like, then (3.1.2) is negative definite, and

$$(-1)^{n-k} \det g^{AB} > 0.$$

As $(-1)^{n-1} g > 0$, it follows from (3.1.3) and $(-1)^k \det g_{\alpha\beta} > 0$ that $(-1)^{n-k-1} \det g^{AB} > 0$, and so Π' is time-like. The other case, of a time-like Π, is settled by a similar argument. As the relation between Π and Π' is symmetric, the lemma follows. $\square$

It remains to prove (3.1.3). This can be done by evaluating the following product of determinants:

$$
\begin{vmatrix}
0 & 0 & . & . & . & . & 0 \\
0 & 1 & 0 & . & . & . & 0 \\
. & . & . & . & . & . & . \\
0 & . & . & 0 & 1 & 0 & 0 \\
g^{1,k+1} & . & . & . & g^{k,k+1} & . & g^{n,k+1} \\
\hdotsfor{7} \\
g^{1,n} & . & . & . & . & . & g^{n,n}
\end{vmatrix}
\begin{vmatrix}
g_{11} & g_{12} & \cdots & g_{1n} \\
\hdotsfor{4} \\
\vdots & & & \\
\vdots & & & \\
\vdots & & & \\
\vdots & & & \\
g_{n1} & g_{n2} & \cdots & g_{nn}
\end{vmatrix}.
$$

The first of these has the value $\det g^{AB}$, while the second is g. But it easily follows from $g^{ik} g_{jk} = \delta^i_j$ that the product determinant is equal to $\det g_{\alpha\beta}$.

It also follows from the argument that if Π is not null, then $\Pi \cap \Pi' = 0$ and the direct sum of Π and Π' spans TM_p. But if Π is null, then $\Pi \cap \Pi'$ is the unique null line contained in Π.

A hypersurface will be called space-like, null, or time-like respectively, if its tangent hyperplane is space-like, null, or time-like respectively. It follows from Lemma 3.1.1 that a necesssary and sufficient condition for a hypersurface to be of a certain type is that the normal should have the complementary type. Suppose in particular that $\Omega \subset M$ is an open set, that $\Sigma \subset \Omega$ is a hypersurface, and that there exists a function $S(p)$ on Ω such that $S(p) = 0$ and $\nabla S \neq 0$ on Σ. As ∇S is normal to Σ, we can assert that Σ is

space-like if $\langle \nabla S, \nabla S \rangle|_{S=0} > 0,$ (3.1.4)

null if $\langle \nabla S, \nabla S \rangle|_{S=0} = 0,$ (3.1.5)

time-like if $\langle \nabla S, \nabla S \rangle|_{S=0} < 0.$ (3.1.6)

Note that this also holds for a C^1 hypersurface.

A C^1 function $S\colon \Omega \to \mathbf{R}$, such that $\nabla S \neq 0$ on Ω, will be called a space-like field, a time-like field, or a null field respectively if the hypersurfaces $\{p;\, p \in \Omega, S(p) = \text{const.}\}$ are all respectively space-like, time-like, or null. By (3.1.4)–(3.1.6), applied to $S(p) - t, t \in \mathbf{R}$, it follows that S is space-like if $\langle \nabla S, \nabla S \rangle > 0$, time-like if $\langle \nabla S, \nabla S \rangle < 0$, and null if S satisfies the first-order partial differential equation

$$\langle \nabla S, \nabla S \rangle = g^{ij}\, \partial_i S\, \partial_j S = 0 \tag{3.1.7}$$

in all coordinate charts that meet Ω.

3.2 Characteristics and bicharacteristics

As explained in the introduction to this chapter, a partial differential equation of the form

$$Pu \equiv \square\, u + \langle a, \nabla u \rangle + bu = f \tag{3.2.1}$$

will be called a (scalar) wave equation on the space–time M. In conformity with the assumption that M has a C^∞ structure, the vector field a and the scalar field b will also be supposed to be C^∞. (Many of the results that will be derived will also hold if a, b and the differentiable structure of M are only in a finite differentiability class.) The unknown u and the second member f may be functions or distributions.

Tensor wave equations are defined similarly. As the metric provides a natural isomorphism between TM and T^*M (the raising and lowering of tensor sub- and superscripts), it will be sufficient to consider tensor fields of type $(0, m)$. A tensor wave equation is then one that can be written, in local coordinates, as

$$(PU)_{i_1 \cdots i_m} = \nabla_j \nabla^j U_{i_1 \cdots i_m} + a_{i_1 \cdots i_m}{}^{jj_1 \cdots j_m} \nabla_j U_{j_1 \cdots j_m}$$
$$+ b_{i_1 \cdots i_m}{}^{j_1 \cdots j_m} U_{j_1 \cdots j_m} = F_{i_1 \cdots i_m}. \tag{3.2.2}$$

The sum of the second-order derivatives of the components of U in $(PU)_{i_1\ldots i_m}$ is $g^{ij}\partial_i\partial_j U_{i_1\ldots i_m}$. For this reason, the theory of tensor wave equations is in most respects a simple extension of the theory of the scalar wave equation, and we shall therefore mainly work with scalar differential operators of the type (3.2.1).

It is natural, in applications, to think of a wave equation as being defined on the whole of a space–time M. But the theory that is the subject of this book is a local one, and it will usually only be assumed that a wave equation (3.2.1) or (3.2.2) is defined on an open set $\Omega \subset M$; in fact, one may as well also suppose that Ω is connected (a domain).

A hypersurface Σ is called a *characteristic* of a differential operator P (of the second order) if the restriction $Pu|_\Sigma$ can be expressed solely in terms of derivatives tangential to Σ of $u|_\Sigma$ and $\nabla u|_\Sigma$. The next theorem states the fundamental property that links wave equations with the metric of the underlying space–time.

Theorem 3.2.1. *The characteristics of scalar and tensor wave equations on M are the null hypersurfaces of M.*

PROOF. Let Σ be a characteristic. As it is a hypersurface, every point of Σ has a coordinate neighbourhood Ω such that

$$\Omega \cap \Sigma = \{x; x^1 = 0\}.$$

The only term in $Pu|_\Sigma$ that is not a tangential derivative of $u|_\Sigma$ or of one of the $D_i u|_\Sigma$ is then $g^{11}D_1^2 u$. Hence Σ will be a characteristic if and only if $g^{11} = 0$ on $x^1 = 0$. But by (3.1.5), applied with $S = x^1$, this means that Σ is null. The proof in the tensor case is exactly the same. $\square$

In the tensor case, the null hypersurfaces are multiple characteristics. For example, when $m = 1$ in (3.2.2), then a routine calculation shows that the condition for a hypersurface that is given locally in the form $\{x; S(x) = 0\}$ to be a characteristic is

$$(\langle \nabla S, \nabla S \rangle)^n = 0.$$

In the general theory of partial differential equations, the appearance of multiple characteristics leads to various difficulties. But the tensor wave equation (3.2.2) is a system of a particularly simple form, and these difficulties will therefore not arise.

On the other hand, there are two important systems in general relativity that are not of the form (3.2.2). The first of these is the set of Maxwell–Einstein equations, for the electromagnetic field. This is an antisymmetric field F_{ij}. It will be recalled that these equations consist of two groups, the first of which can be summarized as

$$d(F_{ij}dx^i \wedge dx^j) = 0.$$

Locally, this implies that the form $F_{ij}dx^i \wedge dx^j$ is exact, so that

$$F_{ij} = \nabla_i A_j - \nabla_j A_i,$$

where A is the vector potential. The second group of equations says that the coderivative of the field is proportional to the current density σ, and one readily finds that the vector potential therefore satisfies the equations

$$\nabla_j \nabla^j A_i + R_i^j A_j - \nabla_i \nabla^j A_j = \sigma_i \quad (i = 1, ..., n).$$

The usual device for reducing this system to the form (3.2.2) is to impose a gauge condition, for example the standard condition $\nabla^j A_j = 0$, which is compatible with the resulting equations.

The second system in question consists of the variational equations of the Einstein field equations. These govern the small perturbations of the metric of a given space–time. Suppose, for simplicity, that both the unperturbed and the perturbed space–times are empty, so that the Einstein equations imply the vanishing of the Ricci tensor. Denoting the unperturbed and the perturbed metric tensor components by g_{ij} and $g_{ij} + \delta g_{ij}$ respectively, one then has the conditions $\delta R_{ij} = 0$. It can be shown that

$$-2\delta R_{ij} = \nabla_k \nabla^k \delta g_{ij} - \nabla_i \nabla^k \delta g_{kj} - \nabla_j \nabla^k \delta g_{ik} + \nabla_i \nabla_j g^{kl} \delta g_{kl} + b_{ij}^{kl} \delta g_{kl},$$

where $$b_{ij}^{kl} = 2R_i^{kl}{}_j - R_i^k \delta_j^l - R_j^k \delta_i^l.$$

The equations $\delta R_{ij} = 0$, which must be considered as differential equations for the δg_{ij}, are thus not of the form (3.2.2). In fact, they belong to an exceptional class, for which the characteristic condition is satisfied identically, so that all hypersurfaces are characteristics. (This is closely related to the covariance of the Einstein field equations.) They can be reduced to the form (3.2.2), however, by imposing gauge conditions, for instance

$$\nabla^k \delta g_{ik} - \tfrac{1}{2}\nabla_i(g^{kl}\delta g_{kl}) = 0 \quad (i = 1, ..., n).$$

One then finds that

$$-2\delta R_{ij} = \nabla_k \nabla^k \delta g_{ij} + b_{ij}^{kl} \delta g_{kl} = 0 \quad (i, j = 1, ..., n),$$

which is of the form (3.2.2), with $m = 2$.

In either case, the choice and imposition of gauge conditions raises a number of questions, both mathematical and physical, which it is not proposed to discuss here.

A characteristic Σ is a null hypersurface. Hence there is, at each point of Σ, a unique null direction that is both normal and tangential to Σ. The curves on Σ that are tangential to this null direction field form a congruence on Σ. They are called the *bicharacteristics*.

Theorem 3.2.2. *The bicharacteristics are null geodesics.*

PROOF. As the assertion is a local one, it is sufficient to prove it in a coordinate neighbourhood Ω and to suppose that

$$\Omega \cap \Sigma = \{x; S(x) = 0\},$$

where $S \in C^\infty(\Omega)$ and $\nabla S \neq 0$. (With a minor modification, the argument remains valid when $S \in C^2(\Omega)$.) As Σ is a characteristic, $\langle \nabla S, \nabla S \rangle$ vanishes when $S = 0$. By introducing S as a local coordinate one can see at once that this implies

$$\langle \nabla S, \nabla S \rangle = AS, \tag{3.2.3}$$

where $A \in C^\infty(\Omega)$. The differential equations

$$\frac{dx^i}{dt} = g^{ij}(x)\,\partial_j S(x) \quad (i = 1, \ldots, n), \tag{3.2.4}$$

where t is a parameter, have a unique C^∞ solution $t \to x(t)$, for given initial values $x(0)$. It follows from (3.2.3) that

$$\frac{d}{dt} S(x(t)) = A(x(t))\,S(x(t)),$$

so that $S(x(0)) = 0$ implies that $S(x(t)) = 0$. As ∇S is normal to a hypersurface $S = \text{const.}$, the bicharacteristics of Σ are obtained by integrating (3.2.4) subject to the subsidiary condition $S(x(0)) = 0$.

Let $t \to x(t)$ be an integral of (3.2.4), and put $p_i(t) = \partial_i S(x(t))$. Then

$$\frac{dp_i}{dt} = (\partial_i \partial_j S)\frac{dx^j}{dt} = (\partial_i \partial_j S)\,g^{jk}\,\partial_k S = \tfrac{1}{2}\partial_i \langle \nabla S, \nabla S \rangle - \tfrac{1}{2}p_j p_k \partial_i g^{jk}.$$

On a bicharacteristic, one has $S = 0$, and so it follows from (3.2.3) that the $2n$ equations

$$\frac{dx^i}{dt} = g^{ij}p_j, \quad \frac{dp_i}{dt} = -\tfrac{1}{2}p_j p_k \partial_i g^{jk} + \tfrac{1}{2}Ap_i \quad (i = 1, \ldots, n), \tag{3.2.5}$$

hold on a bicharacteristic. Now one can define a function $t \to r(t)$ by

$$\frac{dr}{dt} = \exp\left(\frac{1}{2}\int_0^t A(x(t'))\,dt'\right), \quad r(0) = 0.$$

This is evidently an admissible parameter transformation. A simple computation shows that if one sets

$$\xi_i = \frac{dt}{dr}p_i = \frac{dt}{dr}\partial_i S(x(t)) \quad (i = 1, \ldots, n), \tag{3.2.6}$$

then the equations (3.2.5) transform to

$$\frac{dx^i}{dr} = g^{ij}\xi_j, \quad \frac{d\xi_i}{dr} = -\tfrac{1}{2}\xi_j\xi_k\partial_i g^{jk} \quad (i = 1, \ldots, n). \tag{3.2.7}$$

But these are the equation of the geodesics, in canonical form. Also, it follows from (3.2.6) and (3.2.3) that ξ is null. Hence the bicharacteristics are null geodesics; the result is obviously independent of the choice of the characteristic Σ, and so the theorem is proved. $\square$

The argument also shows that if Σ is a characteristic, and q is a point on Σ, then Σ contains the null geodesic Λ that has the direction of the normal to Σ at q. Also, Λ is normal to Σ at all of its points. We therefore have the following:

Corollary 3.2.1. *If two characteristics touch at a point, then they touch along a common bicharacteristic through this point. Hence an envelope of a family of characteristics that is a hypersurface is also a characteristic.* $\square$

3.3 The initial value problem for characteristics and null fields

A characteristic contains a congruence of bicharacteristics, which are null geodesics. Conversely, characteristics can be constructed as loci of null geodesics. In particular, one can give a submanifold of M, of co-dimension 2, and ask whether there are any characteristics that contain it. This may be called the initial value problem for characteristics. If Σ is a characteristic which contains σ, and p is a point of σ, then the tangent space to σ at p is contained in the tangent space to Σ at p. It is therefore normal to the bicharacteristic of Σ that goes through p. As the bicharacteristic is a null curve, it follows from Lemma 3.1.1 that every tangent vector of σ is space-like or null, so that σ is space-like or null. For a reason that will become clear presently, we shall suppose that σ is space-like. The first step in the solution of the initial value problem is the construction of a normal cross-section of the normal bundle of σ. We begin by considering this problem locally.

Theorem 3.3.1. *If σ is a space-like sub-manifold of M, of co-dimension 2, then every point of σ has a neighbourhood Ω in which there are just two characteristics that contain $\sigma \cap \Omega$.*

PROOF. Let p be a point of σ; then there is a coordinate neighbourhood Ω_1 of p such that $\Omega_1 \cap \sigma = \{x; x \in \Omega_1, x^{n-1} = x^n = 0\}$. Let us denote a generic point of $\Omega_1 \cap \sigma$ by $\lambda = (\lambda^1, ..., \lambda^{n-2}, 0, 0)$. The normal to $\Omega_1 \cap \sigma$ at λ is the set of all vectors $\eta \in TM_\lambda$ whose first $n-2$ covariant components vanish. By Lemma 3.1.1, with $k = 2$, it is time-like. The null vectors normal to σ at λ satisfy

$$g^{AB}(\lambda)\,\eta_A\eta_B = 0, \tag{3.3.1}$$

where A and B take only the values $n-1$ and n. Let ζ be an auxiliary time-like C^∞ vector field on $\Omega_1 \cap \sigma$, and normalize η by the condition

$$\langle \zeta, \eta \rangle = 1. \tag{3.3.2}$$

Note that, by Lemma 3.1.1, the scalar product of a time-like vector and a null vector cannot vanish, so that this condition is consistent. Also, as the first member of (3.3.1) is an indefinite quadratic form,

$$g_{nn}(\lambda)\,g_{n-1,\,n-1}(\lambda) - (g_{n,\,n-1}(\lambda))^2 < 0,$$

and so the equations (3.3.1) and (3.3.2) have two distinct solutions (η'_{n-1}, η'_n) and (η''_{n-1}, η''_n) which are C^∞ functions of λ on $\Omega_1 \cap \sigma$.

One can now solve the geodesic equations with the initial values

$$\left.\begin{aligned} x|_{r=0} &= \lambda = (\lambda^1, ..., \lambda^{n-2}, 0, 0), \\ \frac{dx}{dr}\bigg|_{r=0} &= \eta' = (g^{1A}(\lambda)\,\eta'_A(\lambda), ..., g^{nA}(\lambda)\,\eta'_A(\lambda)). \end{aligned}\right\} \tag{3.3.3}$$

Let $x(r, \lambda)$ denote the solution of this initial value problem. (It is just the exponential map $\exp_\lambda(r\eta')$.) One may suppose that $\lambda = 0$ at the given point $p \in \sigma$. For $r = 0$, the vectors $\partial x/\partial \lambda^\alpha, \alpha = 1, ..., n-2$ are the first $n-2$ coordinate vectors in TM_p, and $\partial x/\partial r = \eta'$. As σ is space-like and η' is null, these $n-1$ vectors are linearly independent. It follows that the rank of the map $(r, \lambda) \to x(r, \lambda)$ is $n-1$ in a neighbourhood of p, and hence, by the Rank Theorem (Dieudonné (1960) p. 273), that p has a neighbourhood $\Omega' \subset \Omega_1$ in which the image of $(r, \lambda) \to x(r, \lambda)$ is a hypersurface Σ'. (This is the step in the argument that could fail if σ were null, rather than space-like, at p.) It remains to be shown that Σ' is null. But this follows from a well-known property of null geodesics.

Let Ω be a coordinate neighbourhood, and let $T\Omega$ denote the corresponding coordinate neighbourhood in the tangent bundle, with the local coordinates (x, ξ). A parametrized curve in $T\Omega$ is then a C^∞ map of an open interval $\{\lambda; \lambda \in \mathbf{R}, |\lambda| < \delta\}$ into $T\Omega$.

Lemma 3.3.1. *Let $\lambda \to (y, \eta)$ be a parametrized curve in $T\Omega$, and suppose that $\langle \eta, \eta \rangle = 0$. Denote by $x(r, \lambda)$ the solution of the geodesic equations with the initial values $x|_{r=0} = y(\lambda)$, $\partial x / \partial r|_{r=0} = \eta(\lambda)$. Then*

$$\left\langle \frac{\partial x}{\partial r}, \frac{\partial x}{\partial \lambda} \right\rangle = \left\langle \eta, \frac{\partial y}{\partial \lambda} \right\rangle. \tag{3.3.4}$$

PROOF. The differential equations of the geodesics in the canonical form (3.2.7), which can be written as

$$\frac{\partial x^i}{\partial r} = \frac{\partial H}{\partial \xi_i}, \quad \frac{\partial \xi_i}{\partial r} = -\frac{\partial H}{\partial x^i} \quad (i = 1, ..., n), \quad H = \tfrac{1}{2} g^{ij}(x) \, \xi_i \xi_j, \tag{3.3.5}$$

have a unique solution for the initial data

$$x^i|_{r=0} = y^i(\lambda), \quad \xi_i|_{r=0} = \eta_i(\lambda) \quad (i = 1, ..., n), \tag{3.3.6}$$

which is a C^∞ function of r and λ. It follows from (3.3.5) that

$$(\partial / \partial r) \, H(x, \xi) = 0$$

for a solution $r \to (x, \xi)$; this is just the familiar result that

$$\langle \partial x / \partial r, \, \partial x / \partial r \rangle$$

is constant along a geodesic. In the present case, one has

$$\langle \xi, \xi \rangle|_{r=0} = \langle \eta, \eta \rangle = 0,$$

and so $H(x, \xi) = 0$. Now

$$\frac{\partial}{\partial r} \left\langle \frac{\partial x}{\partial r}, \frac{\partial x}{\partial \lambda} \right\rangle = \frac{\partial}{\partial r} \left(\xi_i \frac{\partial x^i}{\partial \lambda} \right) = \frac{\partial \xi_i}{\partial r} \frac{\partial x^i}{\partial \lambda} + \xi_i \frac{\partial^2 x^i}{\partial r \, \partial \lambda}$$

$$= -\frac{\partial H}{\partial x^i} \frac{\partial x^i}{\partial \lambda} + \xi_i \frac{\partial}{\partial \lambda} \frac{\partial H}{\partial \xi_i} = \frac{\partial}{\partial \lambda} \left(\xi_i \frac{\partial H}{\partial \xi_i} - H \right).$$

But $\xi_i(\partial H/\partial \xi_i) = 2H$ and $H = 0$, and so $(\partial/\partial r)\langle \partial x/\partial r, \partial x/\partial \lambda \rangle = 0$. This proves (3.3.4). $\square$

We can now complete the proof of Theorem 3.3.1. The lemma can be applied to the initial value problem (3.3.3), taking

$$\lambda = \lambda^\alpha, \alpha = 1, \ldots, n-2,$$

in turn. As η' is null and normal to σ, by construction, it follows from (3.3.4) that

$$\left\langle \frac{\partial x}{\partial r}, \frac{\partial x}{\partial r} \right\rangle = 0, \quad \left\langle \frac{\partial x}{\partial r}, \frac{\partial x}{\partial \lambda^\alpha} \right\rangle = 0 \quad (\alpha = 1, \ldots, n-2), \quad (3.3.7)$$

holds for $\{(r, \lambda); x(r, \lambda) \in \Omega'\}$. As $\partial x/\partial r, \partial x/\partial \lambda^1, \ldots, \partial x/\partial \lambda^{n-2}$ span the tangent hyperplane of Σ', Σ' is a null hypersurface, and therefore a characteristic. It should be noted that $\Sigma' \cap \sigma = \Omega' \cap \sigma$.

A second characteristic Σ'' can be derived in the same way, starting with the other null field η'', in a neighbourhood Ω'' of p. Taking $\Omega = \Omega' \cap \Omega''$, one thus obtains a neighbourhood of p in which there are two characteristics Σ' and Σ'' that contain $\Omega \cap \sigma$. There are no others. For if $\Sigma \subset \Omega$ is a characteristic that contains $\Omega \cap \sigma$, then it must be tangent to one of the two null fields η', η'' at σ, and so it follows from Corollary 3.2.1 that either $\Sigma = \Sigma'$ or $\Sigma = \Sigma''$. This completes the proof. The result is also valid if σ is of class C^2. $\square$

With a small modification, the proof of Theorem 3.3.1 yields a local existence theorem for the initial value problem for null fields. It has already been pointed out that a null field $S(p)$ is a solution of the first-order partial differential equation (3.1.7),

$$\langle \nabla S, \nabla S \rangle = 0.$$

The initial value problem for this is to find a solution that reduces to a prescribed function S_0 on a given hypersurface Λ. The hypersurfaces $S = $ const. are characteristics which meet Λ in the $(n-2)$-dimensional sub-manifolds $\{p; p \in \Lambda, S_0(p) = \text{const.}\}$. These characteristics can be constructed by the method that has just been developed. It is evident that a necessary condition for the existence of solutions is that these sub-manifolds must be space-like or null; we shall assume them to be space-like.

Theorem 3.3.2. *Let Λ be a hypersurface, and let $S_0 \in C^\infty(\Lambda)$ be a function on Λ such that the subsets $\{p; p \in \Lambda, S_0(p) = \text{const.}\}$ of Λ are space-like sub-manifolds.*

(i) *If Λ is not null at any of its points, then every point of Λ has a neighbourhood Ω in which there are just two null fields S' and S'' such that $S' = S'' = S_0$ on $\Omega \cap \Lambda$.*

(ii) *If Λ is a characteristic, then every point of Λ has a neighbourhood Ω in which there is a unique null field S such that $S = S_0$ on $\Omega \cap \Lambda$.*

PROOF. The hypothesis implies that S_0 has no critical points. Hence, given a point $p \in \Lambda$, there is a coordinate neighbourhood Ω_1 of p such that $\Omega_1 \cap \Lambda = \{x: x^n = 0\}$ and $S_0 = x^{n-1}$. Define a vector field η on Λ by

$$\left. \begin{array}{l} \eta_1 = \eta_2 = \ldots = \eta_{n-2} = 0, \, \eta_{n-1} = 1, \\ \langle \eta, \eta \rangle = g^{nn}(\lambda, \sigma) \eta_n^2 + 2g^{n,\,n-1}(\lambda, \sigma) \eta_n + g^{n-1,\,n-1}(\lambda, \sigma) = 0, \end{array} \right\} \quad (3.3.8)$$

where $(\lambda, \sigma) = (\lambda^1, \ldots, \lambda^{n-1}, \sigma)$, and $x = (\lambda, \sigma, 0)$ is a generic point of $\Lambda \cap \Omega_1$. In case (i), one has $g^{nn} \neq 0$ on $\Lambda \cap \Omega_1$. Also, as

$$\{x; x^{n-1} = \sigma, \, x^n = 0\}$$

is by hypothesis space-like, the quadratic equation has two distinct solutions η'_n and η''_n both of which are C^∞ functions of (λ, σ). Let $x(r, \lambda, \sigma)$ be the unique solution of the geodesic equations that is determined by the initial conditions

$$\left. \begin{array}{l} x^\alpha \big|_{r=0} = \lambda^\alpha, \quad \alpha = 1, \ldots, n-2, \quad x^{n-1} \big|_{r=0} = \sigma, \quad x^n \big|_{r=0} = 0, \\ \dfrac{\partial x^i}{\partial r} \bigg|_{r=0} = g^{ij}(\lambda, \sigma) \eta'_j \quad (i = 1, \ldots, n). \end{array} \right\} \quad (3.3.9)$$

It follows from the choice of local coordinates, and from the conditions (3.3.8), that, for $r = 0$,

$$\left\langle \frac{\partial x}{\partial r}, \frac{\partial x}{\partial r} \right\rangle = 0, \quad \left\langle \frac{\partial x}{\partial r}, \frac{\partial x}{\partial \lambda^\alpha} \right\rangle = 0 \quad (\alpha = 1, \ldots, n-2), \quad \left\langle \frac{\partial x}{\partial r}, \frac{\partial x}{\partial \sigma} \right\rangle = 1. \tag{3.3.10}$$

As $\partial x / \partial r$ is null, Lemma 3.3.1 therefore implies that (3.3.10) holds for $\{(r, \lambda, \sigma); x(r, \lambda, \sigma) \in \Omega_1\}$. When $r = 0$, the Jacobian

$$D(x^1, \ldots, x^n) / D(r, \lambda^1, \ldots, \lambda^{n-2}, \sigma)$$

reduces to
$$\eta'^n = g^{nn} \eta'_n + g^{n,\,n-1}.$$

But as it has been assumed that $g^{nn} \neq 0$ on $\Lambda \cap \Omega_1$, and the discriminant of the quadratic equation in (3.3.8) is different from zero, it is easily seen that $\eta'^n \neq 0$, in $\Omega_1 \cap \Lambda$. Hence, by the inverse function theorem, there is a neighbourhood $\Omega' \subset \Omega$ of p in which $(r, \lambda, \sigma) \to x(r, \lambda, \sigma)$ is a diffeomorphism. In particular, one then has $\sigma = S'(x)$ in Ω', and S' is a null field. For it follows from (3.3.10) and

$$dx = \frac{\partial x}{\partial r} dr + \frac{\partial x}{\partial \lambda^\alpha} d\lambda^\alpha + \frac{\partial x}{\partial \sigma} d\sigma$$

that

$$\left\langle \frac{\partial x}{\partial r}, dx \right\rangle = d\sigma = dS',$$

whence

$$\frac{\partial S'}{\partial x^i} = g_{ij} \frac{\partial x^j}{\partial r} \quad (i = 1, ..., n).$$

As $\partial x / \partial r$ is null, this gives $\langle \nabla S', \nabla S' \rangle = 0$. Also,

$$S'|_\Lambda = S'|_{x^n = 0} = \sigma = S_0.$$

A second null field S'' can be constructed in the same way, starting with the null field η'', in a neighbourhood Ω'' of p. So $\Omega = \Omega' \cap \Omega''$ is a neighbourhood of p in which there are two null fields S' and S'' which reduce to S_0 on Λ. Also, if t is in the range of the function $S_0|_{\Omega \cap \Lambda}$, then $S' = t$ and $S'' = t$ are distinct characteristics that contain

$$\{p;\, p \in \Omega \cap \Lambda,\, S(p) = t\}.$$

As there cannot be any other characteristics with this property, S' and S'' are the only solutions of the initial value problem in Ω.

In case (ii), one has $g^{nn} = 0$ when $x^n = 0$. Then (3.3.8) has a unique solution $\eta_n(\lambda, \sigma)$. A null field S such that $S = S_0$ on Λ is then constructed as in case (i), in a neighbourhood of p. Uniqueness again follows from Theorem 3.3.1. □

Remark. Let S be one of the null fields in case (i), or the unique null field in case (ii). In either case, one can introduce r, σ and the λ^α as local coordinates in Ω, say

$$\sigma = \tilde{x}^1,\, r = \tilde{x}^2,\, \lambda^\alpha = \tilde{x}^{\alpha+2} \quad (\alpha = 1, ..., n-2).$$

Then it follows from (3.3.10) and the law of contravariance that

$$\tilde{g}^{11} = 0,\, \tilde{g}^{12} = 1,\, \tilde{g}^{1i} = 0 \quad (i = 3, ..., n), \qquad (3.3.11)$$

and as $\tilde{g}^{1k} \tilde{g}_{ik} = \delta_i^1$ this implies that

$$\tilde{g}_{21} = 1,\, \tilde{g}_{2i} = 0 \quad (i = 2, ..., n). \qquad (3.3.12)$$

Given a null field S, such coordinates can be introduced locally; one may call them coordinates adapted to S.

There is another useful consequence of Theorem 3.3.2.

Corollary 3.3.1. *A characteristic Σ can be imbedded locally in a null field $S(p)$ that vanishes on Σ.*

PROOF. Every point $p \in \Sigma$ has a neighbourhood Ω that contains a space-like hypersurface $\Lambda \ni p$. As Σ cannot be tangential to Λ, one can introduce local coordinates such that Σ is $x^{n-1} = 0$ and Λ is $x^n = 0$, shrinking Ω, if necessary. Then the field $S_0 = x^{n-1}$ is space-like on Λ in some neighbourhood Ω' of p. By Theorem 3.3.2, there is a neighbourhood $\Omega'' \subset \Omega'$ in which there are two null fields each of which reduces to x^{n-1} on $x^n = 0$, and one of these must have the required property. $\square$

Returning to the initial value problem for characteristics, we prove a simple global version of Theorem 3.3.1

Theorem 3.3.3. *Let σ be a compact $(n-2)$-dimensional sub-manifold of M, without boundary. If σ is space-like, then it has a neighbourhood in which there are just two characteristics that contain σ.*

PROOF. Let M be given a canonical orientation. As σ is compact, it can be covered by a finite number of orientation-preserving coordinate charts $(\Omega_1, \pi_1), \ldots, (\Omega_N, \pi_N)$ such that $\pi_j(\sigma \cap \Omega_j) = \{x ; x^{n-1} = x^n = 0\}$. As σ is orientable, one can also suppose that the restrictions of these charts to σ are an oriented covering of σ permuting the coordinates, if necessary. Let σ have the orientation determined by this covering. Suppose that (Ω, π) and $(\tilde{\Omega}, \tilde{\pi})$ are two charts of the covering, such that $\Omega \cap \tilde{\Omega}$ is not empty; put $x = \pi q, \tilde{x} = \tilde{\pi} q$ for $q \in \Omega \cap \tilde{\Omega}$. Then $\tilde{x}^{n-1} = \tilde{x}^n = 0$ when $x^{n-1} = x^n = 0$, whence

$$\left.\frac{\partial \tilde{x}}{\partial x^\alpha}\right|_{x^n = x^{n-1} = 0} = 0 \quad (\alpha = 1, \ldots, n-2), \qquad (3.3.13)$$

and so
$$\frac{D(\tilde{x}^1, \ldots, \tilde{x}^n)}{D(x^1, \ldots, x^n)} = \frac{D(\tilde{x}^1, \ldots, \tilde{x}^{n-2})}{D(x^1, \ldots, x^{n-2})} \frac{D(\tilde{x}^{n-2}, \tilde{x}^n)}{D(x^{n-1}, x^n)}$$

holds when $x^n = x^{n-1} = 0$. By the choice of the local coordinates, the first member, and the first factor in the second member, are positive. Hence

$$\left.\frac{D(\tilde{x}^{n-1}, \tilde{x}^n)}{D(x^{n-1}, x^n)}\right|_{x^n = x^{n-1} = 0} > 0. \qquad (3.3.14)$$

Again, if $v \in TM_q$ is normal to σ at q, and its covariant components in the two charts are $(0, \ldots, 0, \eta_{n-1}, \eta_n)$ and $(0, \ldots, 0, \tilde{\eta}_{n-1}, \tilde{\eta}_n)$, then

$$\eta_A = \left.\tilde{\eta}_A \frac{\partial \tilde{x}^B}{\partial x^A}\right|_{x^n = x^{n-1} = 0} \quad (A = n-1, n). \qquad (3.3.15)$$

By (3.3.14), this transformation (in the fibre of the normal cotangent bundle of σ over $\Omega \cap \tilde{\Omega}$) is orientation preserving.

Let ζ be an auxiliary time-like vector field on σ. In each of Ω and $\tilde{\Omega}$, one can determine two null fields v' and v'', orthogonal to σ, and such that $\langle v', \zeta \rangle = \langle v'', \zeta \rangle = 1$. If one denotes their respective covariant components by η' and η'' in Ω, and by $\tilde{\eta}'$ and $\tilde{\eta}''$ in $\tilde{\Omega}$, then the transformation (3.3.15) either gives $\tilde{\eta}' \to \eta'$, $\tilde{\eta}'' \to \eta''$, or $\tilde{\eta}' \to \eta''$, $\tilde{\eta}'' \to \eta'$. But one can order the couple (v', v'') by the rule that, if $(v_{(1)}, ..., v_{(n-2)})$ is a basis of the tangent space of σ at q, then $(v_{(1)}, ..., v_{(n-2)}, v', v'')$ is to have the canonical orientation of M. As (3.3.15) is orientation preserving, one then always has $\tilde{\eta}' \to \eta'$, $\tilde{\eta}'' \to \eta''$. Hence this construction yields two null cross-sections of the normal bundle of σ. A straightforward argument then shows that the null geodesics through σ that are tangent to v' generate a characteristic Σ' ,and the null geodesics through σ that are tangent to v'' generate a characteristic Σ'', each in a neighbourhood of σ, and the existence of two characteristics through σ is thus assured in the intersection Ω of these neighbourhoods. Uniqueness is proved as before, by appealing to Corollary 3.2.1. So the theorem is proved. $\square$

There is another method of constructing Σ' and Σ''. The null cone with vertex q, which will be denoted by $C(q)$, is the set of all points of M that can be reached along null geodesics from q. These null geodesics are its null geodesic generators. The null cone is not a hypersurface, as it is singular at q, and may in general also have other singularities (caustics, which will be discussed in the next section). But it is still a characteristic, in the extended sense that will be introduced presently; restricted to a normal neighbourhood of q, $C(q) \backslash (q)$ consists of two null hypersurfaces. If $q \in \sigma$, then $C(q)$ touches both Σ' and Σ'', as it has the tangent vectors v' and v'' in common with these two characteristics. Hence Σ' and Σ'' are the envelopes of the family $\{C(q): q \in \sigma)$ of null cones, restricted to Ω. This is Huygens' construction of wave fronts in geometrical optics.

3.4. Caustics

It is an obvious consequence of Theorem 3.3.1 that a characteristic can be parametrized locally as

$$x = x(r, \lambda) \equiv x(r, \lambda^1, ..., \lambda^{n-2}), \qquad (3.4.1)$$

where, say $|r| < \delta$, and $\lambda \in \Lambda$, an open set in $\mathbf{R}^{n-2}$, and where the curves $r \to x(r, \lambda)$ are the bicharacteristics. The vectors

$$\partial x/\partial r, \partial x/\partial \lambda^1, \ldots, \partial x/\partial \lambda^{n-2}$$

are linearly independent, as they span the tangent hyperplane of the characteristic, and satisfy (3.3.7),

$$\left\langle \frac{\partial x}{\partial r}, \frac{\partial x}{\partial r} \right\rangle = 0, \quad \left\langle \frac{\partial x}{\partial r}, \frac{\partial x}{\partial \lambda^\alpha} \right\rangle = 0, \quad \alpha = 1, \ldots, n-2. \quad (3.4.2)$$

Assuming that the space–time M is geodesically complete, one can extend the bicharacteristics indefinitely in both senses, and obtain a map $\mathbf{R} \times \Lambda \to M$ which reduces to (3.4.1) when $|r| < \delta$. The range Σ of this map may not be a hypersurface, but will still be called a characteristic. It is evident from the proof of Lemma 3.3.1 that the equations (3.4.2) hold everywhere on Σ, in local coordinates.

Let Σ_0 denote the set of all points of Σ at which the vectors

$$\partial x/\partial r, \partial x/\partial \lambda^1, \ldots, \partial x/\partial \lambda^{n-2}$$

are linearly independent; it is characterized by the fact that the matrix, whose rows are the $\partial x^i/\partial r$ and the $\partial x^i/\partial \lambda^\alpha, \alpha = 1, \ldots, n-2$, has rank $n-1$. It is obvious that Σ_0 is open in Σ, and that each component of Σ_0 is a connected null hypersurface. The exceptional set $\Sigma_1 = \Sigma \backslash \Sigma_0$ will be called the *caustic* of the (extended) characteristic Σ. A point $p(r, \lambda)$ is on the caustic if, in some coordinate chart at p, there are real numbers $\zeta^1, \ldots, \zeta^{n-2}$ and ζ^0 such that

$$\zeta^\alpha \frac{\partial x}{\partial \lambda^\alpha} = \zeta^0 \frac{\partial x}{\partial r}. \quad (3.4.3)$$

Note that $\partial x/\partial r$ cannot vanish on a geodesic, by the uniqueness theorem for the solutions of ordinary differential equations, so that the ζ^α cannot all be zero; but one may have $\zeta^0 = 0$. The points at which a bicharacteristic meets the caustic are evidently the *focal points*, where it is touched – or met – by neighbouring bicharacteristics. There is a simple condition characterizing the caustic.

Theorem 3.4.1. *Let* $(r, \lambda) \to p(r, \lambda)$ *be a characteristic* Σ. *Then* p *is on the caustic of* Σ *if and only if*

$$\Delta(r, \lambda) = \det \left\langle \frac{\partial x}{\partial \lambda^\alpha}, \frac{\partial x}{\partial \lambda^\beta} \right\rangle = 0 \quad (3.4.4)$$

in one, and hence every, coordinate chart at p.

PROOF. If p is on the caustic, then (3.4.3) holds, and so it follows from (3.4.2) that

$$\left\langle \frac{\partial x}{\partial \lambda^\alpha}, \frac{\partial x}{\partial \lambda^\beta} \right\rangle \zeta^\beta = 0 \quad (\alpha = 1, ..., n-2).$$

As this system of equations has, by hypothesis, a non-trivial solution, it follows that $\Delta = 0$.

Conversely, suppose that $\Delta = 0$. Then the quadratic form

$$\left\langle \frac{\partial x}{\partial \lambda^\alpha} \xi^\alpha, \frac{\partial x}{\partial \lambda^\beta} \xi^\beta \right\rangle = \left\langle \frac{\partial x}{\partial \lambda^\alpha}, \frac{\partial x}{\partial \lambda^\beta} \right\rangle \xi^\alpha \xi^\beta \qquad (3.4.5)$$

is degenerate. Now the $\partial x/\partial \lambda^\alpha$ are all orthogonal to the null vector $\partial x/\partial r$, and so are either space-like, or null. The same is true of any linear combination of the $\partial x/\partial \lambda^\alpha$, and so the quadratic form (3.4.5) cannot vanish unless there are numbers $\zeta^\alpha, \alpha = 1, ..., n-2$, such that $\zeta^\alpha(\partial x/\partial \lambda^\alpha)$ is a null vector. But this is (3.4.3), and so $p(r, \lambda)$ is on the caustic. The theorem is proved. $\square$

As an example, we shall now consider characteristics and caustics in a Minkowskian space–time. We write $x = (t, X)$, where

$$X = (X_1, ..., X_{n-1}) \in \mathbf{R}^{n-1},$$

and denote the usual scalar product in $\mathbf{R}^{n-1}$ by $X . Y$. The metric of the space–time is given the line element

$$\langle dx, dx \rangle = dt^2 - dX^2, \quad dX^2 \equiv dX . dX. \qquad (3.4.6)$$

The geodesics are straight lines, and so an $(n-2)$-parameter family of null geodesics is given by

$$t = t_0(\lambda) + r, \quad X = Y(\lambda) + r\theta(\lambda),$$

where $\theta \in \mathbf{S}^{n-2}$ is a unit $(n-1)$-vector. As one can replace r by $r - t_0$ and Y by $Y + t_0\theta$, one can assume from the outset that

$$t = r, \quad X = Y(\lambda) + r\theta(\lambda), \quad \theta^2 \equiv \theta . \theta = 1. \qquad (3.4.7)$$

This family of null geodesics will be the bicharacteristics of a characteristic Σ if (3.4.2) holds. The first of these conditions is satisfied, because $(1, \theta)$ is null. The others become

$$\theta . (dY + r\,d\theta) = 0.$$

But $\theta . d\theta = 0$, as θ is a unit vector in $\mathbf{R}^{n-1}$. Hence (3.4.2) holds if

$$\theta . dY = 0. \qquad (3.4.8)$$

4-2

Hypersurfaces in $\mathbf{R}^{n-1}$ will be called surfaces. The condition (3.4.8) means that θ is a unit vector to the 'initial' surface S_0: $X = Y(\lambda)$. Assume S_0 to be orientable, and choose θ to be one of the two unit normals on S_0. The equations (3.4.7) then show that the level surfaces of the characteristic Σ, $r = t = $ const., are the surfaces S_t: $X = Y + t\theta$ parallel to S_0 and at a (signed) distance t from it. They are a family of wave fronts, in the sense of geometrical optics. The projections of the bicharacteristics of Σ on $\mathbf{R}^{n-1}$ are the common normals of the wave fronts; they are the rays of the wave front system S_t.

The map $(r, \lambda) \to Y(\lambda) + r\theta(\lambda)$ has a unique inverse if there is a unique normal to S_0 that goes through X. The foot of this normal is $Y(\lambda)$, and $r = |X - Y|$ is the least distance of X from S_0: this is Fermat's principle. Suppose that X is close enough to S_0 for such a unique inverse to exist, and denote the function $X \to r$ by $\tau(X)$. Since

$$dX = dY + r\,d\theta + \theta\,dr, \qquad (3.4.9)$$

and $\theta \cdot d\theta = \theta \cdot dY = 0$, it follows that $dr = \theta \cdot dX$, whence

$$\operatorname{grad}_X \tau = \theta, \quad (\operatorname{grad}_X \tau)^2 = 1. \qquad (3.4.10)$$

The second of these equations is the eikonal equation of geometrical optics. Note that $(t, X) \to t - \tau(X)$ is a null field, which reduces to t on the hypercylinder parallel to the t-axis whose base is S_0.

To determine the caustic of the characteristic (3.4.7), one can use (3.4.3), which now becomes

$$\zeta^\alpha \left(0, \frac{\partial Y}{\partial \lambda^\alpha} + r \frac{\partial \theta}{\partial \lambda^\alpha} \right) = \zeta^0 (1, \theta).$$

Hence $\zeta^0 = 0$ and, in classical notation,

$$dY + r\,d\theta = 0. \qquad (3.4.11)$$

These n equations, compatible because of $\theta \cdot d\theta = \theta \cdot dY = 0$, determine $n-1$ directions $d\lambda$ on S_0, tangent to the curvature lines of S_0. The corresponding eigenvalues of r are principal radii of curvature $R_1, \ldots, R_{n-1}$ of S_0. (They are independent of the parametrization of S_0 because they are the stationary values of the quotient of the second fundamental form $dY \cdot d\theta$ and the first fundamental form $dY \cdot dY$ of S_0.) So the projection of the caustic on $\mathbf{R}^{n-1}$ consists in general of $n-1$ sheets,
$$X = Y(\lambda) + R_\nu(\lambda)\,\theta(\lambda) \quad (\nu = 1, \ldots, n-1). \qquad (3.4.12)$$

This set of points in $\mathbf{R}^{n-1}$, which may not be a surface, is usually also called the caustic of the wave front family S_t. To obtain the caustic in space–time, (3.4.12) must be supplemented by $t = R_\nu(\lambda), \nu + 1, ..., n - 1$, respectively. The points $r = R_\nu, \nu = 1, ..., n - 1$ on a ray are the focal points.

In the case of 'physical' Minkowskian space–time, $n = 4$, one can make a further simplification by using curvature line parameters λ_1 and λ_2. These are such that

$$\frac{\partial Y}{\partial \lambda_1} + R_1 \frac{\partial \theta}{\partial \lambda_1} = 0, \quad \frac{\partial Y}{\partial \lambda_2} + R_2 \frac{\partial \theta}{\partial \lambda_2} = 0. \qquad (3.4.13)$$

Note that the curvature lines are orthogonal, if $R_1 \neq R_2$. (Points such that $R_1 = R_2$ are umbilical points, and are singularities of the differential equations of the curvature lines.) It now follows from (3.4.9) that

$$dX = \left(1 - \frac{r}{R_1}\right)\frac{\partial Y}{\partial \lambda_1} d\lambda_1 + \left(1 - \frac{r}{R_2}\right)\frac{\partial Y}{\partial \lambda_2} d\lambda_2 + \theta \, dr, \qquad (3.4.14)$$

which makes it obvious that the characteristic is degenerate when either $r = R_1$ or $r = R_2$.

Consider one of the sheets of the caustic, say

$$t = R_1, \quad X = Y + R_1 \theta. \qquad (3.4.15)$$

For a tangent vector, this gives

$$dt = \frac{\partial R_1}{\partial \lambda_1} d\lambda_1 + \frac{\partial R_1}{\partial \lambda_2} d\lambda_2,$$

$$dX = \left(1 - \frac{R_1}{R_2}\right)\frac{\partial Y}{\partial \lambda_2} d\lambda_2 + \left(\frac{\partial R_1}{\partial \lambda_1} d\lambda_1 + \frac{\partial R}{\partial \lambda_2} d\lambda_2\right)\theta.$$

A simple computation shows that (3.4.15) is a two-dimensional surface at all points where both $R_1 \neq R_2$ and $\partial R_1 / \partial \lambda_1 \neq 0$. When either of these exceptional cases arises, the caustic itself is degenerate. There are a number of possibilities which will not be listed here.

3.5   The propagation of discontinuities

The characteristics play an important part in the theory of the wave equation, essentially because they can be wave fronts. In this section, we shall consider solutions of a wave equation which have discontinuities of a simple type. We begin with a classical result.

Let M be a space–time, and let

$$Pu = \Box u + \langle a, \nabla u \rangle + bu = 0$$

be a homogeneous wave equation on M. Let $\Omega \subset M$ be a connected open set, and let $\Sigma \subset \Omega$ be a hypersurface, such that $\Omega\backslash\Sigma$ consists of two components, Ω_1 and Ω_2. Suppose that $u \in C^{k-1}(\Omega)$, where $k \geqslant 2$, and that both $u \in C^k(\Omega_1 \cup \Sigma)$ and $u \in C^k(\Omega_2 \cup \Sigma)$. This means that u is C^k in $\Omega\backslash\Sigma$, and that the one-sided derivatives of u, or order k, exist on Σ, and are continuous on Σ. Let $[\partial^\alpha u]$ denote the discontinuity of $\partial^\alpha u$ on Σ, say the difference between $\partial^\alpha u$ evaluated in $\Omega_2 \cup \Sigma$ and in $\Omega_1 \cup \Sigma$. We shall say that u has a *discontinuity of order k* on Σ if

$$\sum_{|\alpha|=k} |[\partial^\alpha u]| \neq 0$$

in one, and hence in all, coverings of Σ by coordinate charts.

Theorem 3.5.1. *If $Pu = 0$ in $\Omega\backslash\Sigma$, and u has a discontinuity of order $k \geqslant 2$ on Σ, then Σ is a characteristic.*

PROOF. As the assertion is a local one, it will be sufficient to prove it when Ω is a coordinate neighbourhood in which Σ is the coordinate hypersurface $x^1 = 0$ and, say, $\Omega_1 = \{x; x^1 < 0\}$ and $\Omega_2 = \{x; x^1 > 0\}$. As u is supposed to be C^k for $x^1 \neq 0$, one has $\partial_1^{k-2} Pu = 0$ for $x^1 \neq 0$, whence

$$[\partial_1^{k-2} Pu] = \lim_{\epsilon \downarrow 0} (\partial_1^{k-2} Pu(\epsilon, x^2, ..., x^n) - \partial_1^{k-2} Pu(-\epsilon, x^2, ..., x^n)) = 0.$$

Now the $\partial^\alpha u$ with $|\alpha| \leqslant k-1$ are continuous, and so are the derivatives, of order k, except for $\partial_1^k u$. Hence

$$[\partial_1^{k-2} Pu] = g^{11}|_{x^1=0}[\partial_1^k u] = 0.$$

But $[\partial_1^k u] \neq 0$, by hypothesis, and so $g^{11} = 0$ when $x^1 = 0$, which proves that Σ is a characteristic. $\square$

This result can be extended to discontinuities of orders 1 and 0, provided that $Pu = 0$ holds in the sense of distributions.

Theorem 3.5.2. *Suppose that $u \in C^2(\Omega_1 \cup \Sigma)$ and $u \in C^2(\Omega_2 \cup \Sigma)$, that $u \in \mathscr{D}'(\Omega)$ satisfies $Pu = 0$, and that $|[u]| + |[\nabla u]| \neq 0$. Then Σ is necessarily a characteristic.*

PROOF. As u is clearly locally integrable,

$$(Pu, \phi) = (u, {}^t P\phi) = \int u\, {}^t P\phi\mu = 0 \qquad (3.5.1)$$

holds for all $\phi \in C_0^\infty(\Omega)$, where μ denotes the invariant volume element on M, and ^tP is the adjoint of P, which is

$$^tP\phi = \Box\,\phi - \operatorname{div}(a\phi) + b\phi.$$

By Theorem 2.6.1, applied to $|g|^{\frac{1}{2}}P$ in local coordinates, (3.5.1) and $u \in C^2(\Omega\backslash\Sigma)$ imply that u is a C^2 solution of $Pu = 0$ in $\Omega\backslash\Sigma$. Hence, in $\Omega\backslash\Sigma$,

$$-u\,^tP\phi = \phi Pu - u\,^tP\phi$$
$$= \nabla_i(\phi\nabla^i u - u\nabla^i\phi + au\phi), \qquad (3.5.2)$$

in local coordinates. Let $\Omega' \subset \Omega$ be a coordinate neighbourhood that meets Σ, such that $\Omega' \cap \Sigma = \{x;\, x \in \Omega', x^1 = 0\}$. Then, for $\phi \in C_0^\infty(\Omega')$, (3.5.1) becomes

$$\int_{x^1>0} u\,^tP\phi\,|g|^{\frac{1}{2}}\,dx + \int_{x^1<0} u\,^tP\phi\,|g|^{\frac{1}{2}}\,dx = 0.$$

By (3.5.2) and the divergence theorem, this gives

$$\int_{x^1=0} (\phi[\nabla^1 u] - [u]\nabla^1\phi + a^1[u]\phi)\,|g|^{\frac{1}{2}}\,dx' = 0, \qquad (3.5.3)$$

for all $\phi \in C_0^\infty(\Omega')$, where $dx' = dx^2 \ldots dx^n$.

Suppose first that $[u] = 0$ on $\Omega' \cap \Sigma$. Then also $[\partial_i u] = 0$ for $i > 1$, but, by hypothesis, $[\partial_1 u] \neq 0$. Hence (3.5.3) reduces to

$$\int_{x^1=0} g^{11}\phi[\partial_1 u]\,|g|^{\frac{1}{2}}\,dx' = 0.$$

Given $\psi(x^2, \ldots, x^n) \in C_0^\infty(\Omega' \cap \Sigma)$, one can find $\phi \in C_0^\infty(\Omega')$ such that $\phi = \psi$ when $x^1 = 0$. Hence it follows that $g^{11}[\partial_1 u]\,|g|^{\frac{1}{2}} = 0$ when $x^1 = 0$, since $[\partial_1 u]$ is continuous, and so $g^{11} = 0$ when $x^1 = 0$; thus Σ is a characteristic.

Again, if $[u] \neq 0$ on $\Omega' \cap \Sigma$, we can, given $\psi \in C_0^\infty(\Omega' \cap \Sigma)$, choose $\phi \in C_0^\infty(\Omega')$ such that $\phi = 0$ and $\partial_1\phi = \psi$ when $x^1 = 0$. Then the $\partial_i\phi, i > 1$, vanish for $x^1 = 0$, and so (3.5.3) becomes

$$\int_{x^1=0} g^{11}\psi[u]\,|g|^{\frac{1}{2}}\,dx' = 0.$$

As $[u]$ is continuous, this gives $g^{11}\psi[u]\,|g|^{\frac{1}{2}} = 0$ on $x^1 = 0$ when again $g^{11} = 0$ for $x' = 0$. As every point of Σ has a neighbourhood of the type Ω', the theorem follows. $\Box$

Remark. The theorem also holds under the slightly weaker hypothesis $u \in C^2(\Omega\backslash\Sigma)$, $u \in C^1(\Omega_1 \cup \Sigma)$, and $u \in C^1(\Omega_2 \cup \Sigma)$. Furthermore, both Theorem 3.5.1 and 3.5.2 can be extended to the inhomogeneous wave equation $Pu = f$ provided that f is continuous in a neighbourhood of Σ.

As characteristics can be generated as envelopes of null cones, these results show that discontinuities propagate with the speed of light. A more detailed examination of (3.5.3) shows that there is a further link between geometrical optics and the laws governing the propagation of discontinuities. Before stating this, we introduce some auxiliary concepts. An $(n-2)$-surface (that is, a sub-manifold of M of co-dimension 2) can be considered as a manifold, with the differentiable structure and metric tensor inherited from M. If σ is space-like, then this induced metric on σ is negative definite, so that σ is, in effect, Riemannian. We thus have an invariant $(n-2)$-surface element on σ, which will be denoted by $d\sigma$. In a local coordinate chart such that $\sigma = \{x \colon x^{n-1} = x^n = 0\}$, this is

$$d\sigma = \left|\det g_{\alpha\beta}\big|_{x^{n-1}=x^n=0}\right|^{\frac{1}{2}} dx^1 \wedge \dots \wedge dx^{n-2}, \qquad (3.5.4)$$

where α, β run from 1 to $n-2$. (It is assumed here that σ is orientable, and has the orientation determined by $d\sigma > 0$.) If σ is the image of an imbedding $\tilde{\sigma} \to M$ on an $(n-2)$-dimensional manifold, then it is given in parametric form locally as

$$x = x(\lambda) \equiv x(\lambda^1, \dots, \lambda^{n-2}), \qquad (3.5.5)$$

where $\lambda \in \Lambda$, an open set in $\mathbf{R}^{n-2}$, and $x \in \Omega$, a coordinate neighbourhood in M; the map $\lambda \to x$ is of rank $n-2$ at every point. The restriction of the line element to σ is then

$$\langle dx, dx \rangle\big|_\sigma = \left\langle \frac{\partial x}{\partial \lambda^\alpha}, \frac{\partial x}{\partial \lambda^\beta} \right\rangle d\lambda^\alpha \, d\lambda^\beta.$$

Hence $\qquad d\sigma = \left|\det \left\langle \frac{\partial x}{\partial \lambda^\alpha}, \frac{\partial x}{\partial \lambda^\beta} \right\rangle\right|^{\frac{1}{2}} d\lambda^1 \wedge \dots \wedge d\lambda^{n-2}. \qquad (3.5.6)$

Suppose now that a characteristic is given as a map $(r, \lambda) \to p(r, \lambda)$ of $\mathbf{R} \times \Lambda$ into M, where $\Lambda \in \mathbf{R}^{n-2}$ is open, and the curves $r \to p(r, \lambda)$ are the bicharacteristics. Suppose further that, for

$$\lambda \in \Lambda \quad \text{and} \quad \delta_1(\lambda) < r < \delta_2(\lambda),$$

this map is injective and of rank $n-1$ everywhere (an imbedding). Then the image of $\{r, \lambda \colon \lambda \in \Lambda, \delta_1(\lambda) < r < \delta_2(\lambda)\}$ is a null hypersurface Σ. (One can think of $r = \delta_1$ and $r = \delta_2$ as adjacent focal points on a bicharacteristic.) Let Λ' be a bounded open subset of Λ, such that $\overline{\Lambda'} \subset \Lambda$. We shall call the restriction of $(r, \lambda) \to p(r, \lambda)$ to

$$\{(r, \lambda)\,;\, \lambda \in \overline{\Lambda'}, \delta_1(\lambda) < r < \delta_2(r)\}$$

a bicharacteristic tube Σ' in Σ. An $(n-2)$-surface σ, with boundary, will be called a cross-section of Σ' if it is met once and only once by each bicharacteristic of Σ', and is transversal to the bicharacteristics. It is then necessarily space-like, and has an equation of the form

$$r = \rho(\lambda) \in C^\infty(\overline{\Lambda'}).$$

Finally, let us say that a solution u of the homogeneous wave equation $Pu = 0$ has a discontinuity of order zero on Σ if every point of Σ has a neighbourhood in which the hypotheses of Theorem 3.5.2 hold, and $[u] \neq 0$ on Σ.

Theorem 3.5.3. *Let Σ be a null hypersurface, and suppose that u is a solution of the self-adjoint homogeneous wave equation $(\Box + b)u = 0$ that has a discontinuity of order zero on Σ. Let $\Sigma' \subset \Sigma$ be a bicharacteristic tube. Then the integral*

$$\int_\sigma [u]^2 d\sigma \qquad (3.5.7)$$

has the same value for all cross-sections σ of Σ'.

PROOF. By Corollary 3.3.1, Σ can be imbedded in a null field, and by the remark preceding this corollary, one can, again locally, introduce coordinates adapted to this null field. Then (3.3.11) and (3.3.12) hold,

$$g^{11} = 0,\, g^{12} = 1,\, g^{1i} = 0 \quad (i = 3, ..., n),\}$$
$$g_{21} = 1,\, g_{2i} = 0 \qquad (i = 2, ..., n).\} \qquad (3.5.8)$$

Let Ω denote a coordinate neighbourhood of a point of Σ which is of this type. Note that $\Omega \cap \Sigma = \{x\,;\, x^1 = 0\}$, that x^2 is the affine parameter, and that the coordinate lines $x^1 = 0$, $x^j = \text{const.}, j = 3, ..., n$ are the bicharacteristics of Σ. One can evidently suppose, shrinking Ω if necessary, that the hypotheses of Theorem 3.5.2 hold in Ω. It follows that (3.5.3) holds for all $\phi \in C_0^\infty(\Omega)$. As $\nabla^1 = \partial_2$, by (3.5.8), and as it has been assumed that $a \equiv 0$, (3.5.3) becomes

$$\int_{x^1=0} (\phi[\partial_2 u] - [u]\,\partial_2\phi)\,|g|^{\frac12} dx^2 ... dx^n = 0, \quad \phi \in C_0^\infty(\Omega).$$

Now $[\partial_2 u] = \partial_2[u]$, as it is a derivative tangential to Σ; in fact, it is a derivative in the direction tangent to the bicharacteristics. One can therefore integrate by parts, and as ϕ has compact support this gives

$$\int_{x^1=0} \phi(2\,\partial_2[u]+[u]\,|g|^{-\frac{1}{2}}\,\partial_2\,|g|^{\frac{1}{2}})\,|g|^{-\frac{1}{2}}\,dx^2\ldots dx^n = 0.$$

But the restriction map $C_0^\infty(\Omega) \to C_0^\infty(\Omega \cap \Sigma)$ is surjective, and the integrand is continuous; hence

$$2\,\partial_2[u]+[u]\,|g|^{-\frac{1}{2}}\,\partial_2\,|g|^{\frac{1}{2}} = 0, \quad \text{if} \quad x^1 = 0. \tag{3.5.9}$$

This is an ordinary differential equation for $[u]$ on each bicharacteristic, which is called a *transport equation*. As it can be put into the form

$$\partial_2(|g|^{\frac{1}{2}}\,[u]^2) = 0,$$

it implies that $[u]^2\,|g|^{\frac{1}{2}}$ is independent of x^2, say

$$[u]^2\,|g|^{\frac{1}{2}}|_{x^1=0} = F(x^3, \ldots, x^n). \tag{3.5.10}$$

In the special coordinate system characterized by (3.5.8), the line element is

$$\langle dx, dx \rangle = g_{11}(dx^1)^2 + 2dx^1 dx^2 + 2\sum_{i=3}^{n} g_{1i}\,dx^1 dx^i + \sum_{i,j=3}^{n} g_{ij}\,dx^i dx^j.$$

Hence if $\sigma = \{x;\; x^1 = 0,\; x^2 = \rho(x^3, \ldots, x^n)\}$ is a cross-section, then

$$\langle dx, dx \rangle|_\sigma = \sum_{i,j=3}^{n} g_{ij}|_\sigma\,dx^i dx^j,$$

and so the $(n-2)$-surface element on σ is

$$d\sigma = |\det(g_{ij})_{3\leqslant i,j\leqslant n}|_\sigma|^{\frac{1}{2}}\,dx^3 \wedge \ldots \wedge dx^n.$$

But, again by (3.5.8),

$$-g = \det(g_{ij})_{3\leqslant i,j\leqslant n}.$$

Hence

$$d\sigma = |g|^{\frac{1}{2}}|_\sigma\,dx^3 \wedge \ldots \wedge dx^n. \tag{3.5.11}$$

It therefore follows from (3.5.10) that $[u]^2\,d\sigma$ is independent of the affine parameter x^2. This conclusion is independent of the choice of local coordinates, and clearly implies (3.5.7). So the theorem is proved. □

Suppose that Σ is given locally in the form (3.4.1),

$$x = x(r,\lambda) \equiv x(r,\lambda^1,\ldots,\lambda^{n-2}), \tag{3.5.12}$$

where $\quad \left\langle \dfrac{\partial x}{\partial r},\dfrac{\partial x}{\partial r}\right\rangle = 0, \quad \left\langle \dfrac{\partial x}{\partial r},\dfrac{\partial x}{\partial \lambda^\alpha}\right\rangle = 0 \quad (\alpha = 1,\ldots,n-2). \tag{3.5.13}$

Then the restriction of the line element to Σ is

$$\langle dx, dx \rangle|_\Sigma = \left\langle \frac{\partial x}{\partial \lambda^\alpha}, \frac{\partial x}{\partial \lambda^\beta} \right\rangle d\lambda^\alpha d\lambda^\beta.$$

If σ is a cross-section of a bicharacteristic tube Σ', then

$$r = \rho(\lambda) \in C^\infty(\overline{\Lambda'}),$$

where Λ' is the base of Σ'. Hence

$$\langle dx, dx \rangle|_\sigma = \left\langle \frac{\partial x}{\partial \lambda^\alpha}, \frac{\partial x}{\partial \lambda^\beta} \right\rangle \Big|_{r=\rho} d\lambda^\alpha d\lambda^\beta,$$

whence $d\sigma = |\Delta(r, \lambda)|^{\frac{1}{2}} d\lambda^1 \wedge \ldots \wedge d\lambda^{n-2},$

where, again,

$$\Delta(r, \lambda) = \det\left\langle \frac{\partial x}{\partial \lambda^\alpha}, \frac{\partial x}{\partial \lambda^\beta} \right\rangle. \qquad (3.5.14)$$

Theorem 3.5.3 then implies that, if $x(r', \lambda)$ and $x(r'', \lambda)$ are points on the same bicharacteristic, then

$$[u(x(r', \lambda))]^2 |\Delta(r', \lambda)|^{\frac{1}{2}} = [u(x(r'', \lambda))]^2 |\Delta(r'', \lambda)|^{\frac{1}{2}}, \qquad (3.5.15)$$

which is an explicit formula for the variation of $[u]^2$ along a bicharacteristic.

It was assumed in the proof of Theorem 3.5.3 that Σ is a (connected) null hypersurface, and so contains no caustics. Hence $\Delta(r, \lambda) \neq 0$ on Σ, by Theorem 3.4.1. But this theorem also stated that $\Delta = 0$ on a caustic. It therefore follows from (3.5.15) that the discontinuity $[u(p(r, \lambda))]$ becomes infinite when $p(r, \lambda)$ approaches a focal point along a bicharacteristic. So the generalized geometrical optics intensity law (3.5.7) breaks down at a caustic. The simple method by which Theorem 3.5.3 has been proved cannot be used to elucidate the character of the singularity near the caustic, or to predict the singularity on an adjacent sheet of the characteristic of which Σ is a part. However, there is a simple rule connecting the singularities on adjacent sheets of a characteristic, which will be stated in the next section, in a more general context.

It is of some interest to write out (3.5.15) for a four-dimensional Minkowskian space–time. As in section 3.4, we write $x = (t, X), X \in \mathbf{R}^3$, and suppose that Σ is given by

$$t = r, \quad X = Y(\lambda_1, \lambda_2) + \theta(\lambda_1, \lambda_2) r,$$

where λ_1 and λ_2 are curvature line parameters on the initial wave front S_0, where $r = 0$. Then (3.4.11) holds,

$$dX = \left(1 - \frac{r}{R_1}\right)\frac{\partial Y}{\partial \lambda_1}d\lambda_1 + \left(1 - \frac{r}{R_2}\right)\frac{\partial Y}{\partial \lambda_2}d\lambda_2 + \theta\,dr. \qquad (3.5.16)$$

Now θ is a unit vector orthogonal to the tangential vectors $\partial Y/\partial\lambda_1$, $\partial Y/\partial\lambda_2$, and these are orthogonal as they are tangent to the curvature lines. Hence

$$dX^2 = E\left(1 - \frac{r}{R_1}\right)^2 d\lambda_1^2 + G\left(1 - \frac{r}{R_2}\right)^2 d\lambda_2^2 + dr^2,$$

where $E = (\partial Y/\partial\lambda_1)^2$, $G = (\partial Y/\partial\lambda_2)^2$, so that $E d\lambda_1^2 + G d\lambda_2^2$ is the line element (the 'first differential form') on S_0. Again,

$$\langle dx, dx\rangle = dt - dX^2 = -E\left(1 - \frac{r}{R_1}\right)^2 d\lambda_1^2 - G\left(1 - \frac{r}{R_2}\right)^2 d\lambda_2^2$$

when $t = r$, and so

$$\Delta(r, \lambda) = EG\left(1 - \frac{r}{R_1}\right)^2\left(1 - \frac{r}{R_2}\right)^2.$$

In the present case, one can consider $[u]$ as a function of r on each ray, rather than on each bicharacteristic. So (3.5.15) becomes

$$[u(r', \lambda)]\,|(R_1 - r')(R_2 - r')|^{\frac{1}{2}} = [u(r'', \lambda)]\,|(R_1 - r'')(R_2 - r'')|^{\frac{1}{2}}. \quad (3.5.17)$$

It follows from (3.5.16) that $R_1 - r$ and $R_2 - r$ are the principal radii of curvature of the wave front S_r at (r, λ). So (3.5.17) means that $[u]$ varies along a ray as the square root of the Gaussian curvature of the wave fronts. Note also that (3.5.16) and (3.5.17) imply the intensity law of geometrical optics in its usual form: the product of $[u]^2$ and of the area of a normal cross-section of an infinitesimal ray tube is constant along the tube.

It has been assumed that P is the self-adjoint operator $\Box + b$. It is not difficult to derive the corresponding results when

$$Pu = \Box u + \langle a, \nabla u\rangle + bu = 0.$$

In the special coordinate system characterized by (3.5.8), the identity (3.5.3) then becomes

$$\int_{x^1 = 0}(\phi[\partial_2 u] - [u]\,\partial_2\phi + a^1[u]\,\phi)\,|g|^{\frac{1}{2}}\,dx^2 \ldots dx^n = 0.$$

Arguing as before, one obtains the following transport equation:

$$2\,\partial_2[u] + (|g|^{-\frac{1}{2}}\partial_2|g|^{\frac{1}{2}} + a^1)\,[u] = 0 \quad \text{and} \quad x^1 = 0. \quad (3.5.18)$$

So (3.5.10) must be replaced by

$$[u]^2 \, |g|^{\frac{1}{2}}|_{x^1=0} \exp\left(\int^{x^2} a^1(0, s, x^3, \ldots, x^n) \, ds \right) = F(x^3, \ldots, x^n),$$

where the integral is taken from some suitable base point on each bicharacteristic. Now $\Sigma = \{x : x^1 = 0, x^2 = r, x^\alpha = \lambda^\alpha, \alpha = 3, \ldots, n\}$ in the notation (3.5.12). Hence

$$a^1 dx^2 = \left\langle a, \frac{\partial x}{\partial r} \right\rangle dr.$$

It is now obvious that (3.5.15) becomes, in the general case,

$$[u(x(r'', \lambda))]^2 \, |\Delta(r'', \lambda)|^{\frac{1}{2}}$$
$$= [u(x(r', \lambda))]^2 |\, \Delta(r', \lambda)|^{\frac{1}{2}} \exp\left(-\int_{r'}^{r''} \left\langle a(x(r, \lambda)), \frac{\partial x(r, \lambda)}{\partial r} \right\rangle dr \right).$$
$$\tag{3.5.19}$$

The results which have been obtained for scalar wave equations can be carried over to the tensor case. As all the arguments are local, and tensor sub- and superscripts can be raised and lowered in local coordinates, it will be sufficient to treat tensor differential operators that are of the form

$$(PU)_{i_1 \ldots i_m} = \nabla_j \nabla^j U_{i_1 \ldots i_m} + a_{i_1 \ldots i_m}{}^{jj_1 \ldots j_m} \nabla_j U_{j_1 \ldots j_m} + b_{i_1 \ldots i_m}{}^{j_1 \ldots j_m} U_{j_1 \ldots j_m}.$$

To simplify the notation, the m-tuplets $(i_1, \ldots, i_m)$ and $(j_1, \ldots, j_m)$ will be denoted by I and J, respectively. Thus

$$(PU)_I = \nabla_j \nabla^j U_I + a_I{}^{jJ} \nabla_j U_J + b_I{}^J U_J. \tag{3.5.20}$$

The adjoint of P is then a tensor differential operator acting on field of the type $(m, 0)$, given locally by

$$({}^tP\phi)^I = \nabla_j \nabla^j \phi^I - \nabla_i(a_J{}^{iI}\phi^J) + b_J{}^I \phi^J. \tag{3.5.21}$$

The two operators P and ${}^tP$ are connected by the differential identity

$$\phi^I(PU)_I - U_I({}^tP\phi)^I = \nabla_i(\phi^I \nabla^i U_I - U_I \nabla^i \phi^I + a_I{}^{iJ} U_J \phi^I).$$

The proof of Theorem 3.5.1 is virtually unaltered, and is left to the reader. To deal with discontinuities of orders zero and one, one proceeds as in the scalar case. Suppose that U is a (tensor-valued) distribution that satisfies $PU = 0$ and has such a discontinuity on a hypersurface Σ. Choosing a coordinate neighbourhood Ω in which Σ is the coordinate hypersurface $x^1 = 0$, one obtains, as in the proof of Theorem 3.5.2, an integral identity (analogous to (3.5.3))

$$\int_{x^1=0} (\phi^I[\nabla^1 U_I] - [U_I]\nabla^1\phi^I + a_I{}^{1J}[U_J]\phi^I)\,|g|^{\frac{1}{2}}\,dx' = 0, \quad (3.5.22)$$

for all tensor fields ϕ of type $(m,0)$ in $C_0^\infty(\Omega)$. The proof that this implies $g^{11} = 0$ when $x^1 = 0$, so that Σ is a characteristic, is exactly as before.

Suppose now that Σ is a characteristic, and that $[U] \neq 0$. To derive the transport equations, one can again work in a coordinate neighbourhood Ω in which Σ is $x^1 = 0$, and $g^{11} = g^{13} = \dots = g^{1n} = 0$, $g^{12} = 1$. Let I_r denote the m-tuplet obtained from $(i_1, \dots, i_m)$ when i_r, $r = 1, \dots, m$, is replaced by j. Then (with $x^1 = 0$)

$$[\nabla^1 U_I] = g^{1i}[\nabla_i U_I] = \nabla_2[U_I] = \partial_2[U_I] - r_{2i_r}^j[U_{I_r}]$$

and

$$\nabla^1\phi^I = g^{1i}\nabla_i\phi^I = \nabla_2\phi^I = \partial_2\phi^I + \Gamma_{2j}^{i_r}\phi^{I_r}.$$

Moreover,

$$[U_I]\,\Gamma_{2j}^{i_r}\phi^{I_r} = [U_{I_r}]\,\Gamma_{2i_r}^j\phi^I,$$

as can be seen by interchanging i_r and j in the first member. Hence (3.5.22) becomes

$$\int_{x^1=0} (\phi^I(\partial_2[U_I] - 2\Gamma_{2i_r}^i[U_{I_r}] + a_I{}^{1J}[U_J]) - [U_I]\,\partial_2\phi^I)\,|g|^{\frac{1}{2}}\,dx' = 0.$$

Here one can integrate by parts, and note again that every $\psi \in C_0^\infty(\Omega \cap \Sigma)$ is the restriction to $\Omega \cap \Sigma$ of some $\phi \in C_0^\infty(\Omega)$. Thus one can conclude that the transport equations

$$2\partial_2[U_I] - \Gamma_{2i_r}^j[U_{I_r}] + a_I{}^{1J}[U_J] + [U_I]\,\partial_2(\log|g|^{\frac{1}{2}}) = 0$$

hold, which can be written as

$$2\nabla_2[U_I] + a_I{}^{1J}[U_J] + [U_I]\,\partial_2(\log|g|^{\frac{1}{2}}) = 0. \quad (3.5.23)$$

Let τ be a solution of the scalar transport equation for the self-adjoint operator $\square$, so that

$$2\partial_2\tau + \tau\,\partial_2(\log|g|^{\frac{1}{2}}) = 0. \quad (3.5.24)$$

One can then put

$$[U_I] = \tau V_I \quad (3.5.25)$$

in (3.5.23), and deduce that

$$2\nabla_2 V_I + a_I{}^{1J} V_J = 0.$$

Finally, $g^{1i} = \xi^i = \partial x^i/\partial r$ is a vector field that is tangent to the bicharacteristics of Σ; so the last equation can be written in covariant form as

$$2\frac{\partial x^i}{\partial r}\nabla_i V_I + \frac{\partial x^i}{\partial r}a_{Ii}{}^J V_J = 0. \tag{3.5.26}$$

These are the equations that govern the variation of $V_I = [U_I]/\tau$ along the bicharacteristics of Σ. They cannot be integrated by quadratures, as in the scalar case. However, for an operator P with $a \equiv 0$ one has the noteworthy conclusion that $[U_I]/\tau$ remains parallel to itself along the bicharacteristics of Σ. This case will be discussed in more detail at the end of the next section.

We conclude the section with the remark that the results that have been obtained can be extended to some other simple types of singularities. Let Σ be a hypersurface, and let Ω be a neighbourhood of Σ in which there exists a field $S \in C^\infty(\Omega)$ such that $\nabla S \neq 0$ on Ω, $S \neq 0$ in $\Omega \backslash \Sigma$, and $S = 0$ on Σ. Suppose that $u \in \mathscr{D}'(\Omega)$ satisfies the homogeneous wave equation $Pu = 0$, and that it is of one of the following forms:

$$vS_+^\lambda,\ vS_-^\lambda,\ vS^{-1},\ v\log|S|, \tag{3.5.27}$$

where $v \in C^\infty(\Omega)$ and $v \neq 0$ on Σ. (Note that these distributions are all of the type that was discussed in section 2.9; the one-dimensional distributions t_+^λ, t_-^λ, (λ complex, $\lambda \neq -1, -2, \ldots$) and t^{-1} were defined in section 2.7.) Then one can show that Σ is necessarily a characteristic, and if one takes S to be a null field, then $v|_\Sigma$ satisfies the same propagation law on the bicharacteristics as the discontinuities which we have discussed already.

3.6   Progressing waves

The distributions (3.5.27) are all of the form

$$u = U(p)f(S(p)), \tag{3.6.1}$$

where U and S are C^∞ functions defined on some open set $\Omega \subset M$, with $\nabla S \neq 0$ on Ω, and $f \in \mathscr{D}'(\mathbf{R})$. Such a distribution will be called a *simple progressing wave*, with phase S, wave form f, and amplitude U. In a flat space–time, the ordinary wave equation has the plane wave solutions

$$u = f(\langle x, \xi \rangle),$$

where ξ is a fixed null vector; these simple progressing waves are solutions of the wave equation for all $f \in \mathscr{D}'(\mathbf{R})$. If $n = 4$, then there is a second class of progressing wave solutions, spherical waves. One can ask whether, in a general space–time, a homogeneous wave equation $Pu = 0$ can have progressing wave solutions with fixed phase and amplitude, and of arbitrary wave form.

To answer this question, we compute $P(Uf(S))$. By Theorem 2.9.2, $f(S)$ can be differentiated by the chain rule. Hence

$$\partial_i(Uf(s)) = f(S)\,\partial_i U + Uf'(S)\,\partial_i S \quad (i = 1,\dots,n),$$

and so

$$\square\,(Uf(S)) = \nabla_i(f(S)\,\nabla^i U + Uf'(S)\,\nabla^i S)$$

$$= f(S)\,\square\,U + (2\,\langle\nabla S,\nabla U\rangle + U\,\square\,S)f'(S)$$

$$+ U\,\langle\nabla S,\nabla S\rangle f''(S).$$

Thus

$$P(Uf(S)) = f(S)\,PU + (2\,\langle\nabla S,\nabla U\rangle + (\square\,S + \langle a,\nabla S\rangle)\,U)f'(S)$$

$$+ U\,\langle\nabla S,\nabla S\rangle f''(S), \quad (3.6.2)$$

an identity that will be used repeatedly in the next chapter. If $P(Uf(S))$ is to vanish for all $f \in \mathscr{D}'(\mathbf{R})$, then the coefficients of f, f' and f'' must all vanish identically. So one has three conditions for the two functions U and S, and in general one cannot expect that they can be satisfied, except trivially. The case of the ordinary wave equation in four dimensions shows that there are exceptions. In fact, all simple progressing wave solutions of the ordinary wave equations can be found explicitly; this will be done in the next section. On the other hand, the Klein–Gordon equation $(\square + \mu^2)\,u = 0$, where $\mu \neq 0$ is a constant, does not have simple progressing wave solutions.

However, it is a general rule that $f'(t)$ is 'worse' than $f(t)$; for instance, if f is of order k, then f' is of order $k + 1$. So, instead of trying to satisfy $Pu = 0$ exactly, one can adopt the more modest aim of making Pu as 'good' as possible, for all f, by an appropriate choice of U and S. First, one can take S to be a null field, so that the term in $f''(S)$ disappears. When a null field has been chosen, one can remove the term in $f'(S)$ by requiring U to be a solution of

$$2\,\langle\nabla S,\nabla U\rangle + (\square\,S + \langle a,\nabla S\rangle)\,U = 0.$$

This is just the transport equation (3.5.18) in general form, as it reduces to (3.5.18) in a coordinate neighbourhood in which $S = x^1$ and $g^{11} = g^{13} = \dots = g^{1n} = 0$, $g^{12} = 1$. One is then left with

$$P(Uf(S)) = P(U)f(S), \quad (3.6.3)$$

and this is the most that can be achieved for a simple progressing wave.

If $f(t) \in \mathscr{D}'(\mathbf{R})$, then there are distributions $f_1 \in \mathscr{D}'(\mathbf{R})$ such that

$f_1' = f$; they are called primitives (or antiderivatives) of f. To prove this, choose an $\alpha(t) \in C_0^\infty(\mathbf{R})$ such that

$$\int \alpha \, dt = 1.$$

Then, if $\phi \in C_0^\infty(\mathbf{R})$, one has

$$\int \left(\phi(s) - \alpha(s) \int \phi \, dt \right) ds = 0.$$

Hence
$$\psi(t) = \int_t^\infty \left(\phi(s) - \alpha(s) \int \phi \, dt \right) ds \qquad (3.6.4)$$

has compact support, so that it is in $C_0^\infty(\mathbf{R})$, and

$$\phi(t) = -\psi'(t) + \alpha(t) \int \phi \, dt. \qquad (3.6.5)$$

One can now define a primitive of f by putting

$$(f_1, \phi) = (f, \psi) + C \int \phi \, dt, \qquad (3.6.6)$$

where $C - (f_1, \alpha)$ is an arbitrary constant.

For it is an obvious consequence of (3.6.4) that $\psi \to 0$ in $C_0^\infty(\mathbf{R})$ when $\phi \to 0$ in $C_0^\infty(\mathbf{R})$, so that (f_1, ϕ) is a distribution. Also, the second member of (3.6.4) becomes $-\phi$ when ϕ is replaced by ϕ', whence

$$(f_1', \phi) = -(f_1, \phi') = (f, \phi).$$

Any other primitive of f differs from f_1 by a constant. For the difference of two primitives is a distribution g such that $g' = 0$. By (3.5.6), one has

$$(g, \phi) = -(g, \psi') + (g, \alpha) \int \phi \, dt = (g', \psi) + (g, \alpha) \int \phi \, dt,$$

and as $g' = 0$ it follows that $(g, \phi) = (C, \phi)$ where $C = (g, \alpha)$ is a constant.

Let us replace f by a primitive f_1 in (3.6.2), and at the same time replace U by another C^∞ function U_1. As S is now supposed to be a null field, this gives

$$P(U_1 f_1(S)) = P(U_1) f_1(S) + (2\langle \nabla S, \nabla U_1 \rangle + (\Box S + \langle a, \nabla S \rangle) U_1) f(S).$$

If we also suppose that U and U_1 are linked by the equation

$$2\langle \nabla S, \nabla U_1 \rangle + (\Box S + \langle a, \nabla S \rangle) U_1 = -PU,$$

and add the identity for $P(U_1 f_1(S))$ to (3.6.2), we obtain

$$P(U f(S) + U_1 f_1(S)) = P(U_1) f_1(S).$$

This is an improvement on (3.6.3), as f_1 is 'better' than f. By repeating this step a finite number of times, one obtains an algorithm for the construction of approximate solutions of $Pu = 0$. Let us write f_0 for f and U_0 for U, and suppose that functions $U_0, U_1, ..., U_N$, all in C^∞, have been constructed so as to satisfy the transport equations

$$2\langle \nabla S, \nabla U_0 \rangle + (\square S + \langle a, \nabla S \rangle) U_0 = 0, \qquad (3.6.7)$$

$$2\langle \nabla S, \nabla U_\nu \rangle + (\square S + \langle a, \nabla S \rangle) U_\nu = - P U_{\nu-1} \quad (\nu = 1, ..., N). \quad (3.6.8)$$

Let $f_1, f_2, ..., f_N$ be a sequence of iterated primitives of f_0, so that

$$f'_\nu(t) = f_{\nu-1}(t) \quad (\nu = 1, 2, ..., N). \qquad (3.6.9)$$

Then
$$u_N = \sum_{\nu=0}^{N} U_\nu(p) f_\nu(S(p)) \qquad (3.6.10)$$

will be called a *progressing wave of order* N. It follows from (3.6.2), (3.6.9), and $\langle \nabla S, \nabla S \rangle = 0$, that

$$P(U_\nu f_\nu(S)) = P(U_\nu)(f_\nu(S) + (2\langle \nabla S, \nabla U_\nu \rangle + (\square S + \langle a, \nabla S \rangle) U_\nu) f_{\nu-1}(S),$$

for $\nu = 0, 1, ..., N$, with $f_{-1} = f'_0$. By (3.6.7) and (3.6.8), these identities become

$$P(U_0 f_0(S)) = P(U_0) f_0(S),$$

$$P(U_\nu f_\nu(S)) = P(U_\nu) f_\nu(S) - P(U_{\nu-1}) f_{\nu-1}(S) \quad (\nu = 1, ..., N).$$

Hence
$$Pu_N = P(U_N) f_N(S), \qquad (3.6.11)$$

and this relation, valid for fixed $S, U_0, ..., U_N$, and for all $f_0 \in \mathscr{D}'(\mathbf{R})$, characterizes progressing waves of order N. If it happens that $PU_N = 0$, then u_N is actually a solution of $Pu = 0$ for all $f \in \mathscr{D}'(\mathbf{R})$. As in the case of simple progressing waves, this is an exceptional case. For the $N+2$ functions $S, U_0, ..., U_N$ have to satisfy $N+3$ equations, namely, $\langle \nabla S, \nabla S \rangle = 0$, the $N+1$ equations (3.6.7), (3.6.8), and $PU_N = 0$.

It has already been pointed out that the transport equation of order zero, (3.6.7), is similar to the transport equation for a discontinuity. The only difference is that (3.6.7) holds on all bicharacteristics of S. (Here, a curve is called a bicharacteristic of the null field S if it is a bicharacteristic of one of the characteristics $S = $ const.) Locally, one can use a coordinate system adapted to S, so that

$$S = x^1, \quad g^{12} = 1, \quad g^{11} = g^{12} = ... = g^{1n} = 0.$$

Then the coordinate curves $x^1 = $ const., $x^3 = $ const., $..., x^n = $ const.

are the bicharacteristics of S, and x^2 is an affine parameter on each of these. The transport equation (3.6.7) becomes

$$2\partial_2 U_0 + (a^1 + \partial_2 \log |g|^{\frac{1}{2}}) U_0 = 0.$$

One can fix U_0 by giving its value on a hypersurface that is transversal to the bicharacteristics of S, say on $x^2 = 0$. Then

$$U_0 = U_0 |g|^{\frac{1}{4}} \Big|_{x^2=0} |g|^{-\frac{1}{4}} \exp\left(-\int_0^{x^2} a^1 \Big|_{x^2=s} ds\right). \qquad (3.6.12)$$

The U_ν can then be computed recursively, provided that their values on $x^2 = 0$ are also prescribed. As it follows from (3.6.7) and (3.6.8) that

$$\langle \nabla S, \nabla(U_\nu/U_0)\rangle = -\frac{PU_{\nu-1}}{2U_0},$$

one finds

$$U_\nu = -\frac{U_\nu}{2U_0}\Big|_{x^2=0} U_0 \int_0^{x^2} \frac{PU_{\nu-1}}{U_0}\Big|_{x^2=s} ds \quad (\nu = 1, ..., N). \qquad (3.6.13)$$

Note that (3.6.12) implies that $U_0 \neq 0$ so that all the U_ν are C^∞ functions, provided that their initial values (on $x^2 = 0$) are C^∞ functions of $x^1, x^3, ..., x^n$.

It is not difficult to write down similar formulae for the U_ν when S has been determined as the solution of an initial value problem (Theorem 3.3.2). One then has a diffeomorphism

$$x = x(r, \sigma, \lambda) \equiv x(r, \sigma, \lambda^1, ..., \lambda^{n-2}) \qquad (3.6.14)$$

such that $S = \sigma(x)$; the curves

$$\sigma = \text{const.}, \quad \lambda^\alpha = \text{const.} \quad (\alpha = 1, ..., n-2),$$

are the bicharacteristics of S, and r is a common affine parameter. Coordinates adapted to S are obtained by putting

$$\tilde{x}^1 = \sigma, \quad \tilde{x}^2 = r, \quad \tilde{x}^{\alpha+2} = \lambda^\alpha \quad (\alpha = 1, ..., n-2).$$

In these coordinates, (3.6.12) and (3.6.13) hold. Also,

$$|g(x)|^{\frac{1}{2}} \left|\frac{Dx}{D\tilde{x}}\right| = |\tilde{g}(\tilde{x})|^{\frac{1}{2}}.$$

Let us put

$$\kappa(r, \lambda, \sigma) = |g(x(r, \sigma, \lambda))|^{-\frac{1}{4}} \left|\frac{D(x^1, ..., x^n)}{D(r, \sigma, \lambda^1, ..., \lambda^{n-2})}\right|^{-\frac{1}{2}}. \qquad (3.6.15)$$

To simplify the notation, the composite functions $U_\nu(x(r, \sigma, \lambda))$ will also be denoted by U_ν. Then

$$
\left.
\begin{aligned}
U_0 &= \frac{U_0}{\kappa}\bigg|_{r=0} \kappa \exp\left(-\int_0^x \left\langle (s, \sigma, \lambda), \frac{\partial}{\partial s} x(s, \sigma, \lambda) \right\rangle ds \right), \\
U_\nu &= -\frac{U_\nu}{2U_0}\bigg|_{r=0} U_0 \int_0^r \frac{PU_{\nu-1}}{U_0}\bigg|_{r=s} ds \quad (\nu = 1, \ldots, N).
\end{aligned}
\right\}
\tag{3.6.16}
$$

The motivation for the construction of progressing waves was that $f_{\nu+1}$ is in some sense 'better' than f_ν. A simple but important example is obtained by taking $f_\nu = (i\omega)^{-\nu} \exp(i\omega t)$, where ω is a fixed real number. Then (3.6.10) becomes

$$
u_N = e^{i\omega S} \sum_{\nu=0}^{N} (i\omega)^{-N} U_\nu.
\tag{3.6.17}
$$

The leading term, which dominates as $|\omega| \to \infty$, can be taken to represent a simple harmonic wave of (circular) frequency ω. As S is a null field, and U_0 varies along the bicharacteristics of S according to a generalized geometrical optics intensity law, this leading term can be interpreted as the geometrical optics approximation to a solution of $Pu = 0$. The other terms are corrections to this geometrical optics field. It follows from (3.6.11) that

$$
Pu_N = (i\omega)^{-N} e^{i\omega S} PU_N.
$$

Suppose that there exists a function v_N such that

$$
Pv_N = -e^{i\omega S} Pu_N,
$$

and that $v_N \to 0$ as $|\omega| \to \infty$. Then

$$
u = u_N + (i\omega)^{-N} v_N
$$

is a solution of $Pu = 0$ that is asymptotic to u in the usual sense,

$$
\lim_{|\omega| \to \infty} |\omega|^N (u - u_N) = 0.
$$

The existence of such a v_N, in a definite mathematical context that is of physical interest, is usually difficult to establish. Nevertheless, progressing waves of this type have been used successfully in applications, especially to obtain high frequency approximations to the solutions of scattering problems.

One can also make precise statements about the f_ν when f_0 is a distribution of finite order. Distributions of finite order were defined at the end of section 2.1. For $\mathscr{D}'(\mathbf{R})$, an equivalent definition is as follows: $f \in \mathscr{D}'(\mathbf{R})$ is of order $k < \infty$ if (i) for every $a > 0$, there is a constant $C = C(a)$ such that

$$|(f, \phi)| \leqslant C \sum_{j=0}^{k} \sup |\phi^{(j)}(t)|, \qquad (3.6.18)$$

for all $\phi \in C_0^\infty(\mathbf{R})$ whose supports are contained in the closed interval $[-a, a]$, and (ii) when $k \geqslant 1$, then there is no a such that a semi-norm estimate of this type holds when k is replaced by l, $0 \leqslant l < k$.

It is easy to prove that if f is of order $k \geqslant 1$ then every primitive f_1 of f is of order $k - 1$. As constants are distributions of order zero, one can take f_1 to be given by (3.6.6) with $C = 0$,

$$(f_1, \phi) = (f, \psi), \quad \psi = \int_t^\infty \left(\phi(s) - \alpha(s) \int \phi\, dt \right) ds,$$

where $\alpha \in C_0^\infty$ and $\int \alpha\, dt = 1$. Let a_0 be such that $\alpha(t) = 0$ when $|t| \geqslant a_0$. Then it follows from supp $\phi \subset [-a, a]$ that supp $\psi \subset [-b, b]$, where $b = \max(a_0, a)$. From this and (3.6.18) one can easily conclude that there is a constant $C_1 = C_1(a)$ such that

$$|(f_1, \phi)| \leqslant C_1 \sum_{j=0}^{k-1} \sup |\phi^{(j)}(t)|,$$

which proves that f_1 is of order $l \leqslant k - 1$. If $k = 1$, then there is nothing more to prove. If $k > 1$ and $l < k - 1$, then $(f, \phi) = (f_1', \phi) = -(f_1, \phi')$ implies that f is of order $l + 1 < k$, which contradicts the hypothesis on f. Hence f_1 is of order $k - 1$.

Suppose now that f_0 is of order k, and that the f_ν, $\nu = 1, 2, \ldots$, are iterated primitives of f_0. Clearly, f_ν is of order $k - \nu$ if $\nu \leqslant k$. As f_k is of order zero, it follows from a well-known theorem, due to F. Riesz, that it is a measure. In one dimension this means that there exists a function $m_k(t)$, of bounded variation in every finite interval, such that

$$(f_k, \phi) = \int \phi(t)\, dm_k(t).$$

This function is determined uniquely by f_k, up to an additive constant. Integration by parts now shows that $f_{k+1} = m_k$. For $\nu \geqslant k + 2$, the f_ν are just primitives of m_k in the usual sense, and so $f_\nu \in C^{\nu-k-2}(\mathbf{R})$ for $\nu \geqslant k + 2$. This also holds when $k = 0$.

These properties of the f_ν are relevant when progressing waves are used to represent a singularity carried on characteristics. There are situations where it is possible to expand a solution u of a homogeneous wave equation as a progressing wave, in the sense that, for all $N \geqslant 0$, one has a decomposition

$$u = \sum_{\nu=0}^{N} U_\nu f_\nu(S) + v_N, \qquad (3.6.19)$$

with a remainder that is smoother than the last term in the sum. A case in point (and the only one which will appear in this book) is the field of a line source, which will be treated in section 5.6.

As in the case of the propagation of simple discontinuities, the progressing wave formalism breaks down at a caustic. For the Jacobian $D(x^1, \ldots, x^n)/D(r, \sigma, \lambda^1, \ldots, \lambda^{n-2})$ vanishes on the union of the caustics of the characteristics $S = \sigma = $ const., and so it can be seen from (3.5.15) and (3.6.16) that all the U_ν become infinite on this set. This is an inherent defect of the method. It can be overcome, but it is beyond the scope of this book to discuss the focusing problem. However, it is easy to state the rule for connecting the leading terms of progressing wave expansions on the two sides of a caustic.

Suppose that, in (3.6.14), $(\sigma, \lambda) \in D$, a connected open set in $\mathbf{R}^{n-1}$, and $0 \leqslant r \leqslant r_1$, where r_1 is a positive constant; denote $[0, r_1] \in \mathbf{R}$ by I. It is of course assumed that (3.3.10) holds on $I \times D$, so that (3.6.14) represents a family of (extended) characteristics $S(x) = \sigma = $ const. Suppose also that $D(x^1, \ldots, x^n)/D(r, \sigma, \lambda^1, \ldots, \lambda^{n-2}) \neq 0$ on $I \times D$, except when $r = \rho(\sigma, \lambda)$, where $0 < \rho < r_1$, and $\rho \in C^\infty(D)$. Thus $x = x(\rho, \sigma, \lambda)$ is the union of the caustics of these characteristics.

Assume that, for $0 \leqslant r < \rho$, u is a solution of the homogeneous wave equation $Pu = 0$ that is given by (3.6.19), where $f_0(t)$ has compact support, and v_N is sufficiently smooth. Then there is a corresponding progressing wave expansion in $\rho < r \leqslant r_1$. The leading term of this is $U_0 \hat{f}_0(\sigma)$, where U_0 is still given by (3.6.16), and $\hat{f}_0$ is the *Hilbert transform* of f_0. This is defined by

$$\hat{f}_0(t) = -\frac{1}{\pi t} * f_0(t).$$

If, more generally, $f_0 \in L_2(\mathbf{R})$ (that is to say if it is square integrable), then $\hat{f}_0$ is a principal value,

$$\hat{f}_0(t) = \frac{1}{\pi} \lim_{\epsilon \downarrow 0} \left(\int_{-\infty}^{t-\epsilon} + \int_{t+\epsilon}^{\infty} \right) \frac{f_0(s)}{s-t} ds,$$

and $f_0 \to \hat{f}_0$ is an isometry of $L_2(\mathbf{R})$,

$$\int |\hat{f}_0(t)|^2 dt = \int |f_0(t)|^2 dt.$$

Generally, the Hilbert transform maps the space of distributions $\mathscr{D}'_{L_2}$, which consists of finite sums of derivatives of L_2 functions, onto itself. One then has $\hat{\hat{f_0}} = -f_0$. If $f_0(t)$ has a jump discontinuity at $t = t_0$, then $\hat{f}$ behaves like

$$\frac{1}{\pi}(f(t_0+)-f_0(t_0-))\log\frac{1}{|t-t_0|}$$

near $t = t_0$; this is relevant when the wave u has a discontinuity of order zero, in the sense defined in the last section, on the characteristic $S = t_0$. Near the union of the caustics of the characteristics

$$S = \sigma = \text{const.},$$

the behaviour of the solution is more complicated, but a uniformly valid approximation can be obtained.

The progressing wave formalism can also be developed for tensor wave equations, and we conclude the section with a brief account of this. Consider, for simplicity, the equation

$$(Pu)_I = \nabla_j \nabla^j u_I + b_I{}^J u_J = 0, \tag{3.6.20}$$

where, as in the last section, $I = (i_1, ..., i_m)$, $J = (j_1, ..., j_m)$, and u is a tensor field of rank m. If S is a null field, then a simple computation gives

$$P(Uf(S))_I = (PU)_I f(S) + (2\nabla^i S \nabla_i U_I + U_I \square S)f'(S).$$

One can now set up a progressing wave expansion of degree N, in the form

$$\sum_{\nu=0}^{N} U_I^{(\nu)}(p)f_\nu(S),$$

where the f_ν are again iterated primitives of some $f_0(t) \in \mathscr{D}'(\mathbf{R})$. Corresponding to the scalar transport equations (3.6.7) and (3.6.8), one then has the transport equations

$$2\nabla^i S \nabla_i U_I^{(0)} + U_I^{(0)} \square S = 0, \tag{3.6.21}$$

$$2\nabla^i S \nabla_i U_I^{(\nu)} + U_I^{(\nu)} \square S = -(PU^{(\nu-1)})_I \quad (\nu = 1, 2, ..., N), \tag{3.6.22}$$

which ensure that

$$(Pu_{(N)})_I = (P\,U^{(N)})_I f_N(S).$$

Let κ again denote a solution of the first scalar transport equation (3.6.7), with $a \equiv 0$. Then, if one sets

$$U_I^{(\nu)} = \kappa V_I^{(\nu)} \quad (\nu = 0, 1, ..., N), \tag{3.6.23}$$

it follows from (3.6.21) and (3.6.22) that

$$\nabla^i S \, \nabla_i \, V_I^{(0)} = 0, \tag{3.6.24}$$

$$\nabla^i S \nabla_i V_I^{(\nu)} = -\tfrac{1}{2}\kappa^{-1}P(\kappa V^{(\nu-1)})_r \quad (\nu = 1, 2, ..., N). \tag{3.6.25}$$

As $\nabla^i S = \partial x^i/\partial r$, where $r \to x(r, \cdot)$ are the bicharacteristics of S, and r is an affine parameter, these equations are ordinary differential equations on each bicharacteristic. In particular, (3.6.24) shows that $V^{(0)}$ is obtained by parallel transport along the bicharacteristics (which are null geodesics). Let $\xi^{(l)}, l = 1, ..., n$ be a set of n covectors defined by parallel transport,

$$\nabla^i S \nabla_i \xi^{(l)} = \frac{\partial \xi_i^{(l)}}{\partial r} - \Gamma_{ik}^j \, \xi_j^{(l)} \frac{\partial x^k}{\partial r} = 0 \quad (i, l = 1, ..., n), \tag{3.6.26}$$

and the initial conditions

$$\xi_i^{(l)}|_{r=0} = \delta_i^l \quad (i, l = 1, ..., n). \tag{3.6.27}$$

Then it is easily verified that, for given initial values of the $V_I^{(0)}$, the solution of (3.6.24) is

$$V_I^{(0)} = V_{i_1, ..., i_m}^{(0)} = V_{j_1 \dots j_m}^{(0)}|_{r=0} \xi_{i_1}^{(j_1)} \dots \xi_{i_m}^{(j_m)}. \tag{3.6.28}$$

The equations (3.6.25) can also be solved for the $V_I^{(\nu)}$ in terms of the $V_I^{(\nu-1)}$, by the method of variation of parameters; the details are straightforward, and are left to the reader.

3.7 Simple progressing wave solutions of the ordinary wave equation

The simple progressing wave solutions of the ordinary wave equation

$$\Box u = (\partial_1^2 - \partial_2^2 - \partial_3^2 - \partial_4^2)\, u = 0 \tag{3.7.1}$$

can be determined explicitly. Such a solution is of the form (3.6.1), which we now write as

$$u = U(x)f(S(x)), \tag{3.7.2}$$

where U and S are C^∞ functions, defined on some open set $\Omega \subset \mathbf{R}^4$, and $f \in \mathscr{D}'(\mathbf{R})$. It follows from (3.6.2) that (3.7.2) will be a solution of (3.7.1) for all f if

$$\langle \nabla S, \nabla S \rangle = 0, \quad 2\langle \nabla S, \nabla U \rangle + U \Box S = 0, \quad \Box U = 0. \tag{3.7.3}$$

It can be shown – by taking $f(s) = \delta(s - \sigma)$, where σ is a constant – that these conditions, which are sufficient, are also necessary.

As in section 3.4, we shall also write $x = (t, X)$, where $X \in \mathbf{R}^3$, and $S(x) = S(t, X)$, $U(x) = U(t, X)$. If $\tau(X)$ is a solution of the eikonal equation $(\operatorname{grad}_X \tau)^2 = 1$, then $S = t - \tau(X)$ is a null field. Taking $S = t - \tau(X)$ and $U = V(X)$ in (3.7.2), one obtains a special class of simple progressing waves,

$$u = V(X) f(t - \tau(X)). \qquad (3.7.4)$$

The conditions (3.7.3) then become

$$(\operatorname{grad}_X \tau)^2 = 1, \quad 2 \operatorname{grad}_X \tau . \operatorname{grad}_X V + V \Delta \tau = 0, \quad \Delta V = 0, \quad (3.7.5)$$

where Δ denotes the Laplacian in $\mathbf{R}^3$.

The general case (3.7.2) can be reduced locally to the special case (3.7.4). As

$$\langle \nabla S, \nabla S \rangle = (\partial_1 S)^2 - (\partial_2 S)^2 - (\partial_3 S)^2 - (\partial_4 S)^2 = 0,$$

$\partial_1 S = 0$ at a point y implies that $\nabla S(y) = 0$, and this case must be excluded as such a point would be on the caustic of the characteristic $S(x) = S(y)$. Hence it must be assumed that $\partial_1 S \neq 0$ on Ω. By the inverse function theorem, every point $y \in \Omega$ therefore has a connected open neighbourhood $\Omega' \subset \Omega$ in which the map

$$x \to \tilde{x} = (S(x), x^2, x^3, x^4)$$

has a unique inverse, say

$$\tilde{x} \to x = (\tau(\tilde{x}), \tilde{x}^2, \tilde{x}^3, \tilde{x}^4),$$

and $x \to \tilde{x}$ is a diffeomorphism between connected open sets in $\mathbf{R}^4$. It follows that if $\sigma \in S(\Omega')$, then $S(x) = \sigma$ can be solved for x^1 by setting $x^1 = \tau(\sigma, X)$. By writing out the conditions (3.7.3) in the coordinates $\tilde{x}^i$, one can prove that, if $V(\sigma, X)$ is defined by

$$V(\sigma, X) = U(\tau(\sigma, X), X) (\partial/\partial\sigma) \tau(\sigma, X),$$

then $X \to \tau(\sigma, X)$ and $X \to V(\sigma, X)$ satisfy the conditions (3.7.5). It is therefore sufficient to determine the simple progressing wave solutions of the wave equation which are of the special form (3.7.4).

Conversely, if $X \to \tau(\sigma, X)$ and $X \to V(\sigma, X)$ are a pair of functions that satisfy (3.7.5), and depend smoothly on a parameter $\sigma \in \mathbf{R}$, then one can construct, locally, functions $S(x)$ and $U(x)$ which satisfy (3.7.3). The function S is obtained by solving $\tau(S, X) = t$ for S, assuming that $(\partial/\partial\sigma) \tau(\sigma, X) \neq 0$; U is then given by

$$U(x) = V(\sigma, X) ((\partial/\partial\sigma) \tau(\sigma, X))^{-1}|_{\sigma = S(x)}.$$

It was shown in section 3.4 that, if $\tau(X)$ is a solution of the eikonal equation, then the surfaces $\{X; \tau(X) = \text{const.}\}$ in $\mathbf{R}^3$ are a family of wave fronts, in the sense of ordinary geometrical optics. These surfaces are parallel to each other; their common normals are the associated system of rays, and they have a common focal surface which consists in general of two components. Now it can be shown that if the equations (3.7.5), which are an over-determined system, are to have a solution, then both components of this focal surface must be curves. The systems of wave fronts with this property are families of certain algebraic surfaces of the fourth order, the cyclides of Dupin. There is a sub-class of third-order surfaces, and plane, cylindrical, conical, spherical and toroidal wave fronts are also included.

The focal curves can, in fact, be determined as follows. The condition implies that the caustic of the characteristic $S = t - \tau(X) = 0$ also degenerates to two curves F' and F''. Let these be given by $x = \phi(\lambda)$ and $x = \psi(\mu)$ respectively, where λ and μ are real parameters. The bicharacteristics are the straight lines in $\mathbf{R}^4$ joining a point of F' to a point of F''. As they must be null lines, the functions ϕ and ψ must satisfy the identity

$$\langle \phi(\lambda) - \psi(\mu), \phi(\lambda) - \psi(\mu) \rangle = 0 \qquad (3.7.6)$$

for all λ, μ in $\mathbf{R}$, or in appropriate open intervals. Differentiation with respect to λ and μ yields two more identities,

$$\langle \phi(\lambda) - \psi(\mu), \phi'(\lambda) \rangle = 0, \quad \langle \psi(\mu) - \phi(\lambda), \psi'(\mu) \rangle = 0.$$

Putting $\mu = \mu_1$ and $\mu = \mu_2$, in turn, in the first of these, and subtracting, one obtains

$$\langle \psi(\mu_2) - \psi(\mu_1), \phi'(\lambda) \rangle = 0.$$

Similarly, the second identity gives

$$\langle \phi(\lambda_1) - \phi(\lambda_2), \psi'(\mu) \rangle = 0.$$

As these identities hold for all (μ_1, μ_2, λ) and all $(\lambda_1, \lambda_2, \mu)$ respectively, it follows that every chord of F' is orthogonal to every tangent of F'', and vice versa. Hence F' and F'' lie in two orthogonal 2-planes, which we denote by Σ' and Σ'' respectively. There are now three possibilities.

(i) Suppose that Σ' is space-like; then Σ'' is time-like, and $\Sigma' \cap \Sigma''$ consists of a single point, which may be taken as the origin of co-ordinates. Then also $\langle \phi(\lambda), \psi(\mu) \rangle = 0$, and so (3.7.6) becomes

$$\langle \phi(\lambda), \phi(\lambda) \rangle + \langle \psi(\mu), \psi(\mu) \rangle = 0.$$

As λ and μ are independent, this can only hold if

$$\langle \phi(\lambda), \phi(\lambda) \rangle = -\alpha^2, \quad \langle \psi(\mu), \psi(\mu) \rangle = \alpha^2,$$

where α is a constant. So F' and F'' are pseudo-circles, with a common centre and the same 'radius' α, lying in orthogonal 2-planes that are respectively time-like and space-like.

(ii) A similar argument shows that, if Σ' and Σ'' are both null, then F' and F'' are again pseudo-circles.

(iii) If one of the caustic curves is at infinity, then the other one is a straight line which is either space-like or null.

The corresponding characteristics will be described in reverse order. In case (iii), one finds that the characteristic in question has the equation
$$\langle x, x \rangle \langle \xi, \xi \rangle = (\langle x, \xi \rangle)^2,$$

where ξ is a constant vector, which is space-like or null. If ξ is null, then this reduces to $\langle x, \xi \rangle = 0$, and the characteristic is a hyperplane; so in this case the wave fronts are plane. If ξ is space-like, then the appearance of the wave fronts depends on the choice of the coordinate system, which is only determined up to a Lorentz transformation; one may say that it depends on the rest frame of the observer. In a coordinate system in which $\xi_0 = 0$ (so that the caustic curve is at rest), they are circular cylinders, with a common axis. In general, they are seen as circular cones with a common axis.

In case (ii), one can choose the coordinate system so that Σ' is

$$x^1 + x^2 = t + X_1 = -p, \quad x^4 = X_3 = 0,$$

and Σ'' is $\qquad x^1 + x^2 = t + X_1 = p, \qquad x^3 = X_2 = 0,$

where p is a constant. One then finds that F' and F'' are parabolas in orthogonal planes which pass through each other's foci,

$$F': t + X_1 = -p, \quad X_2^2 = 4p(X_1 + \tfrac{1}{2}p), \quad X_3 = 0,$$
$$F'': t + X_1 = p, \quad X_3^2 = 4p(X_1 + \tfrac{1}{2}p), \quad X_2 = 0.$$

The characteristic is an algebraic surface of the third order,

$$(t + X_1)(t^2 - X_1^2 - X_2^2 - X_3^2) = p(X_2^2 - X_3^2) + p^2(t - X_1).$$

Finally, in case (i), the focal curves in $\mathbf{R}^3$ are an ellipse and a hyperbola, lying in orthogonal planes, and passing through each other's foci, and the characteristic is a fourth-order surface. However, if the coordinate system is chosen so that Σ' is at rest, say

$\Sigma' = \{(t, X); t = X_3 = 0\}$, then F' is the circle $X_1^2 + X_2^2 = \alpha^2$, and F'' is the hyperbola $\{(t, X); t^2 - X_3^2 = \alpha^2, X_2 = X_3 = 0\}$. The wave fronts are concentric tori,

$$t^2 = ((X_1^2 + X_2^2)^{\frac{1}{2}} - \alpha)^2 + X_3^2,$$

which intersect themselves when $|t| \geqslant \alpha$. For $\alpha = 0$, one has spherical waves. The general case can be deduced from this special one by a Lorentz transformation.

It can be shown that, for each of these types of characteristic, there is a non-trivial amplitude V which satisfies (3.7.5). A distinctive feature of the result is that in most cases V is many-valued, so that the resulting simple progressing wave solutions should be considered as solutions of the wave equation on a suitable covering space–time. For cylindrical waves, one can put

$$X_1 = r \cos \lambda, \quad X_2 = r \sin \lambda, \quad X_3 = z.$$

One then finds that

$$\tau(X) = r, \quad V(X) = \frac{(Az + B)(C \cos \tfrac{1}{2}\lambda + D \sin \tfrac{1}{2}\lambda)}{r^{\frac{1}{2}}},$$

provided that $r \neq 0$; A, B, C and D are arbitrary constants. For toroidal waves, put

$$X_1 = (\alpha + \nu \cos \lambda) \cos \mu, \quad X_2 = (\alpha + \nu \cos \lambda) \sin \mu, \quad X_3 = \nu \sin \lambda.$$

Then

$$\tau(X) = \nu, \quad V(X) = \frac{(A \cos \tfrac{1}{2}\lambda + B \sin \tfrac{1}{2}\lambda)(C \cos \tfrac{1}{2}\mu + D \sin \tfrac{1}{2}\mu)}{|\nu(\alpha + \nu \cos \lambda)|^{\frac{1}{2}}},$$

where A, B, C and D are arbitrary constants, and $\nu \neq 0, \alpha + \nu \cos \lambda \neq 0$. The focal lines are the X_3-axis and the circle $X_1^2 + X_2^2 = \alpha^2, t = X_3 = 0$.

NOTES

Sections 3.1–3.4 The general definition of characteristics is as follows. Let $\Omega \subset \mathbf{R}^n$, where $n > 1$, be an open set, and let

$$u \to P(x, D) u = \sum_{|\alpha| \leqslant m} a_\alpha(x) D^\alpha u$$

be a linear partial differential operator of order m, defined on Ω, say with C^∞ coefficients. Let

$$P_m(x, D) u = \sum_{|\alpha| = m} a_\alpha(x) D^\alpha u$$

be the principal part of P, consisting of the terms that contain derivatives of order m. Let Σ be a surface, defined locally by an equation $S(x) = 0$, where $\operatorname{grad} S \neq 0$. Then Σ is a characteristic of P if

$$P_m(x, \operatorname{grad} S) = 0,$$

when $S = 0$. So one can associate, with a characteristic, a cross-section $\Lambda\colon x \to (x, \xi)$ of the cotangent bundle $T^*\Omega$ for which

$$P_m(x, \xi) = 0,$$

and on which the restriction of the canonical symplectic form

$$\sum_{j=1}^{n} dx^j \wedge d\xi_j$$

of $T^*\Omega$ vanishes. This definition generalizes at once to a linear partial differential operator defined on a manifold.

The bicharacteristics of P are the projections on Ω (or on the manifold) of the curves in $T^*\Omega$ (called bicharacteristic strips) that are the integral curves of the Hamiltonian system of ordinary differential equations

$$\frac{dx^j}{dt} = \frac{\partial P_m(x, \xi)}{\partial \xi_j}, \quad \frac{d\xi_j}{dt} = -\frac{\partial P_m(x, \xi)}{\partial x^j} \quad (j = 1, \ldots, n).$$

The theory of characteristics and bicharacteristics belongs in effect, therefore, to the theory of first-order partial differential equations. The classical reference is Carathéodory (1935); a sketch of the geometric approach that has just been outlined will be found in Hörmander (1971a) Chapter III. See also Courant and Hilbert (1962).

Sections 5.5 and 5.6 The material in these sections only touches the fringe of the subject of singularities of solutions of linear partial differential equations. The recognition of the connection between the propagation of discontinuities and geometrical optics, and of the relevance of the progressing wave formalism, goes back to Luneberg (Kline and Kay (1965), see also Friedlander (1958)). A comprehensive extension of the progressing wave formalism to systems with analytic coefficients is given in Gårding, Kotake and Leray (1964). Starting from the progressing wave formalism, and from classical methods used in wave mechanics, a considerable body of work on singularities has come into being; see Lax (1957) and Ludwig (1960), Hörmander (1971a,b) and Duistermaat and Hörmander (1972). For the problem of caustics, see Ludwig (1966), and Guillemin and Schaeffer (1973).

Section 3.7 See Friedlander (1946) and M. Riesz (1957). The elegant proof that the two focal curves are Lorentz circles is due to M. Riesz.

4

Fundamental solutions

This chapter and the next deal with wave equations on four-dimensional space–times. Section 4.1 recapitulates well-known results in the theory of the ordinary wave equation which serve as a guide to the investigation of the general case. The ordinary wave equation has two fundamental solutions that are particularly important. They are, respectively, the values at the origin of the advanced and retarded potentials of a test function; each is a measure, and their supports are the future and the past null semi-cones with vertex O respectively. Corresponding to the first of these, the general scalar wave equation in four dimensions has, locally, a fundamental solution $G_q^+(p)$ which is the sum of a singular part and a regular part. The singular part is a measure supported on the future null semi-cone with vertex q, $C^+(q)$: this is determined in section 4.2. The construction can only be carried out in a geodesically convex domain Ω as it depends on the existence of a unique geodesic joining q and p. The construction of the regular part (the 'tail') is carried out in two steps. In effect, the derivatives of all orders of this term on $C^+(q)$ can be computed by solving recurrence equations which are similar to the transport equations of the progressing wave formalism. One can then form a C^∞ parametrix. This is a distribution $\tilde{G}_q^+$ whose support is contained in the future emission $J^+(q)$ (the set of all points in Ω that can be reached along future-directed causal geodesics from q), and such that $P\tilde{G}_q^+ - \delta_q \in C^\infty(\Omega)$. It is derived in section 4.3.

In order to complete the construction of the forward fundamental solution G_q^+, a further restriction on the neighbourhood in which it is to be defined is needed, which is of a causal nature. Some restriction of this kind is always made in the literature, sometimes tacitly. In this book, an explicit condition has been chosen, which is a simplified version of Leray's condition of global hyperbolicity, appropriate for connected open subsets of a geodesically convex domain. Domains that satisfy this condition are here called causal domains; they are introduced in section 4.4. Finally, the construction of G_q^+ in a causal domain (and of the analogous backward fundamental solution), which is carried out in section 4.5, turns out to be quite simple. It involves the solution of an integral equation by means of successive approximations. The last section contains an example (space–times of constant curvature), and some remarks on conformal space–times.

The application of the fundamental solutions obtained in this chapter to the derivation of existence, uniqueness, and representation theorems, for the Cauchy problem and certain characteristic initial value problems, will be found in the next chapter.

4.1 The ordinary wave equation

In this section, we shall discuss the ordinary wave equation

$$\Box u = (\partial_1^2 - \partial_2^2 - \partial_3^2 - \partial_4^2)\, u = f. \tag{4.1.1}$$

Strictly speaking, this is a wave equation on a Minkowskian space–time. But in a fixed coordinate system, one can consider it as a partial differential equation on $\mathbf{R}^4$. As in section 3.4, we shall write $x = (t, X)$, where $X \in \mathbf{R}^3$; the Euclidean norm of X will be denoted by r.

A fundamental solution of (4.1.1) is a distribution G such that $\Box G = \delta$. As the differential operator $\Box$ is self-adjoint, this means that $(G, \Box \phi) = \phi(0)$ for all $\phi \in C_0^\infty(\mathbf{R}^4)$. There are two fundamental solutions that are particularly important. They are, respectively, the advanced and retarded potentials of ϕ, evaluated at the origin.

Theorem 4.1.1. *The two distributions*

$$G^+ = \frac{\delta(t-r)}{4\pi r}, \quad G^- = \frac{\delta(t+r)}{4\pi r}, \tag{4.1.2}$$

which are defined by

$$(G^+, \phi) = \int \frac{\phi(r, X)}{4\pi r}\, dX, \quad (G^-, \phi) = \int \frac{\phi(-r, X)}{4\pi r}\, dX, \tag{4.1.3}$$

are fundamental solutions of the wave equation (4.1.1).

PROOF. It will be sufficient to prove this for G^+, as the argument for G^- is similar. Note first that G^+, as defined by (4.1.3), is a distribution, as $1/r$ is locally integrable in $\mathbf{R}^3$, so that $(G^+, \phi) \to 0$ when $\phi \to 0$ in $C_0^\infty(\mathbf{R}^4)$. For $r \neq 0$, G^+ is equal to the distribution $\delta(t-r)/4\pi r$, defined in accordance with Theorem 2.9.1.

If $\psi(t) \in C_0^\infty(\mathbf{R})$, then the function $(t, X) \to \psi(t-r)/r$ is locally integrable, and so can be identified with the distribution

$$\left(\frac{\psi(t-r)}{r}, \phi\right) = \int \frac{\psi(t-r)}{r} \phi(t, X) \, dt \, dX. \tag{4.1.4}$$

The d'Alembertian of this distribution is

$$\left(\Box \frac{\psi(t-r)}{r}, \phi\right) = \left(\frac{\psi(t-r)}{r}, \Box \phi\right) = \lim_{\epsilon \downarrow 0} \int_{r \geqslant \epsilon} \frac{\psi(t-r)}{r} \Box \phi \, dt \, dX.$$

As $\Box(\psi(t-r)/r) = 0$ in $r \geqslant \epsilon$, the last integral is equal to

$$\int \left\{ \frac{\partial}{\partial t} \left(\frac{\psi(t-r)}{r} \frac{\partial \phi}{\partial t} - \phi \frac{\psi'(t-r)}{r} \right) \right.$$
$$\left. - \operatorname{div}_X \left(\frac{\psi(t-r)}{r} \operatorname{grad}_X \phi - \phi \operatorname{grad}_X \frac{\psi(t-r)}{r} \right) \right\} dt \, dX,$$

where div_X and grad_X refer to the Euclidean space $\mathbf{R}^3$. The integral of the $\partial/\partial t$ term is obviously zero, and the remaining terms are, by the divergence theorem, equal to

$$\int dt \int_{r=\epsilon} \left[\frac{\psi(t-r)}{r} \frac{\partial \phi}{\partial r} - \phi \frac{\partial}{\partial r} \frac{\psi(t-r)}{r} \right] r^2 \, d\omega,$$

where $d\omega$ is the surface element on $\mathbf{S}^2$. This, in turn, is evidently

$$\int dt \int \phi(t, \epsilon\theta) \psi(t-\epsilon) \, d\omega + O(\epsilon), \quad \theta \in \mathbf{S}^2,$$

and tends to

$$4\pi \int \psi(t) \phi(t, 0) \, dt = 4\pi(\psi(t) \otimes \delta(X), \phi(t, X))$$

when $\epsilon \downarrow 0$. Hence

$$\Box \frac{\psi(t-r)}{r} = 4\pi \psi(t) \otimes \delta(X). \tag{4.1.5}$$

Now take $\psi(t)$ such that $\psi \in C_0^\infty$ and $\psi \geqslant 0$,

$$\operatorname{supp} \psi \subset \{t; |t| \leqslant 1\}, \int \psi(t) \, dt = 1;$$

then, by Lemma 2.1.1., $\psi_\nu(t) = \nu\psi(\nu t)$, where ν is a positive number, tends to $\delta(t)$ in $\mathscr{D}'(\mathbf{R})$ when $\nu \to \infty$. Replace ψ by ψ_ν. We have, from (4.1.4),

$$\left(\frac{\psi_\nu(t-r)}{r}, \phi \right) = \int \psi_\nu(t) \, dt \int \frac{\phi(t+r, X)}{r} \, dX,$$

whence, by Lemma 2.1.1,

$$\lim_{\nu \to \infty} \left(\frac{\psi_\nu(t-r)}{r}, \phi \right) = \int \frac{\phi(r, X)}{r} dX = 4\pi(G^+, \phi).$$

On the other hand, $\psi_\nu(t) \otimes \delta(X) \to \delta(t) \otimes \delta(X) = \delta(x)$ when $\nu \to \infty$, and so it follows from (4.1.5) that $\Box\, G^+ = \delta$, as asserted. $\Box$

The support of G^+ is the future null semi-cone with vertex $(0, 0)$, which will be denoted by $C^+(0)$:

$$C^+(0) = \{(t, X); t = r = |X|\}. \tag{4.1.6}$$

The support of G^- is the past null semi-cone with vertex $(0, 0)$, which will be denoted by $C^-(0)$. The two distributions G^+ and G^- will be called the forward and backward fundamental solutions of $\Box$ respectively. One can use them to construct solutions of (4.1.1). We shall only discuss G^+, as the corresponding results deduced from G^- can then be derived by reversing the time orientation.

Theorem 4.1.2. *Suppose that $f(t, X) \in \mathscr{D}'(\mathbf{R}^4)$, and that*

$$\mathrm{supp} f \subset \{(t, X); t \geqslant 0\}.$$

Then the convolution $\qquad u = G^+ * f \tag{4.1.7}$

exists, and is a solution of (4.1.1), with support in $\{(t, X); t \geqslant 0\}$. Moreover, it is the only solution whose support is contained in $\{(t, X); t \geqslant 0\}$.

PROOF. By Theorem 2.5.3, $G^+ * f$ will exist if, for every positive constant c, and $(t', X') \in \mathrm{supp}\, G^+$, $(t'', X'') \in \mathrm{supp} f$, such that

$$|t' + t''| + |X' + X''| < c, \tag{4.1.8}$$

there exist positive real numbers a and b such that $|t'| + |X'| < a$, $|t''| + |X''| < b$. Now as both the supports of f and of G^+ are in $t \geqslant 0$, (4.1.8) implies that $0 \leqslant t' < c$, $0 \leqslant t'' < c$, and as also $t' = |X'| = r'$, one has $|t'| + |X'| < 2c$. Again,

$$|X''| = |X'' + X' - X'| \leqslant |X'' + X'| + |X'| < 2c,$$

whence $|t''| + |X''| < 3c$. So $G^+ * f$ exists, and

$$\Box u = \Box (G^+ * f) = \Box G^+ * f = \delta * f = f.$$

Furthermore, by Theorem 2.5.3,

$$\mathrm{supp}\, u \subset \mathrm{supp}\, G^+ + \mathrm{supp} f \subset \{(t, X); t \geqslant 0\}. \tag{4.1.9}$$

If $\square u = f$ had two solutions that vanish for $t < 0$, then their difference v would satisfy $\square v = 0$ and $\operatorname{supp} v \subset \{(t, X); t \geqslant 0\}$. By what has just been proved, $G^+ * \square v$ exists, and equals $\square G^+ * v = v$. Hence $v = G^+ * \square v = 0$, and this proves uniqueness. $\square$

Remark. The argument shows that G^+, as a convolution operator on the class of distributions supported in $t \geqslant 0$, is both a right inverse and a left inverse of the d'Alembertian.

One can, of course, obtain more precise information about the support of u from (4.1.7). Let $C^+(t, X)$ denote the future null semi-cone with vertex (t, X),

$$C^+(t, X) = (t, X) + C^+(0) = \{(t', X'); t' = t + |X - X'|\}. \quad (4.1.10)$$

Also, if F is a set in $\mathbf{R}^4$, put

$$C^+(F) = \bigcup_{(t, X) \in F} C^+(t, X). \quad (4.1.11)$$

Then (4.1.9) just states that

$$\operatorname{supp} u \subset C^+(\operatorname{supp} f). \quad (4.1.12)$$

This is the causality principle for the wave equation (4.1.1). A more general form of this will be discussed in detail in chapter 5.

To compute $G^+ * f$, one can use (2.5.4), which gives, for $\phi \in C_0^\infty(\mathbf{R}^4)$,

$$(G^+ * f, \phi) = (f(t', X'), (G^+(t'', X''), \phi(t' + t'', X' + X''))).$$

So it follows from (4.1.3) that

$$(G^+ * f, \phi) = \left(f(t', X'), \int \frac{\phi(t' + |X''|, X' + X'')}{4\pi |X''|} dX'' \right). \quad (4.1.13)$$

If f is a function that is bounded and measurable (for instance, piecewise continuous), then this becomes

$$(G^+ * f, \phi) = \int f(t', X') \, dt' \, dX' \int \frac{\phi(t' + |X''|, X' + X'')}{4\pi |X''|} dX'',$$

and the order of integration can be inverted, by Fubini's theorem. If one also puts $X' + X'' = X$ and $t' + |X''| = t$, one then obtains

$$(G^+ * f, \phi) = \int \phi(t, X) \, dt \, dX \int \frac{f(t - |X - X'|, X')}{4\pi |X - X'|} dX'.$$

Hence
$$G^+ * f = \int \frac{f(t - |X - X'|, X')}{4\pi |X - X'|} dX', \quad (4.1.14)$$

which is the retarded potential of f; note that, as it was supposed that $f = 0$ for $t < 0$, the integral is over $\{X'; |X - X'| \leqslant t\}$. One can also write (4.1.14) as

$$G^+ * f = \int \frac{f(t - |x'|, X + X')}{4\pi |x'|} \, dX',$$

and this shows that $G^+ * f \in C^2(\mathbf{R}^4)$ if $f \in C^2(\mathbf{R}^4)$. One can therefore conclude from Theorem 2.6.1 that $u = G^+ * f$ is then a classical solution of $\square u = f$. One can check easily that both u and $\partial_1 u$ vanish for $t = 0$. So the retarded potential of a function $f \in C^2(\mathbf{R}^4)$ such that $f = 0$ when $t < 0$ solves the initial value problem

$$\square u = f, \quad t > 0, \quad u = \partial_1 u = 0, \quad t = 0 \qquad (4.1.15)$$

in $t > 0$. To obtain the solution of the corresponding initial value problem with non-zero initial data, one can argue as in section 2.6. The computations made there can also be carried out as follows.

Suppose that $u \in C^2(\mathbf{R}^4)$ is a given function; one can then define a distribution u^+, supported in $t \geqslant 0$, by setting

$$u^+ = u(t, X) H(t), \qquad (4.1.16)$$

where $H(t)$ is the Heaviside unit function, $H(t) = 1$ if $t \geqslant 0$, and $H(t) = 0$ if $t < 0$. Then (as $t \equiv x^1$)

$$\partial_1 u^+ = (\partial_1 u) H(t) + u \delta(t) = (\partial_1 u) H(t) + u|_{t=0} \delta(t),$$

$$\partial_i u^+ = (\partial_i u) H(t) \quad (i = 2, 3, 4).$$

Differentiating again, and putting

$$\square u = f, \quad u|_{t=0} = u_0(X), \quad \partial_1 u|_{t=0} = u_1(X), \qquad (4.1.17)$$

one finds that

$$\square u^+ = f H(t) + \delta(t) \otimes u_1(X) + \delta'(t) \otimes u_0(X). \qquad (4.1.18)$$

The terms in the second member in u_0 and u_1 are a monopole layer and a dipole layer on $t = 0$ respectively.

As the supports of both u^+ and of the second member of (4.1.18) are contained in $t \geqslant 0$, it follows from Theorem 4.1.2 that

$$u^+ = G^+ * (f H(t) + \delta(t) \otimes u_1(X) + \delta'(t) \otimes u_0(X)). \qquad (4.1.19)$$

As f, u_0 and u_1 are here defined by (4.1.17), this is a representation of u, in $t > 0$, in terms of the restriction of $\square u$ to $t > 0$ and of $u(0, X)$ and $\partial_1 u(0, X)$. To obtain the second member explicitly, we note first that the term in f is the retarded potential (4.1.14). The contribution due to $\delta(t) \otimes u_1$ is, by (4.1.13),

$$\int u_1(X') \, dX' \int \frac{\phi(|X''|, X'+X'')}{4\pi |X''|} \, dX''.$$

As u_1 is continuous and ϕ has compact support, the order of integration can be inverted. Also, one can set $X' = X - X''$, so that the integral becomes

$$\frac{1}{4\pi} \iint u_1(X - X'') \, \phi(|X''|, X) \frac{dX \, dX''}{|X''|}.$$

Now put $X'' = -t\theta$, where $t \geqslant 0$ and $\theta \in \mathbf{S}^2$, so that $dX'' = t^2 dt \, d\omega_\theta$, $d\omega_\theta$ being the surface element on $\mathbf{S}^2$. Then

$$(G^+ * \delta(t) \otimes u_1, \phi) = \frac{1}{4\pi} \int \phi(t, X) \, t \, dt \, dX \int u_1(X + t\theta) \, d\omega_\theta,$$

whence, obviously,

$$G^+ * \delta(t) \otimes u_1 = \frac{t}{4\pi} \int u_1(X + t\theta) \, d\omega_\theta.$$

The right-hand side is the spherical mean of u_1 over the sphere with centre X and radius t, multiplied by t. The contribution form $\delta'(t) \otimes u_0$ is deduced from this if one replaces u_1 by u_0 and differentiates with respect to t. Hence (4.1.19) gives, for $t > 0$,

$$u(t, X) = \int_{|X - X'| \leqslant t} \frac{f(t - |X - X'|, X')}{4\pi |X - X'|} \, dX'$$

$$+ \frac{t}{4\pi} \int u_1(X + t\theta) \, d\omega_\theta + \frac{\partial}{\partial t} \left(\frac{t}{4\pi} \int u_0(X + t\theta) \, d\omega_\theta \right), \quad (4.1.20)$$

where f, u_0 and u_1 are defined by (4.1.17).

Now suppose that f, u_0 and u_1 are given functions, such that f is C^2 in $t \geqslant 0$, and $u_0 \in C^3(\mathbf{R}^3)$, $u_1 \in C^2(\mathbf{R}^3)$. Then the second member of (4.1.18), formed with these functions, is a distribution whose support is contained in $t \geqslant 0$, and so, by Theorem 4.1.2, (4.1.18) has a unique solution $u \in \mathscr{D}'(\mathbf{R}^4)$, also supported in $t \geqslant 0$, which is given by (4.1.19). As we have just seen, this implies that (4.1.20) holds in $t > 0$, and in view of the hypotheses on f, u_0 and u_1, it is evident that u is C^2 in $t > 0$. By Theorem 2.6.1, it is therefore a classical solution of the wave equation for $t > 0$. It is also easily verified that $u \to u_0$ and $\partial_t u \to u_1$ when $t \downarrow 0$. So (4.1.20) is then the solution of the classical initial value problem, in $t \geqslant 0$. As a matter of fact, it is also the solution of this problem in $t \leqslant 0$, with the domain of integration in the retarded potential term replaced by $\{X'; |X - X'| \leqslant -t\}$.

Formally, a Cauchy problem for one of the other coordinate hyper-planes, say $x^4 = 0$, can be treated in the same way. If $u \in C^2(\mathbf{R}^4)$, then

$$\Box(uH(x^4)) = H(x^4)\,\Box\,u - \delta'(x^4) \otimes u\big|_{x^4=0} - \delta(x^4) \otimes \partial_4 u\big|_{x^4=0}.$$

If one supposes, for simplicity, that $u = 0$ for $x^4 < 0$, then the second member has its support contained in $x^4 \geqslant 0$, and one can again obtain a representation of u in terms of the second member of the identity. But one can no longer convert this into a solution of the Cauchy problem, by replacing $\Box\,u\,H(x^4)$ and $u\big|_{x^4=0}, \partial_4 u\big|_{x^4=0}$ by 'arbitrary' functions. We shall return to this point presently. But first it will be shown that the Cauchy problem with data on $x^4 = 0$ is improperly posed.

A boundary value problem for a partial differential equation is said to be *properly posed* if it has a unique solution, and if this solution depends continuously on the data. The continuity requirement must be made precise by stating appropriate topologies for the data and the solutions, which must be considered as members of a space of functions. If one or more of these conditions is not satisfied, the problem is said to be *improperly posed*.

Theorem 4.1.2 shows that, for all $f \in \mathscr{D}'(\mathbf{R}^4)$ supported in $t \geqslant 0$, the problem
$$\Box\,u = f;\ u \in \mathscr{D}'(\mathbf{R}^4),\, \text{supp}\,u \subset \{(t, X);\, t \geqslant 0\}$$

is properly posed, if one considers f as the data, and uses the distribution topology for both f and u. For it has been shown that $u = G^+ * f$ is a solution, and that it is the only solution. Also, if $f_\nu \to f$ in $\mathscr{D}'(\mathbf{R}^4)$, and $f_\nu = 0$ for $t < 0$, then $G^+ * f_\nu \to G^+ * f$, by Theorem 2.5.5.

Likewise, the classical initial value problem is properly posed. We have just shown that the solution exists; uniqueness follows from the fact that it is also a distribution solution, or directly from the representation (4.1.20). Again, (4.1.20), considered now as the solution of the problem for given f, u_0 and u_1, maps

$$(f, u_0, u_1) \in C^2(\{(t, X);\, t \geqslant 0\}) \times C^3(\mathbf{R}^3) \times C^2(\mathbf{R}^3)$$

continuously to $u \in C^3(\{t, X;\, t \geqslant 0\})$. There are other important continuity properties, connected with energy norms, which will not be discussed here.

On the other hand, the Cauchy problem for $x^4 = 0$ is improperly posed, even for C^∞ data. This can be shown by means of a simple example. If A and λ are real numbers, and $\lambda > 0$, then

$$u = A\,e^{-\lambda^{\frac{1}{2}}} \cos(\lambda x^3) \cosh(\lambda x^4)$$

is a plane harmonic function of (x^3, x^4), and so also satisfies $\square\, u = 0$. On $x^4 = 0$, one has estimates of the type

$$\left| \partial^\alpha u \right|_{x^4=0} \leqslant C_\alpha \lambda^{|\alpha|} e^{-\lambda^{\frac{1}{2}}}$$

for all multi-indices α, where the C_α do not depend on λ. Hence all the derivatives of u, restricted to $x^4 = 0$, tend to zero when $\lambda \to \infty$. But for $x^4 \neq 0$, the factor $\cosh(\lambda x^4)$ swamps the exponential, and $|u| \to \infty$ when $\lambda \to \infty$. So the Cauchy problem for the wave equation, relative to $x^4 = 0$, is highly unstable. (One can also deduce from this and the closed graph theorem that there exist C^∞ Cauchy data for which this Cauchy problem has no solution.)

The crucial difference between the two cases is that the 'good' hyperplane $t \equiv x^1 = 0$ is space-like, and the 'bad' hyperplane $x^4 = 0$ is time-like. Generally, a Cauchy problem for a space-like hypersurface Σ is properly posed, at least locally. (If, for every $x \in \Sigma$ the null cone with vertex x meets Σ at no point other than x, then the Cauchy problem for Σ is also properly posed globally.) On the other hand, the Cauchy problem for a time-like hypersurface is always improperly posed.

This fact is relevant when one considers another well-known representation theorem for the wave equation, Kirchhoff's formula. This can be derived by an argument similar to that which led to (4.1.20), but it is technically easier to use a different method, based on a direct computation of the restriction of $\square\, u$, where $u \in C^2(\mathbf{R}^4)$, to a past null semi-cone.

If $F(t, X) \in C^2(\mathbf{R}^4)$, then one can define a function $Y \to [F]$ on $\mathbf{R}^3$ by setting
$$[F] = F(t - R, Y), \quad R = |X - Y|. \tag{4.1.21}$$

This is the restriction of F to $C^-(t, X)$, with $C^-(t, X)$ parametrized as $t' = t - R, X' = Y$; it is in $C^2(\mathbf{R}^3 \backslash (X))$. For the derivatives of $Y \to [F]$ one has, if $Y \neq X$,

$$\operatorname{grad}_Y [F] = [\operatorname{grad}_X F] - \left[\frac{\partial F}{\partial t}\right] \operatorname{grad}_Y R. \tag{4.1.22}$$

Replacing F by u, and differentiating again, one obtains, as $\Delta_Y R = 2/R$ and $(\operatorname{grad}_Y R)^2 = 1$,

$$[\Delta_X u] = \Delta_Y [u] - 2 \operatorname{grad}_Y R \cdot \left[\operatorname{grad}_X \frac{\partial u}{\partial t}\right] + \left[\frac{\partial^2 u}{\partial t^2}\right] - \frac{2}{R}\left[\frac{\partial u}{\partial t}\right].$$

Again, by (4.1.22) with $F = \partial u/\partial t$,

$$\operatorname{grad}_Y \left[\frac{\partial u}{\partial t}\right] = \left[\operatorname{grad}_X \frac{\partial u}{\partial t}\right] - \left[\frac{\partial^2 u}{\partial t^2}\right] \operatorname{grad}_Y R.$$

By combining these two identities, one finds that

$$[\Box u] + \Delta_Y[u] + 2\,\mathrm{grad}_Y\,R\,.\,\mathrm{grad}_Y\left[\frac{\partial u}{\partial t}\right] + \frac{2}{R}\left[\frac{\partial u}{\partial t}\right] = 0.$$

As $\Delta_Y(1/R) = 0$ for $Y \neq X$, and

$$\mathrm{div}_Y\,(R^{-1}\mathrm{grad}_Y\,R) = \Delta_Y\log R = \frac{1}{R},$$

the last identity becomes, after multiplication by $1/R$,

$$\frac{1}{R}\,\Box\,u + \mathrm{div}_Y\left(\frac{1}{R}\,\mathrm{grad}_Y\,[u] - [u]\,\mathrm{grad}_Y\,\frac{1}{R} + \frac{2}{R}\left[\frac{\partial u}{\partial t}\right]\mathrm{grad}_Y\,R\right) = 0,$$

or, by (4.1.22) with $F = u$,

$$\frac{1}{R}[\Box\,u] + \mathrm{div}_Y\left(\frac{[\mathrm{grad}_X\,u]}{R} - [u]\,\mathrm{grad}_Y\,\frac{1}{R} + \frac{1}{R}\left[\frac{\partial u}{\partial t}\right]\mathrm{grad}_Y\,R\right) = 0.$$
$$(4.1.23)$$

So $\Box u/R$ is an intrinsic divergence on $C^-(t, X)\backslash(t, X)$; this result is closely related to Theorem 4.1.1. It may be noted that the representation (4.1.20) can be obtained by taking $r > 0$, integrating (4.1.23) over $\{Y; \epsilon \leqslant R \leqslant t\}$, and then making $\epsilon \downarrow 0$. By the divergence theorem, the integral of the divergence term is the sum of an integral over the sphere $R = t$ and an integral over the sphere $R = \epsilon$; the latter is

$$-\int_{R=\epsilon}\left(\frac{1}{R}\left[\frac{\partial u}{\partial R}\right] - [u]\frac{\partial}{\partial R}\frac{1}{R} + \frac{1}{R}\frac{\partial u}{\partial t}\right)R^2\,d\omega. \qquad (4.1.24)$$

When $\epsilon \downarrow 0$, this obviously tends to $-4\pi u(t, X)$. The simple proof that the integral over $R = t$ gives the initial value terms in (4.1.20) is left to the reader.

To derive Kirchhoff's formula, we consider a bounded domain $D \in \mathbf{R}^3$ with a piece-wise smooth boundary. Let $\Omega = \{(t, X); X \in D\}$ be the hypercylinder with base D and generators parallel to the t-axis. Let u be a given function in $C^2(\bar{\Omega})$, so that it can be extended as a C^2 function to $\mathbf{R}^4$. Then (4.1.23) holds for $Y \in \bar{D}\backslash(X)$, and for all (t, X). Suppose that $X \in D$. Then one can integrate (4.1.23) over

$$\{Y; Y \in D, |Y - X| \geqslant \epsilon\},$$

where ϵ is a positive number, less than the distance of X from ∂D. The

divergence term contributes an integral over ∂D, and the integral
(4.1.24). Making $\epsilon \downarrow 0$, one therefore obtains the identity

$$4\pi u(t, X) = \int_D \frac{[\square u]}{R} \, dY$$

$$+ \int_{\partial D} \left(\frac{[\nu_Y \cdot \mathrm{grad}_X u]}{R} + \left(\frac{[u]}{R^2} + \frac{1}{R}\left[\frac{\partial u}{\partial t}\right] \right) \right) \nu_Y \cdot \mathrm{grad}_Y R \right) dS_Y,$$

$$(4.1.25)$$

where ν_Y is the Euclidean unit normal to ∂D, drawn away from D, and dS_Y is the surface element on ∂D.

This identity gives a representation of the restriction $u|_\Omega$ of u to Ω, in terms of $\square u|_\Omega$, $u|_{\partial\Omega}$ and $\mathrm{grad}_X u|_{\partial\Omega}$. It can be used to relate the properties of u in Ω to those of $\square u|_\Omega$, $u|_{\partial\Omega}$ and $\mathrm{grad}_X u|_{\partial\Omega}$. For instance, (4.1.25) shows that if $v \in C^2(\bar{\Omega})$, such that $\square v = \square u$ in Ω, and $v = u$, $\mathrm{grad}_X v = \mathrm{grad}_X u$ on $\partial\Omega$, then $v = u$ in Ω. This uniqueness theorem is the logical basis for the familiar procedure by which the field due to sources outside Ω is replaced by a field produced by monopoles and dipoles distributed on $\partial\Omega$.

But it must be emphasized that (4.1.25) does not furnish the solution of a Cauchy problem for $\square u = f$ with data on $\partial\Omega$, when $\square u$, $u|_{\partial\Omega}$ and $\mathrm{grad}_X u|_{\partial\Omega}$ are replaced by smooth, but otherwise arbitrary, functions. The point is that such data cannot be given freely. If one takes $X \in \mathbf{R}^4 \backslash \bar{D}$, then the function $Y \to R$ is bounded away from zero in $\bar{D}$, and (4.1.23) can be integrated over $\bar{D}$; the term $-4\pi u(t, X)$ is then absent, and one obtains

$$\int_D \frac{[\square u]}{R} \, dY + \int_{\partial D} \left(\frac{[\nu_Y \cdot \mathrm{grad}_X u]}{R} \right.$$

$$+ \left. \left(\frac{[u]}{R^2} + \frac{1}{R}\left[\frac{\partial u}{\partial t}\right] \right) \nu_Y \cdot \mathrm{grad}_Y R \right) dS_Y = 0. \quad (4.1.26)$$

If one fixes $\square u = f$, for example, then (4.1.26) gives an infinite set of subsidiary conditions – one for each $X \in \mathbf{R}^4 \backslash \bar{D}$ – that must be satisfied by the Cauchy data on $\partial\Omega$. (These are automatically satisfied if the data are in fact the restrictions of u and $\mathrm{grad}_X u$ to $\partial\Omega$, where u is a C^2 solution of $\square u = f$.) One can see that these subsidiary conditions arise because $\partial\Omega$ is time-like. For (4.1.26) is obtained by integrating $\square u/R$ over $C^-(t, X) \cap \Omega$, which is not empty. In the case of the initial value problem, the integration is over the portion of $C^-(t, X)$ that is in $t > 0$, which is empty when $t < 0$; hence there are no subsidiary conditions. So one can say that, if the Cauchy data are replaced by layers of monopoles and dipoles on a Cauchy hypersurface Σ, then the fields of the constituent point sources do not interfere with each other when Σ is space-like, but do when Σ is time-like.

So far, the wave equation has been treated as a partial differential equation on $\mathbf{R}^4$. We conclude this section by deriving alternative expressions for the two fundamental solutions G^+ and G^- which are manifestly invariant under proper Lorentz transformations. (The proper Lorentz transformations preserve the time-orientation.) In this form, Theorem 4.1.1 will apply to the wave equation when it is considered on a four-dimensional Minkowskian space–time M_4.

We can work with the collection of coordinate systems that are related to each other by proper Lorentz transformations. The time-orientation is then defined by the vector field $(1, 0, 0, 0)$, at any point of M_4. The square of the geodesic distance of a point $x = (t, X)$ from the origin is

$$\gamma(x) = \eta_{ij}x^ix^j = t^2 - |X|^2. \qquad (4.1.27)$$

Let $D^+(0)$ denote the interior of the future null semi-cone $C^+(0)$,

$$D^+(0) = \{x = (t, X); t > |X|\}.$$

As $\operatorname{grad}\gamma \neq 0$ in $D^+(0)$, we can, given a positive number ϵ, define a distribution $\delta_+(\gamma - \epsilon) \in \mathcal{D}'(D^+(0))$ by means of Theorem 2.9.1. It is

$$(\delta_+(\gamma - \epsilon), \phi) = \int_{\gamma = \epsilon} \phi\mu_\gamma, \quad \phi \in C_0^\infty(D^+(0)),$$

where μ_γ is a Leray form such that $d\gamma \wedge \mu_\gamma = dx$. The support of $\delta_+(\gamma - \epsilon)$ is the upper sheet of the hyperboloid $\{x; \gamma(x) = \epsilon\}$, which will be denoted by Σ_ϵ^+; it is oriented by $\mu_\gamma > 0$. Now one can obviously extend $\delta_+(\gamma - \epsilon)$ to $\mathcal{D}'(M_4)$ by setting it equal to zero in

$$\{x = (t, X); t < (|X|^2 + \epsilon)^{\frac{1}{2}}\},$$

so that

$$(\delta_+(\gamma - \epsilon), \phi) = \int_{\Sigma_\epsilon^+} \phi\mu_\gamma, \quad \phi \in C_0^\infty(M_4). \qquad (4.1.28)$$

Similarly, we can define a distribution $\delta_-(\gamma - \epsilon)$ whose support is the lower sheet Σ_ϵ^- of $\gamma(x) = \epsilon$,

$$(\delta_-(\gamma - \epsilon), \phi) = \int_{\Sigma_\epsilon^-} \phi\mu_\gamma, \quad \phi \in C_0^\infty(M_4). \qquad (4.1.29)$$

Both $\delta_+(\gamma - \epsilon)$ and $\delta_-(\gamma - \epsilon)$ are evidently invariant under proper Lorentz transformations.

It follows from (4.1.27) and $d\gamma \wedge \mu_\gamma = dx$ that one can take $\mu_\gamma = dX/2t$. Hence

$$(\delta_+(\gamma - \epsilon), \phi) = \frac{1}{2} \int \frac{\phi((|X|^2 + \epsilon)^{\frac{1}{2}}, X)}{(|X|^2 + \epsilon)^{\frac{1}{2}}} \, dX. \qquad (4.1.30)$$

For $\delta_-(\gamma - \epsilon)$ one has, by the orientation rule,

$$(\delta_-(\gamma - \epsilon), \phi) = \frac{1}{2} \int \frac{\phi(-(|X|^2 + \epsilon)^{\frac{1}{2}}, X)}{(|X|^2 + \epsilon)^{\frac{1}{2}}} \, dX. \qquad (4.1.31)$$

By dominated convergence, one can make $\epsilon \downarrow$, in each of these, under the integral sign, and comparison with (4.1.3) then shows that

$$G^+ = \frac{1}{2\pi} \delta_+(\gamma) = \lim_{\epsilon \downarrow 0} \frac{1}{2\pi} \delta_+(\gamma - \epsilon) \qquad (4.1.32)$$

and

$$G^- = \frac{1}{2\pi} \delta_-(\gamma) = \lim_{\epsilon \downarrow 0} \frac{1}{2\pi} \delta_-(\gamma - \epsilon), \qquad (4.1.33)$$

where both the limits are in the distribution topology. These are the required invariant forms of G^+ and G^-.

One can now establish a relation that will be needed in the next section. For $\epsilon > 0$, $\delta_+(\gamma - \epsilon)$ can be differentiated by the chain rule (Theorem 2.9.2). So

$$\Box \delta_+(\gamma - \epsilon) = \partial_i(\eta^{ij} \partial_j \delta_+(\gamma - \epsilon)) = \partial_i(2x^i \delta'_+(\gamma - \epsilon)),$$

whence

$$\Box \delta_+(\gamma - \epsilon) = 4\gamma \delta''_+(\gamma - \epsilon) + 8\delta'_+(\gamma - \epsilon).$$

Let us write this as

$$\Box \delta_+(\gamma - \epsilon) = 4\epsilon \delta''_+(\gamma - \epsilon) + 4(\gamma - \epsilon) \delta''_+(\gamma - \epsilon) + 8\delta'_+(\gamma - \epsilon).$$

In $\mathscr{D}'(\mathbf{R})$, one has $t\delta(t) = 0$, whence $t\delta''(t) + 2\delta(t) = 0$, by differentiation. Hence also $(\gamma - \epsilon)\delta''_+(\gamma - \epsilon) + 2\delta'_+(\gamma - \epsilon) = 0$, and so one is left with

$$\Box \delta_+(\gamma - \epsilon) = 4\epsilon \delta''_+(\gamma - \epsilon).$$

As the differentiation of distributions is a continuous operation, one can now make $\epsilon \downarrow 0$. By (4.1.32), this gives

$$\lim_{\epsilon \downarrow 0} 4\epsilon \delta''_+(\gamma - \epsilon) = 2\pi \delta(x). \qquad (4.1.34)$$

Applied to a test function $\phi \in C_0^\infty(M^4)$, this means that

$$\lim_{\epsilon \downarrow 0} \epsilon \left(\frac{\partial}{\partial \epsilon} \right)^2 \int \frac{\phi((|X|^2 + \epsilon)^{\frac{1}{2}}, X)}{(|X|^2 + \epsilon)^{\frac{1}{2}}} \, dX = \phi(0),$$

an identity which gives $\phi(0)$ in terms of 'hyperbolic means' of ϕ.

By a similar argument, one also obtains

$$\lim_{\epsilon \downarrow 0} 4\epsilon \delta''_-(\gamma - \epsilon) = 2\pi\delta(x). \tag{4.1.35}$$

4.2 The singular part of the fundamental solutions

Let us now consider a general four-dimensional space–time M_4, and a scalar differential operator with C^∞ coefficients

$$Pu = \square + \langle a, \nabla u \rangle + bu, \tag{4.2.1}$$

that is defined on a connected open set $\Omega \subset M_4$. A fundamental solution of P is a distribution $(G_q(p), \phi(p))$ in $\mathscr{D}'(\Omega)$, which is a function of q, such that $PG_q = \delta_q$. As explained at the end of section 2.8, this means that

$$(PG_q, \phi) = (G_q, {}^t P\phi) = \phi(q), \quad \phi \in C_0^\infty(\Omega), \tag{4.2.2}$$

where ${}^t P$ is the adjoint of P,

$$^t P\phi = \square \phi - \mathrm{div}\,(a\phi) + b\phi. \tag{4.2.3}$$

In this section, we shall begin the construction of two fundamental solutions of P that are similar to the fundamental solutions G^+ and G^- of the ordinary wave equation. For the time being, it will be assumed that Ω is geodesically convex. A further restriction, which is a causality postulate, will be made in section 4.4.

As Ω is geodesically convex, it is time-orientable. For if q is a point in Ω, then one can give the tangent space TM_q a time-orientation, and this can then be carried to any other point $p \in \Omega$ by parallel transport along the unique geodesic $\widehat{qp} \subset \Omega$. It will therefore be assumed that Ω has been given a time-orientation.

Definition 4.2.1. *Let Ω be a geodesically convex domain, and let q be a point in Ω. The future null semi-cone $C^+(q)$ is the set of all points $p \in \Omega$ that can be reached along future-directed null geodesics from q; the future dependence domain $D^+(q)$ is the set of all points $p \in \Omega$ that can be reached along future-directed time-like geodesics from q. The past null semi-cone $C^-(q)$ and the past dependence domain $D^-(q)$ are defined similarly.*

Remark. The exponential map at q is a diffeomorphism between an open set $(\exp_q)^{-1}\Omega \subset TM_q$ and Ω. It maps the tangent null cone at q to the null cone $C(q)$ with vertex q. The future-directed null vectors at q

are mapped to $C^+(q)$, and the future-directed time-like vectors at q are mapped to $D^+(q)$. Obviously, $C^+(q) = \partial D^+(q)$. Similarly,

$$C^-(q) = \partial D^-(q).$$

Let $(x^1, ..., x^4)$ be a local coordinate system that is normal and Minkowskian at q, and such that the vector $(1, 0, 0, 0) \in TM_q$ is future-directed. Then q has the coordinates $x = 0$, and $g_{ij}(0) = \eta_{ij}$. By (1.2.16) one therefore has

$$g_{ij}(x)\, x^j = \eta_{ij} x^j \quad (i = 1, ..., 4). \tag{4.2.4}$$

Note that this coordinate system is defined on all of Ω, as Ω is geodesically convex. In this coordinate system, the geodesics through q are the straight lines through the origin, and

$$D^+(q) = \{x;\ x^1 > ((x^2)^2 + (x^3)^2 + (x^4)^2)^{\frac{1}{2}}\},$$

$$D^-(q) = \{x;\ x^1 < -((x^2)^2 + (x^3)^2 + (x^4)^2)^{\frac{1}{2}}\}.$$

As Ω is geodesically convex, the square of the geodesic distance of two points p and q in Ω can be defined. Its properties are summarized in Theorem 1.2.3. It will again be denoted by $\Gamma(p, q) = \Gamma(q, p)$. By (1.2.17), one has

$$\Gamma(p, q) = \gamma(x) = \eta_{ij} x^i x^j, \tag{4.2.5}$$

in the special coordinate system that has just been introduced. It is obvious that $\{p;\ p \in \Omega,\ \Gamma(p, q) > 0\}$ is $D^+(q) \cup D^-(q)$.

If $f(t) \in \mathscr{D}'(\mathbf{R})$ has its support contained in $t > 0$, then one can define a distribution $f_+(\Gamma) \in \mathscr{D}'(D^+(q))$ by means of Theorem 2.9.1, as $\operatorname{grad}_p \Gamma \neq 0$ in $D^+(q)$. (This distribution acts on test functions $\phi(p)$, and is also a function of q.) As $f_+(\Gamma)$ vanishes in a neighbourhood Σ of $C^+(q)$, it can be extended to $\mathscr{D}'(\Omega)$ by setting it equal to zero in $(\Omega \backslash D^+(q)) \cup \Sigma$. By (2.9.7), this extended distribution, which will still be denoted by $f_+(\Gamma)$, is

$$(f_+(\Gamma), \phi) = \left(f(t), \int_{\Sigma_t^+} \phi \mu_\Gamma(p) \right), \quad \phi \in C_0^\infty(\Omega), \tag{4.2.6}$$

where $\Sigma_t^+ = \{p;\ p \in D^+(q),\ \Gamma(p, q) = t > 0\}$ and μ_Γ is a Leray form such that $d_p \Gamma(p, q) \wedge \mu_\gamma = \mu(p)$. (It will be recalled that $\mu(p)$ denotes the invariant volume element.) Similarly, a distribution $f_-(\Gamma)$, whose support is contained in $D^-(q)$, is defined by

$$(f_-(\Gamma), \phi) = \left(f(t), \int_{\Sigma_t^-} \phi(p) \mu_\Gamma(p) \right), \quad \phi \in C_0^\infty(\Omega), \tag{4.2.7}$$

where $\qquad\qquad \Sigma_t^- = \{p;\ p \in D^-(q),\ \Gamma(p, q) = t > 0\}.$

In general, these distributions need not exist if one only supposes that $\operatorname{supp} f \subset \{t; t \geqslant 0\}$, as $\operatorname{grad}_p \Gamma(p,q) = 0$ when $p = q$. But in special cases, (4.2.6) and (4.2.7) may still have a meaning. In particular, it follows from (4.2.5) that the results obtained in the last section can be applied in a coordinate system that is normal and Minkowskian at q. Hence the distributions

$$\delta_+(\Gamma) = \lim_{\epsilon \downarrow 0} \delta_+(\Gamma - \epsilon), \quad \delta_-(\Gamma) = \lim_{\epsilon \downarrow 0} \delta_-(\Gamma - \epsilon) \qquad (4.2.8)$$

exist. By analogy with the Minkowskian case, one may expect that P will have fundamental solutions that behave like $\delta_+(\Gamma)$ and $\delta_-(\Gamma)$ near $C(q)$. But the discussion of the propagation of singularities in the last chapter shows that $\delta_+(\Gamma)$ and $\delta_-(\Gamma)$ will have to be multiplied by a smooth function U which satisfies an appropriate transport equation. Even so, $U\delta_+(\Gamma)$ and $U\delta_-(\Gamma)$ will not (in general) be fundamental solutions of P; the situation is similar to that which arises when one considers simple progressing wave solutions of P. However, it will turn out that these two distributions are the singular parts of the two basic fundamental solutions of P, and so the next theorem is a key step in the argument.

Theorem 4.2.1. *There exists a function* $U(p,q) \in C^\infty(\Omega \times \Omega)$ *such that*

$$P(U\delta_+(\Gamma)) = P(U)\,\delta_+(\Gamma) + 2\pi\delta_q \qquad (4.2.9)$$

and
$$P(U\delta_-(\Gamma)) = P(U)\,\delta_-(\Gamma) + 2\pi\delta_q, \qquad (4.2.10)$$

where $U\delta_+(\Gamma) = U(p,q)\,\delta_+(\Gamma(p,q))$ *and* $U\delta_-(\Gamma) = U(p,q)\,\delta_-(\Gamma(p,q))$ *are distributions that act on test functions* $\phi(p) \in C_0^\infty(\Omega)$.

PROOF. It will be sufficient to establish (4.2.9), as the proof of (4.2.10) is similar. We first consider $U\delta_+(\Gamma - \epsilon)$, where ϵ is a positive number. This distribution is well defined, and can be differentiated by the chain rule (Theorem 2.9.2). Hence, by (3.6.2), and taking into account that $\langle \nabla\Gamma, \nabla\Gamma \rangle = 4\Gamma$,

$$P(U\delta_+(\Gamma - \epsilon)) = P(U)\,\delta_+(\Gamma - \epsilon) + 4\Gamma U\delta_+''(\Gamma - \epsilon)$$
$$+ (2\langle \nabla\Gamma, \nabla U \rangle + (\square\,\Gamma + \langle a, \nabla\Gamma \rangle)\,U)\,\delta_+'(\Gamma - \epsilon).$$

Now it follows from $t\delta''(t) + 2\delta'(t) = 0$ that

$$(\Gamma - \epsilon)\,\delta_+''(\Gamma - \epsilon) = -2\delta_+'(\Gamma - \epsilon),$$

whence
$$4\Gamma\delta_+''(\Gamma - \epsilon) = 4\epsilon\delta_+''(\Gamma - \epsilon) - 8\delta_+'(\delta - \epsilon).$$

Hence

$$P(U\delta_+(\Gamma-\epsilon)) = P(U)\,\delta_+(r-\epsilon) + 4\epsilon U\delta_+''(\Gamma-\epsilon)$$
$$+ (2\langle\nabla\Gamma,\nabla U\rangle + (\square\,\Gamma + \langle a,\nabla\Gamma\rangle - 8)\,U)\,\delta_+'(\Gamma-\epsilon).$$
$$(4.2.11)$$

We shall now show that one can determine U by solving the equation

$$2\langle\nabla\Gamma,\nabla U\rangle + (\square\,\Gamma + \langle a,\nabla\Gamma\rangle - 8)\,U = 0, \qquad (4.2.12)$$

subject only to the condition $U(q,q) = 1$. Note that this is a first order equation whose characteristic curves are the geodesics through q. The reason why it has solutions that remain bounded when $p \to q$ is that the coefficient of U vanishes at q. To see this, and to obtain the integral, one can work with local coordinates that are normal at q. Then the coordinates of q are $x = 0$, and, by (1.2.16) and (1.2.17),

$$g_{ij}(x)\,x^j = g_{ij}(0)\,x^j \ (i = 1, ..., 4), \quad \Gamma = g_{ij}(0)\,x^i x^j.$$

So

$$\nabla^i\Gamma = 2g^{ij}(x)\,g_{jk}(0)\,x^k = 2g^{ij}(x)\,g_{jk}(x)\,x^k = 2x^i \quad (i = 1, ..., 4),$$

whence

$$\square\,\Gamma = \nabla_i\nabla^i\Gamma = |g|^{-\frac{1}{2}}\,\partial_i(2\,|g|^{\frac{1}{2}}\,x^i) = 8 + 2x^i\,\partial_i \log|g|^{\frac{1}{2}}.$$

Thus (4.2.12) becomes, after multiplication by $|g|^{\frac{1}{4}}$,

$$x^i\,\partial_i(|g|^{\frac{1}{4}}\,U) + \tfrac{1}{2}a_i x^i\,|g|^{\frac{1}{4}}\,U = 0.$$

If x is now replaced by rx, $0 \leqslant r \leqslant 1$, then this implies

$$\frac{\partial}{\partial r}\,(|g|^{\frac{1}{4}}\,U|_{rx}) = -\tfrac{1}{2}a_i(rx)\,x^i(|g|^{\frac{1}{4}}U|_{rx}).$$

So the solution of (4.1.12), such that $U(0) = 1$, is

$$U(x) = \left|\frac{g(0)}{g(x)}\right|^{\frac{1}{4}} \exp\left(-\frac{1}{2}\int_0^1 a_i(rx)\,x^i dr\right). \qquad (4.2.13)$$

The second member is a C^∞ function of x; as the exponential map depends smoothly on q, it is clear that $U \in C^\infty(\Omega \times \Omega)$.

With this choice of U, (4.2.11) reduces to

$$P(U\delta_+(\Gamma-\epsilon)) = P(U)\,\delta_+(\Gamma-\epsilon) + 4\epsilon U\delta_+''(\Gamma-\epsilon).$$

It has already been noted that $\delta_+(\Gamma-\epsilon)$ tends to a limit $\delta_+(\Gamma)$, in $\mathscr{D}'(\Omega)$, when $\epsilon \downarrow 0$. By the continuity of the map $P: \mathscr{D}'(\Omega) \to \mathscr{D}'(\Omega)$, one also has $P(U\delta_+(\Gamma-\epsilon)) \to P(U\delta_+(\Gamma))$. Finally, it follows from (4.2.5) that

$$4\epsilon U\delta_+''(\Gamma-\epsilon) = 4\epsilon U\delta_+''(\gamma-\epsilon)$$

in the special coordinate system used there. Hence

$$(4\epsilon U\delta''_+(\Gamma-\epsilon), \phi) = 4\epsilon(\delta''_+(\gamma(x)-\epsilon), U(x)\,\phi(x)\,|g(x)|^{\frac{1}{2}}).$$

By (4.1.34), this tends to $2\pi U(0)\,\phi(0)\,|g(0)|^{\frac{1}{2}}$ when $\epsilon \downarrow 0$. But $U(0) = 1$, by construction, and $g(0) = 1$, as the special coordinate system is Minkowskian at q. So one obtains (4.2.9) in the limit. The proof of (4.2.10) is similar; the only point to note is that U was in effect obtained by integrating along geodesics from q, and so is defined on all of $\Omega \times \Omega$ by (4.2.13). This completes the proof of the theorem. $\square$

The reason for choosing U as a solution of (4.2.12) is that $\delta'_+(\Gamma-\epsilon)$ does not tend to a limit when $\epsilon \downarrow 0$. It would, in fact, have been sufficient to require that (4.2.12) should hold only on $C(q)$, and that U should be C^∞ in a neighbourhood of $C(q)$. One can then obtain the identities (4.2.9) and (4.2.10) again, as follows.

Suppose that

$$2\langle\nabla\Gamma, \nabla U'\rangle + (\square\,\Gamma + \langle a, \nabla\Gamma\rangle - 8)\,U' = A\Gamma, \quad U'(q,q) = 1,$$

where A is C^∞ in a neighbourhood of $\{(p,q); (p,q)\in\Omega\times\Omega, \Gamma(p,q) = 0\}$. Then, as $\operatorname{grad}_p \Gamma$ is tangent to the bicharacteristics (the null geodesic generators) of $C(q)$, it is clear that $U' = U$ on $C(q)$. Thus

$$U'\delta_+(\Gamma) = U'|_{\Gamma=0}\delta_+(\Gamma) = U\delta_+(\Gamma). \tag{4.2.14}$$

Now (4.2.11) becomes, with U' instead of U,

$$P(U'\delta_+(\Gamma-\epsilon)) = P(U')\,\delta_+(\Gamma-\epsilon) + A\Gamma\delta'_+(\Gamma-\epsilon) + 4\epsilon U'\delta''_+(\Gamma-\epsilon).$$

It follows from $t\delta'(t) + \delta(t) = 0$ that

$$\Gamma\delta'_+(\Gamma-\epsilon) + (\Gamma-\epsilon)\,\delta'_+(\Gamma-\epsilon) + \epsilon\delta'_+(\Gamma-\epsilon) = \epsilon\delta'_+(\Gamma-\epsilon) - \delta_+(\Gamma-\epsilon).$$

Hence

$$P(U'\delta_+(\Gamma-\epsilon)) = (P(U') - A)\,\delta_+(\Gamma-\epsilon) + A\epsilon\delta'_+(\Gamma-\epsilon) + 4\epsilon U'\delta''_+(\Gamma-\epsilon).$$

Now we shall prove presently that

$$\lim_{\epsilon\downarrow 0}\epsilon\delta'_+(\Gamma-\epsilon) = 0. \tag{4.2.15}$$

Assuming this, and making $\epsilon \downarrow 0$, one obtains, arguing as before,

$$P(U'\delta_+(\Gamma)) = (P(U') - A)\,\delta_+(\Gamma) + 2\pi\delta_q.$$

Because of (4.2.14), this must be identical with (4.2.9), whence

$$(P(U') - A)|_{\Gamma=0} = P(U)|_{\Gamma=0}.$$

This can also be verified by a simple computation. There is a similar argument for the backward case.

It remains to prove (4.2.15), which will also be needed in the next section. Let us again choose local coordinates that are normal and Minkowskian at q, with the x^1-axis future-directed. Writing $x = (t, X)$, $X \in \mathbf{R}^3$, and $|X| = r$, one has $\Gamma = t^2 - r^2$, by (4.2.5). Hence one can take, on $\Sigma_\epsilon^+ = \{(t, X); t = (r^2 + \epsilon)^{\frac{1}{2}}\}$,

$$\mu_\Gamma = \frac{1}{2} \frac{|g|^{\frac{1}{2}} dX}{(r^2 + \epsilon)^{\frac{1}{2}}}.$$

Then, by (4.2.6),

$$(\epsilon \delta'_+(\Gamma - \epsilon), \phi) = -\epsilon \frac{d}{d\epsilon}(\delta_+(\Gamma - \epsilon), \phi) = -\tfrac{1}{2}\epsilon \frac{d}{d\epsilon} \int \frac{\psi((r^2 + \epsilon)^{\frac{1}{2}}, X)}{(r^2 + \epsilon)^{\frac{1}{2}}} dX,$$

where $\psi = \phi |g|^{\frac{1}{2}}$, or

$$(\epsilon \delta'_+(\Gamma - \epsilon), \phi) = \tfrac{1}{4}\epsilon \int \left(\frac{\psi((r + \epsilon^2)^{\frac{1}{2}}, X)}{(r^2 + \epsilon)^{\frac{3}{2}}} + \frac{\partial_1 \psi((r^2 + \epsilon)^{\frac{1}{2}}, X)}{r^2 + \epsilon} dX \right).$$

As $\psi = \phi |g|^{\frac{1}{2}} \in C_0^\infty(\Omega_0)$, there are positive constants a and b such that $\psi = 0$ for $r > a$, and $|\psi| \leqslant b, |\partial_1 \psi| \leqslant b$. Hence

$$|(\epsilon \delta'_+(\Gamma - \epsilon), \phi)| \leqslant \tfrac{1}{4}b\epsilon \int_{r<a} \left(\frac{1}{(r^2 + \epsilon)^{\frac{3}{2}}} + \frac{1}{r^2 + \epsilon} \right) dX$$

$$= \pi b \epsilon \int_0^a \left(\frac{r^2}{(r^2 + \epsilon)^{\frac{3}{2}}} + \frac{r^2}{r^2 + \epsilon} \right) dr.$$

But $r \leqslant (r^2 + \epsilon)^{\frac{1}{2}}$, and so

$$|(\epsilon \delta'_+(\Gamma - \epsilon), \phi)| \leqslant \pi b \epsilon \int_0^a \left(1 + \frac{r}{r^2 + \epsilon} \right) dr = \pi b \epsilon \left(a + \tfrac{1}{2} \log \frac{a^2 + \epsilon}{\epsilon} \right),$$

which goes to zero when $\epsilon \downarrow 0$, and (4.2.15) is proved.

As $U\delta_+(\Gamma)$ and $U\delta_-(\Gamma)$ are the singular parts of the fundamental solutions of P which will be constructed, we interrupt the argument to examine the function U in more detail. Let (Ω, π) be a coordinate chart covering Ω, and set $\pi p = x, \pi q = y, \pi_* v = \theta$, where $v \in TM_q$. Let $r \to z(r) = h(y, r\theta)$ denote the solutions of the differential equations of the geodesics, with the initial values $z(0) = y, (\partial/\partial r) z(0) = \theta$. Then $(y, \theta) \to x = h(y, \theta)$ is the exponential map $\exp_q v$, and (as Ω is a geodesically convex domain) can be inverted to give θ as a function of x and y. Also, the θ^i are local coordinates, admissible in Ω, and normal at q; let us write $\theta = \tilde{x}$. Then, by (4.2.13),

$$U = \left| \frac{\tilde{g}(0)}{\tilde{g}(\tilde{x})} \right|^{\frac{1}{4}} \exp\left(-\frac{1}{2} \int_0^1 \tilde{a}_i(r\tilde{x}) \, \tilde{x}^i \, dr \right).$$

This obviously implies that

$$U(x, y) = \kappa(x, y) \exp\left(-\frac{1}{2}\int_0^1 \left\langle a(z(r)), \frac{\partial z}{\partial r}\right\rangle dr\right), \quad (4.2.16)$$

where $\kappa \in C^\infty(\Omega \times \Omega)$ satisfies

$$2\langle\nabla\Gamma, \nabla\kappa\rangle + (\square\,\Gamma - 8)\kappa = 0, \quad \kappa|_{p=q} = 1. \quad (4.2.17)$$

Also $\qquad |\tilde{g}(\tilde{x})|^{\frac{1}{2}} = |g(x)|^{\frac{1}{2}} |Dx/D\tilde{x}|^{-1},$

where the Jacobian determinant $Dx/D\tilde{x}$ is expressed as a function of x and y. Hence, as $Dx/D\tilde{x} = Dx/D\theta = 1$, when $x = y$,

$$\kappa(x, y) = \left|\frac{g(y)}{g(x)}\right|^{\frac{1}{4}}\left|\frac{Dx}{D\theta}\right|^{-\frac{1}{2}}. \quad (4.2.18)$$

This expression is similar to the solution of the transport equation of order zero of the progressing wave formalism, (3.6.15). It has already been pointed out that only the restrictions of U and PU to $C(q)$ figure in (4.2.9) and (4.2.10). It can be seen from (4.2.16) and (4.2.18) that they become infinite on a caustic of $C(q)$. In the present context, this cannot happen, as Ω is supposed to be a geodesically convex domain.

Another interesting expression for κ can be derived from (4.2.18). The second equation of (1.2.15) of Theorem 1.2.3 is, in the present notation, $\quad (\partial/\partial y^i)\,\Gamma(x, y) = -2g_{ij}(y)\,\theta^j \quad (i = 1, \ldots, 4).$

Differentiating with respect to the x^i, and evaluating the determinants, one finds $\quad |\det \partial^2\Gamma/\partial x^i\,\partial y^j| = 16\,|g(y)|\,|D\theta/Dx|_p.$

Hence (4.2.18) can also be written as

$$\kappa(x, y) = \frac{|\det \partial^2\Gamma/\partial x^i\,\partial y^j|^{\frac{1}{2}}}{4\,|g(x)\,g(y)|^{\frac{1}{4}}}. \quad (4.2.19)$$

As $\Gamma(x, y) = \Gamma(y, x)$, this implies that κ is a symmetric function of its arguments, $\qquad \kappa(p, q) = \kappa(q, p). \quad (4.2.20)$

The function $p \to \kappa(p, q)$ satisfies the transport equation (4.2.17), which is similar to the transport equation for the amplitude of a simple progressing wave, (3.6.7). As the null cone $C(q)$ is a characteristic, it is to be expected that the variation of κ along the bicharacteristics of $C(q)$ will be related to the generalized geometrical optics intensity law (Theorem 3.5.3). We shall now show that this is so, and derive a 'geometric' form of the restriction of κ to $C(q)$. It will obviously be sufficient to do this on $C^+(q)$ only.

One must first introduce a special family of parametrizations of $C^+(q)\backslash(q)$, of the form

$$\mathbf{R}^+ \times \mathbf{S}^2 \ni (r, \xi) \to p(r, \xi) \in C^+(q)\backslash(q),$$

such that ξ is constant on a bicharacteristic, and r is an affine parameter, for constant ξ. (Here, $\mathbf{R}^+$ is the positive real line, and $\mathbf{S}^2$ is the unit sphere in $\mathbf{R}^3$.) Let (Ω, π) be a coordinate chart that is normal and Minkowskian at q, and such that the vector $(1, 0, 0, 0) \in TM_q$ is future-directed. Let $\xi = (\xi_1, \xi_2, \xi_3)$ be a unit vector in $\mathbf{R}^3$, and let $\xi \to \xi_0(\xi)$ be a positive real-valued C^∞ function on S^2. Then the vector $v \in TM_q$ such that $\Pi_* v = (\xi_0, \xi_0 \xi_1, \xi_0 \xi_2, \xi_0 \xi_3)$ is null and future-directed. As Ω is geodesically convex, it follows that, if $p \in C^+(q)\backslash(q)$, then the equation $p = \exp_q(rv)$ has a unique inverse $r = r(p), v = v(p)$, and that

$$(r, \xi) \to p = \exp_q(r, v)$$

is a diffeomorphism between $C^+(q)\backslash(q)$ and a subset of $\mathbf{R}^+ \times \mathbf{S}^2$ that is of the form $\{(r, \xi); \xi \in \mathbf{S}^2, 0 < r < r_0(\xi)\}$, where $\xi \to r_0(\xi)$ is a C^∞ function on $\mathbf{S}^2$. This parametrization of $C^+(q)\backslash(q)$ has the required properties. It is obvious that, if $(\tilde{r}, \tilde{\xi}) \to p(\tilde{r}, \tilde{\xi})$ is another parametrization of the same type, then $\xi \to \tilde{\xi}$ is a diffeomorphism of $\mathbf{S}^1$ onto itself, and $\tilde{r} = \rho(\xi) r$, where $0 < \rho \in C^\infty(\mathbf{S}^2)$.

Theorem 4.2.2. *Let* $\mathbf{R}^+ \times \mathbf{S}^2 \ni (r, \xi) \to p(r, \xi)$ *be a parametrization of* $C^+(q)\backslash(q)$ *such that, on each bicharacteristic of* $C^+(q)$, ξ *is constant, and* r *is an affine parameter. Let* Σ *be a smooth 2-surface, contained in* $C^+(q)\backslash(q)$, *which is met at most once, and transversally, by any bicharacteristic. Let* $d\sigma$ *denote the invariant 2-surface element on* Σ, *determined by the induced metric on* Σ. *Then* $d\sigma = r^2 k(r, \xi) d\omega_\xi$, *where* $d\omega_\xi$ *is the surface element on* $\mathbf{S}^2$, *and* $k(0, \xi) \in C^\infty(\mathbf{S}^2)$. *Also,*

$$\kappa(r, q)\big|_{C^+(q)\backslash(q)} = \left(\frac{k(0, \xi)}{k(r, \xi)}\right)^{\frac{1}{2}} = r\left(\frac{d\sigma_0}{d\sigma}\right)^{\frac{1}{2}}, \qquad (4.2.21)$$

with
$$d\sigma_0 = k(0, \xi) d\omega_\xi = \lim_{r \downarrow 0} \frac{d\sigma}{r^2}. \qquad (4.2.22)$$

PROOF. We first compute the induced metric on $C^+(q)\backslash(q)$, in a coordinate system that is normal at q, so that $\pi p = x, \pi q = 0$, and, by (1.3.16),

$$g_{ij}(x) x^j = g_{ij}(0) x^j \quad (i = 1, \dots, 4). \qquad (4.2.23)$$

In these coordinates, one has $x = r\theta$, where $\theta = \theta(\xi)$ is a null vector. So

$$\langle dx, dx \rangle = g_{ij}(\theta^i dr + r d\theta^i)\,(\theta^j dr + r d\theta^j)$$
$$= g_{ij}(r\theta)\,\theta^i\theta^j\,dr^2 + 2rg_{ij}(r\theta)\,\theta^i d\theta^j\,dr + r^2 g_{ij}(r\theta)\,d\theta^i d\theta^j.$$

By (4.2.23)

$$g_{ij}(r\theta)\,\theta^i\theta^j = g_{ij}(0)\,\theta^i\theta^j = 0,$$
$$2g_{ij}(r\theta)\,\theta^i d\theta^j = 2g_{ij}(0)\,\theta^i d\theta^j = d(g_{ij}(0)\,\theta^i\theta^j) = 0. \qquad (4.2.24)$$

Hence
$$\langle dx, dx \rangle = r^2 g_{ij}(r\theta)\,d\theta^i d\theta^j.$$

Locally, one can introduce coordinates (λ^1, λ^2) on $\mathbf{S}^2$, and so

$$\langle dx, dx \rangle = r^2 h_{AB}\,d\lambda^A d\lambda^B, \quad h_{AB} = g_{ij}(r\theta)\frac{\partial\theta^i}{\partial\lambda^A}\frac{\partial\theta^j}{\partial\lambda^B}, \quad (4.2.25)$$

where A and B take only the values 1 and 2. A 2-surface $\Sigma \subset C^+(q)\backslash(q)$ of the type envisaged has an equation of the form $r = r(\lambda^1, \lambda^2)$, where $r(\lambda^1, \lambda^2)$ is a C^∞ function on some open set in $\mathbf{R}^2$. Hence (4.2.25) is already the induced metric on Σ, and

$$d\sigma = r^2(h_{11}h_{22} - h_{12}^2)^{\frac{1}{2}}\,d\lambda^1 d\lambda^2. \qquad (4.2.26)$$

But
$$d\omega_\xi = \xi_1 d\xi_2 \wedge d\xi_3 + \xi_2 d\xi_3 \wedge d\xi_1 + \xi_3 d\xi_1 \wedge d\xi_2,$$

and so it is clear that (4.2.26) can be written as $d\sigma = r^2 k(r, \xi)\,d\omega_\xi$, and that $k(0, \xi) \in C^\infty(\mathbf{S}^2)$.

We shall now obtain another expression for $(h_{11}h_{22} - h_{12}^2)^{\frac{1}{2}}$. If ξ is a future-directed time-like vector at q, then $\langle \xi, \theta \rangle > 0$, and so, if one sets $\eta = \xi/\langle \xi, \theta \rangle$, one obtains a vector $\eta(\xi)$ such that $\langle \eta, \theta \rangle = 1$. Put

$$a_1^i = \frac{\partial\theta^i}{\partial\lambda^1}, \quad a_2^i = \frac{\partial\theta^i}{\partial\lambda^2}, \quad a_3^i = \theta^i, \quad a_4^i = \eta^i \quad (i = 1, ..., 4). \quad (4.2.27)$$

By the rule for the multiplication of determinants, one has

$$\det\,(g_{kl}(r\theta)\,a_i^k a_j^l) = g(r\theta)\,(\det a_j^i)^2.$$

Now by (4.2.26) and (4.2.27),

$$g_{kl}a_A^k a_B^l = h_{AB}.$$

Again, one has

$$g_{kl}a_3^k a_3^l = g_{kl}(r\theta)\,\theta^k\theta^l = 0, \quad g_{kl}a_3^k a_A^l = g_{kl}(r\theta)\,\theta^k\frac{\partial\theta^l}{\partial\lambda^A} = 0,$$

by (4.2.24); and, by construction,

$$g_{kl}(r\theta)\,a_3^k a_4^l = g_{kl}(r\theta)\,\theta^k\eta^l = g_{kl}(0)\,\theta^k\eta^l = 1.$$

Hence $\qquad -g(r\theta)\,(\det a_j^i)^2 = h_{11}h_{22} - h_{12}^2,$

and so it follows from (4.2.26) that, with $x = r\theta$,

$$d\sigma = r^2\,|g(x)|^{\frac{1}{2}}\,|\det a_j^i|\,d\lambda^1 d\lambda^2. \qquad (4.2.28)$$

But by (4.2.18), and as the coordinates are normal at q, one also has

$$\kappa = \left|\frac{g(0)}{g(x)}\right|^{\frac{1}{4}}.$$

Hence (4.2.28) implies (4.2.21), and so the theorem is proved. $\square$

4.3 The parametrix

We now return to the main argument. Let $H(t)$ denote the Heaviside function, $H(t) = 1$ if $t \geqslant 0$ and $H(t) = 0$ if $t < 0$. Then the distribution $H_+(\Gamma - \epsilon)$, where $\epsilon > 0$, which is given by (4.2.6) with $f(t) = H(t)$, tends to the characteristic function of $\overline{D^+(q)}$ when $\epsilon \downarrow 0$. This set will be called the *future emission* of q, and be denoted by $J^+(q)$; the *past emission* $J^-(q)$ of q is the closure of $D^-(q)$. So

$$H_+(\Gamma) = \begin{cases} 1 & \text{if} \quad p \in J^+(q), \\ 0 & \text{if} \quad p \notin J^+(q). \end{cases} \qquad (4.3.1)$$

Similarly, $H_-(\Gamma)$ is the characteristic function of $J^-(q)$.

Lemma 4.3.1. *There exists a function $V_0(p,q) \in C^\infty(\Omega \times \Omega)$ such that, if $W \in C^\infty(\Omega \times \Omega)$ and $W = V_0$ on $C(q)$, then*

$$P(U\delta_+(\Gamma) + WH_+(\delta)) = P(W)\,H_+(\Gamma) + 2\pi\,\delta_q, \qquad (4.3.2)$$

$$P(U\delta_-(\Gamma) + WH_-(r)) = P(W)\,H_-(\Gamma) + 2\pi\delta_q, \qquad (4.3.3)$$

where U is the function of Theorem 4.2.1.

PROOF. As $H'(t) = \delta(t)$, and $\delta_+'(\sigma)$ does not exist, we first compute $P(WH_+(\delta - \epsilon))$, where $\epsilon > 0$. By (3.6.2), and as $\langle \nabla\Gamma, \nabla\Gamma \rangle = 4\Gamma$, this is

$P(W)\,H_+(\Gamma - \epsilon)$
$\qquad + (2\langle \nabla\Gamma, \nabla W \rangle + (\Box\,\Gamma + \langle a, \nabla\Gamma \rangle)\,W)\,\delta_+(\Gamma - \epsilon) + 4W\Gamma\delta_+'(\Gamma - \epsilon).$

It follows from (4.2.15) that

$$\lim_{\epsilon \downarrow 0} \Gamma\delta_+'(\Gamma - \epsilon) = \lim_{\epsilon \downarrow 0} (\epsilon\delta_+'(\Gamma - \epsilon) - \delta_+(\Gamma - \epsilon)) = -\delta_+(\Gamma).$$

Hence

$$P(WH_+(\Gamma))$$
$$= P(W)H_+(\Gamma) + (2\langle \nabla\Gamma, \nabla W\rangle + (\Box\,\Gamma + \langle a, \nabla\Gamma\rangle - 4)\,W)\,\delta_+(\Gamma). \tag{4.3.4}$$

We shall now show that there is a $V_0 \in C^\infty(\Omega \times \Omega)$ such that

$$2\langle \nabla\Gamma, \nabla V_0\rangle + (\Box\,\Gamma + \langle a, \nabla\Gamma\rangle - 4)\,V_0 = -PU. \tag{4.3.5}$$

It follows from the transport equation for U, (4.2.12), that (4.3.5) can be put into the form

$$2\left\langle \nabla\Gamma, \nabla\frac{V_0}{U}\right\rangle + 4\frac{V_0}{U} = -\frac{PU}{U}.$$

(Note that $U \neq 0$ in $\Omega \times \Omega$, by (4.3.13).) Consider this equation in local coordinates, with the notation that was used in the proof of (4.2.18) and (4.2.19). On the geodesic $\widehat{qp}$: $[0, 1] \ni r \to z(r) = h(y, \theta(x, y)\,r)$ one has, by (1.2.14),

$$\nabla^i\Gamma\big|_{x=z(r)} = 2r\frac{\partial z^i}{\partial r} \quad (i = 1, \ldots, 4),$$

where the ∇^i act at x. Hence

$$\frac{\partial}{\partial r}\left(\frac{rV_0}{U}\bigg|_{x=z(r)}\right) = -\frac{1}{4}\frac{PU}{U}\bigg|_{x=z(r)}.$$

It is obvious that this has one and only one solution that remains bounded when $r \to 0$, namely

$$V_0\big|_{x=z(r)} = -\frac{1}{4r}U\bigg|_{x=z(r)}\int_0^r \frac{PU}{U}\bigg|_{x=z(s)}ds.$$

As $z(1) = x$, one therefore finds that (4.3.5) has a unique solution which remains bounded when $x \to y$,

$$V_0(x, y) = -\tfrac{1}{4}U(x, y)\int_0^1 \frac{PU}{U}\bigg|_{x=z(s)}ds. \tag{4.3.6}$$

Here $Z(s) = h(y, \theta(x, y)\,s)$, where $\theta(x, y)$ is the inverse of the diffeomorphism $\theta \to x = h(y, \theta)$. Hence $V_0 \in C^\infty(\Omega \times \Omega)$.

Now

$$(2\langle \nabla\Gamma, \nabla W\rangle + (\Box\,\Gamma + \langle a, \nabla\Gamma\rangle - 4)\,W)\,\delta_+(\Gamma)$$
$$= (2\langle \nabla\Gamma, \nabla W\rangle + (\Box\Gamma + \langle a, \nabla\Gamma\rangle - 4)\,W)\big|_{C^+(q)}\delta_+(\Gamma).$$

As $\langle \nabla\Gamma, \nabla W\rangle$ is an intrinsic derivative on $C(q)$, it follows that, if $W = V_0$ on $C^+(q)$, then, by (4.3.4) and (4.3.5),

$$P(WH_+(\Gamma)) = P(W)H_+(\Gamma) - P(U)\delta_+(\Gamma).$$

When this is added to (4.2.9), one obtains (4.3.2). The other identity, (4.3.3), is proved in the same way, or can be deduced by reversing the time-orientation. So the lemma is proved. □

Lemma 4.3.1 shows that one can obtain a fundamental solution of P, whose support is contained in $J^+(q)$, if one can solve the characteristic initial value problem

$$PV^+ = 0, \, p \in D^+(q); \; V^+ = V_0, p \in C^+(q). \tag{4.3.7}$$

For with $W = V^+$, (4.3.2) implies that

$$G_q^+ = \frac{1}{2\pi}(U\delta_+(\Gamma) + V^+ H_+(\Gamma))$$

is a fundamental solution of P. Likewise, if the characteristic initial value problem

$$PV^- = 0, p \in D^-(q); \; V^- = V_0, p \in C^-(q) \tag{4.3.8}$$

has a solution, then

$$G_q^- = \frac{1}{2\pi}(U\delta_-(\Gamma) + V^- H_+(\Gamma))$$

is a fundamental solution of P whose support is contained in $J^-(q)$. Both (4.3.7) and (4.3.8) are properly posed; this will become clear in the next chapter.

Formally, the combined characteristic initial value problem

$$PV = 0, \quad V = V_0 \quad \text{on} \quad C(q), \tag{4.3.9}$$

can be solved by a series

$$\sum_{\nu=0}^{\infty} V_\nu \frac{\Gamma^\nu}{\nu!}, \tag{4.3.10}$$

where V_0 is the function just determined, if the V_ν, $\nu \geqslant 1$, satisfy a certain set of transport equations. For, by (3.6.2) and $\langle \nabla\Gamma, \nabla\Gamma \rangle = 4\Gamma$,

$$P\left(V_\nu \frac{\Gamma^\nu}{\nu!}\right) = P(V_\nu)\frac{\Gamma^\nu}{\nu!} + (2\langle \nabla\Gamma, \nabla V_\nu \rangle + (\Box\,\Gamma + \langle a, \nabla\Gamma \rangle + 4\nu - 4)\,V_\nu)\frac{\Gamma^{\nu-1}}{(\nu-1)!}.$$

Suppose that the V_ν can be chosen so that

$$2\langle \nabla\Gamma, \nabla V_\nu \rangle + (\Box\,\Gamma + \langle a, \nabla\Gamma \rangle + 4\nu - 4)\,V_\nu = -PV_{\nu-1} \quad (\nu = 1, 2, \ldots). \tag{4.3.11}$$

Then $\quad P\left(V_\nu \frac{\Gamma^\nu}{\nu!}\right) = P(V_\nu)\frac{\Gamma^\nu}{\nu!} - P(V_{\nu-1})\frac{\Gamma^{\nu-1}}{(\nu-1)!} \quad (\nu = 1, 2, \ldots). \tag{4.3.12}$

So the series (4.3.10) is then a formal solution of the homogeneous wave equation $Pu = 0$, and reduces to V_0 on $C(q)$, where $\Gamma = 0$.

It is easy to see that the recurrence equations (4.3.11) determine a unique sequence of functions V_ν, if one adds the condition that each V_ν is to remain bounded when $p \to q$. For these equations can be treated like (4.3.5) (which is just (4.3.11) with $\nu = 0$ and $V_{-1} = U$). On the geodesic $\widehat{qp}$: $[0, 1] \ni r \to z(r) = h(y, \theta(x, y)\, r)$ one obtains

$$r\frac{\partial}{\partial r}\left(\frac{V_\nu}{U}\Big|_{x=z(r)}\right) + (\nu+1)\frac{V_\nu}{U}\Big|_{x=z(r)} = -\frac{1}{4}\frac{PV_{\nu-1}}{U}\Big|_{x=z(r)},$$

or, after multiplication by r^ν,

$$\frac{\partial}{\partial r}\left(r^{\nu+1}\frac{V_\nu}{U}\Big|_{x=z(r)}\right) = -\tfrac{1}{4}r^\nu\frac{PV_{\nu-1}}{U}\Big|_{x=z(r)}.$$

Assuming that $V_{\nu-1}$ has already been determined, this can be solved for V_ν; there is one and only one solution that remains bounded when $r \to 0$,

$$\frac{V_\nu}{U}\Big|_{x=z(r)} = -\frac{1}{4r^{\nu+1}}\int_0^r \frac{PV_{\nu-1}}{U}\Big|_{x=z(s)}\, s^\nu\, ds.$$

One can now put $r = 1$, to conclude that

$$V_\nu(x, y) = -\tfrac{1}{4}U(x, y)\int_0^1 \frac{PV_{\nu-1}}{U}\Big|_{x=z(s)}\, s^\nu\, dS \quad (\nu = 1, 2, \ldots). \quad (4.3.13)$$

It is apparent that the V_ν determined recursively by these equations are all in $C^\infty(\Omega \times \Omega)$.

There is one important case where the series (4.3.10), with the coefficients which have just been determined, converges to a solution of the boundary value problem (4.3.9), for sufficiently small $|\Gamma(p, q)|$. Before stating this, some preliminary remarks are required. A function $u(x)$, defined on an open set $E \subset \mathbf{R}^n$, is called a real analytic function if every point $x_0 \in E$ has a neighbourhood in which u can be expanded as a convergent power series,

$$u(x) = \sum_{\alpha \geq 0} u_\alpha(x_0)\frac{(x - x_0)^\alpha}{\alpha!}.$$

This is then, of course, a Taylor series, and $u_\alpha = \partial^\alpha u(x_0)$. Such a function can be extended to a complex analytic function on a neighbourhood of $\{z = x+iy;\, x \in E, y = 0\}$ in $\mathbf{C}^n$, as the power series also converges for sufficiently small $|z - x_0|$ when x is replaced by $z = x+iy$. An analytic structure for an n-dimensional manifold M is defined like a differentiable structure, except that the coordinate transformations are required to be analytic. A function u on M is then analytic if, in every coordinate chart (Ω, π), $u \circ \pi^{-1}(x)$ is analytic on $\pi\Omega$. Analytic vector and tensor fields are defined similarly as analytic cross-sections of the appropriate bundles over M. A space–time is analytic if it has an analytic structure and an analytic metric. A differential operator $P = \Box + \langle a, \nabla \rangle + b$ defined on such a space–time is analytic if, in addition, the vector field a and the scalar field b are analytic. We shall refer to this as the analytic case. It is evident that, in this case, $\Gamma(p, q)$ and the functions $U, V_0, V_1, \ldots$ are all analytic.

Theorem 4.3.1. *In the analytic case, every point $q \in \Omega$ has a neighbourhood $\omega_q \subset \Omega$ in which the series (4.3.10) converges to an analytic function $V(p, q)$ which is a solution of the boundary value problem (4.3.9). The distributions*

$$G_q^+ = \frac{1}{2\pi}(U\delta_+(\Gamma) + VH_+(\Gamma)) \tag{4.3.14}$$

and

$$G_q^- = \frac{1}{2\pi}(U\delta_-(\Gamma) + VH_-(\Gamma)) \tag{4.3.15}$$

are then fundamental solutions of P, on ω_q, whose supports are contained in $J^+(q)$ and $J^-(q)$ respectively. $\Box$

The proof will be omitted. It is a consequence of Hadamard's classic theory of linear hyperbolic differential equations of the second order. In fact, our function V is, apart from a numerical factor, the coefficient of the logarithmic term in Hadamard's 'elementary solution'. The convergence domain ω_q depends on the singularities of the metric and of the fields a and b in the complex extension of Ω, and may therefore be a proper subset of Ω. So Theorem 4.3.1 may be inadequate even in the analytic case, and the theory which will be developed for the C^∞ case is still relevant.

In the C^∞ case, the series (4.3.10) does not converge. However, by inserting suitable factors, one can convert it into a convergent series whose sum is a C^∞ function $\tilde{V}$ such that $\tilde{V} = V_0$ on $C(q)$, and $P\tilde{V}$ vanishes to all orders on $C(q)$. This can then be used to construct C^∞ parametrices for P. The convergence factors are defined in the following lemma.

Lemma 4.3.2. *Let $\sigma(t) \in C_0^\infty(\mathbf{R})$ be such that $0 \leqslant \sigma(t) \leqslant 1$, that $\sigma(t) = 1$ for $|t| \leqslant \frac{1}{2}$, and that $\sigma(t) = 0$ for $|t| \geqslant 1$. Then there exists a sequence of* *positive numbers* $k_1, k_2, \ldots,$ *strictly increasing and tending to infinity, such that the series*

$$\sum_{\nu=1}^{\infty} V_\nu \frac{\Gamma^\nu}{\nu!} \sigma(k_\nu \Gamma) \tag{4.3.16}$$

converges to a function $S(p, q) \in C^\infty(\Omega \times \Omega)$, *uniformly on every compact subset of* $\Omega \times \Omega$. *Also,* $S \to 0$ *when* $\Gamma \to 0$, *and*

$$\lim_{\Gamma \to 0} \Gamma^{-\mu} \left(S - \sum_{\nu=1}^{\mu} V_\nu \frac{\Gamma^\nu}{\nu!} \right) = 0 \quad (\mu = 1, 2, \ldots). \tag{4.3.17}$$

PROOF. If $\{k_\nu\}$ is strictly increasing and $k_\nu \to \infty$, then the series is a finite sum for $\Gamma \neq 0$, as the $\sigma(k_\nu \Gamma)$ vanish when $k_\nu |\Gamma| \geqslant 1$. Its sum is then a C^∞ function $S(p, q)$; the only point at issue is the behaviour of S when $\Gamma \to 0$. Let Φ be a compact subset of $\Omega \times \Omega$. Then the V_ν, and each of their derivatives in a fixed coordinate chart covering Ω, are bounded on Φ. Now in Φ,

$$\left| V_\nu \frac{\Gamma^\nu}{\nu!} \sigma(k_\nu \Gamma) \right| \leqslant \frac{1}{\nu! k_\nu^\nu} \sup_\Phi |V_\nu|,$$

as $\sigma(k_\nu \Gamma) = 0$ for $|\Gamma| \geqslant 1/k_\nu$. Hence the series (4.3.16) converges uniformly on Φ if the k_ν are chosen such that

$$\sum_{\nu=1}^{\infty} \frac{1}{\nu! k_\nu^\nu} \sup_\Phi |V_\nu| < \infty. \tag{4.3.18}$$

If $\{k_\nu\}$ is a sequence for which this holds, and $\{k_\nu'\}$ is another sequence, such that $k_\nu \leqslant k_\nu'$ for $\nu = 1, 2, \ldots,$ then (4.3.18) also holds when the k_ν are replaced by the k_ν'.

One can now repeat this reasoning with the derivatives of the series, computed in some fixed coordinate chart covering Ω. By the remark just made, it follows that one can therefore determine a set of such sequences, $\{k_\nu^{(\mu)}\}_{1 \leqslant \nu < \infty}$, with the following properties: (i) if the k_ν in (4.3.16) are replaced by the $k_\nu^{(\mu)}$, then the series, together with its derivatives with respect to the local coordinates of p and q of order not exceeding $\mu - 1$, converges uniformly on Φ; (ii) for all μ and ν, $k_\nu^{(\mu)} \leqslant k_\nu^{(\mu+1)}$. If one now sets $k_\nu = k_\nu(\Phi) = k_\nu^{(\nu)}$ then (4.3.16), together with its derivatives of all orders, converges uniformly on Φ, and so $S \in C^\infty(\Phi)$. Now let $\Phi_1, \Phi_2, \ldots$ be an expanding sequence of compact subsets of $\Omega \times \Omega$ whose union covers $\Omega \times \Omega$. For each $\lambda = 1, 2, \ldots$ one can find a sequence $\{k_\nu(\Phi_\lambda)\}_{1 \leqslant \nu < \infty}$ of the type just determined; also, one can arrange that $k_\nu(\Phi_\lambda) \leqslant k_\nu(\Phi_{\lambda+1})$ for all ν, λ. So if one finally sets $k_\nu = k_\nu(\Phi_\nu)$, then one obtains a series that has the required behaviour.

To prove (4.3.17), we use the fact that $\sigma(t) = 1$ for $|t| \leqslant \frac{1}{2}$. It is obvious that $S \to 0$ when $\Gamma \to 0$. Generally, if $|\Gamma| \leqslant \frac{1}{2}k_\mu$, then $\sigma(k_\nu \Gamma) = 1$ for $\nu = 1, \ldots, \mu$. So

$$S - \sum_{\nu=1}^{\mu} V_\nu \frac{\Gamma^\nu}{\nu!} = \sum_{\nu=\mu+1}^{\infty} V_\nu \frac{\Gamma^\nu}{\nu!} \sigma(k_\nu \Gamma), \quad |\Gamma| \leqslant \frac{1}{2k_\mu}.$$

As the series on the right-hand side, and the series deduced from it by term-by-term differentiation in local coordinates, converge uniformly on every compact subset of $\Omega \times \Omega$, (4.3.17) follows. This completes the proof. $\square$

Remark. The equation (4.3.17) shows that the series (4.3.16), with the convergence factors $\sigma(k_\nu \Gamma)$ omitted, is an asymptotic expansion of S, valid when $\Gamma \to 0$. To indicate this, we write

$$S \sim \sum_{\nu=1}^{\infty} V_\nu \frac{\Gamma^\nu}{\nu!}. \qquad (4.3.19)$$

A C^∞ *parametrix* of P is a distribution $\tilde{G}_q$ such that

$$P\tilde{G}_q - \delta_q \in C^\infty(\Omega).$$

Theorem 4.3.2. *Let U be the function of Theorem 4.2.1, and $V_0, V_1, \ldots,$ be the functions given by (4.3.6) and (4.3.13); let $\sigma(t)$ and the k_ν be as in Lemma 4.3.2. Put*

$$\tilde{V} = V_0 + S = V_0 + \sum_{\nu=1}^{\infty} V_\nu \frac{\Gamma^\nu}{\nu!} \sigma(k_\nu \Gamma). \qquad (4.3.20)$$

Then
$$\tilde{G}_q^+ = \frac{1}{2\pi}(U\delta_+(\Gamma) + \tilde{V}H_+(\Gamma)) \qquad (4.3.21)$$

is a C^∞ parametrix of P whose support is contained in $J^+(q)$, and

$$\tilde{G}_q^- = \frac{1}{2\pi}(U\delta_-(\Gamma) + \tilde{V}H_-(\Gamma)) \qquad (4.3.22)$$

is a C^∞ parametrix of P whose support is contained in $J^-(q)$.

PROOF. The support of $\tilde{G}_q^+$ is contained in $J^+(q)$, by construction. Since $\tilde{V} = V_0$ on $C^+(q)$, it follows from Lemma 4.3.1 that

$$P\tilde{G}_q^+ = \frac{1}{2\pi}P(\tilde{V})H_+(\Gamma) + \delta_q. \qquad (4.3.23)$$

As $P(\tilde{V}) \in C^\infty(\Omega \times \Omega)$, $P\tilde{G}_q - \delta_q$ is C^∞ for $p \in D^+(q)$, and vanishes for $p \notin D^+(q)$. To show that $P\tilde{G}_q^+ - \delta_q \in C^\infty(\Omega \times \Omega)$ one therefore only has to show that $P(\tilde{V})$ and its derivatives of all orders vanish on $C^+(q)$, that is to say when $\Gamma = 0$. Now, for $\nu \geqslant 1$,

$$P\left(V_\nu \frac{\Gamma^\nu}{\nu!} \sigma(k_\nu \Gamma)\right) = P\left(V_\nu \frac{\Gamma^\nu}{\nu!}\right) \sigma(k_\nu \Gamma) + 2\left\langle \nabla\left(V_\nu \frac{\Gamma^\nu}{\nu!}\right), \nabla\sigma(k_\nu \Gamma) \right\rangle$$
$$+ V_\nu \frac{\Gamma^\nu}{\nu!} (\Box + \langle a, \nabla\rangle) \sigma(k_\nu \Gamma).$$

The V_ν were constructed so as to satisfy (4.3.3),

$$P\left(V_\nu \frac{\Gamma^\nu}{\nu!}\right) = P(V_\nu)\frac{\Gamma^\nu}{\nu!} - P(V_{\nu-1})\frac{\Gamma^{\nu-1}}{(\nu-1)!}.$$

Also, as $\langle \nabla\Gamma, \nabla\Gamma \rangle = 4\Gamma$,

$$\left\langle \nabla\left(V_\nu \frac{\Gamma^\nu}{\nu!}\right), \nabla\sigma(k_\nu \Gamma) \right\rangle = k_\nu(\langle \nabla v_\nu, \nabla\Gamma\rangle + 4\nu V_\nu)\frac{\Gamma^\nu}{\nu!} \sigma'(k_\nu \Gamma)$$

and

$$(\Box + \langle a, \nabla\rangle) \sigma(k_\nu \Gamma) = 4k_\nu^2 \Gamma\sigma''(k_\nu \Gamma) + k_\nu(\Box\Gamma + \langle a, \nabla\Gamma\rangle) \sigma'(k_\nu \Gamma).$$

Hence

$$P\left(V_\nu \frac{\Gamma^\nu}{\nu!} \sigma(k_\nu \Gamma)\right) = \left(P(V_\nu)\frac{\Gamma^\nu}{\nu!} - P(V_{\nu-1})\frac{\Gamma^{\nu-1}}{(\nu-1)!}\right) \sigma(k_\nu \Gamma) + W_\nu \frac{\Gamma^\nu}{\nu!},$$

where the W_ν only involve σ_ν and σ'_ν; in detail,

$$W_\nu = 4k_\nu^2 \Gamma\sigma''(k_\nu \Gamma)$$
$$+ k_\nu(2\langle \nabla V_\nu, \nabla\Gamma\rangle + (\Box\Gamma + \langle a, \nabla\Gamma\rangle + 8\nu) V_\nu) \sigma'(k_\nu \Gamma).$$

Thus
$$P\tilde{V} = (1 - \sigma(k_1 \Gamma)) PV_0 + \sum_{\nu=1}^{\infty} W_\nu \frac{\Gamma^\nu}{\nu!}. \qquad (4.3.24)$$

As $\sigma(t) = 1$ and $\sigma'(t) = 0$ for $|t| \leqslant \frac{1}{2}$, it follows that $1 - \sigma(k_1 \Gamma)$ and the W_ν, and their derivatives of all orders, vanish for $\Gamma = 0$. So it follows by uniform convergence that $P\tilde{V}$ and its derivatives of all orders vanish when $\Gamma = 0$. This proves that $\tilde{G}_q^+$ is a C^∞ parametrix. The argument for $\tilde{G}_q^-$ is similar. $\Box$

Remark. It follows from (4.3.17) and (4.3.20) that

$$\tilde{V} \sim \sum_{\nu=0}^{\infty} V_\nu \frac{\Gamma^\nu}{\nu!} \qquad (4.3.25)$$

holds when $\Gamma \to 0$; this means that

$$\lim_{\Gamma\to 0} \Gamma^{-\mu}\left(\tilde{V} - \sum_{\nu=0}^{\mu} V_\nu \frac{\Gamma^\nu}{\nu!}\right) = 0 \qquad (\mu = 0, 1, \ldots), \qquad (4.3.26)$$

with uniform convergence in every compact subset of $\Omega \times \Omega$ that meets $\{(p,q); \Gamma(p,q) = 0\}$. The series (4.3.25) will be called the *Hadamard series*.

Neither parametrix is determined uniquely. If $F \in C^\infty(\Omega \times \Omega)$ and $F \sim 0$ when $\Gamma \to 0$, then $\tilde{V} + F = V_0$ on $C(q)$, and $P(\tilde{V} + F) \sim 0$ when $\Gamma \to 0$. Hence one can replace $\tilde{V}$ by $\tilde{V} + F$ in each of (4.3.22) and (4.3.23), to obtain another pair of parametrices of P in Ω.

4.4 Causal domains

The starting point for the construction of fundamental solutions is Lemma 4.3.1. Applied to a test function $\phi \in C_0^\infty(\Omega)$, the identity (4.3.2) becomes

$$\phi(q) + \int_{J^+(q)} K(p,q)\,\phi(p)\,\mu(p)$$

$$= \frac{1}{2\pi} \int_{C^+(q)} U(p,q)\,^tP\phi(p)\,\mu_\Gamma(p) + \frac{1}{2\pi} \int_{J^+(q)} W(p,q)\,^tP\phi(p)\,\mu(p),$$

$$(4.4.1)$$

where $K = (\tfrac{1}{2}\pi)\,PW(p,q)$, W being a function such that $W = V_0$ when $p \in C^+(q)$. One can look upon (4.4.1) as an integral equation for ϕ. If it can be solved, then one obtains a linear form $^tP\phi \to \phi(q)$, q being regarded as a fixed point, which will be a fundamental solution of P if it is a distribution. A characteristic feature of this integral equation is that the integral in the first member is over the variable domain $J^+(q)$. The equation is therefore a multi-dimensional analogue of the Volterra equation of the second kind in one dimension. Volterra equations can be solved by the method of successive approximation. Likewise, (4.4.1) can be solved in this way, if an appropriate restriction is placed on the domain in which it is considered.

Definition 4.4.1. *A connected open set Ω will be called a causal domain if*
  (i) *there is a geodesically convex domain Ω_0 such that $\Omega \subset \Omega_0$, and*
  (ii) *for all pairs of points p, q in Ω, $J^+(q) \cap J^-(p)$ is a compact subset of Ω, or void.*

Note that $J^+(q) \cap J^-(p)$ may be void; this will happen if $p \notin J^+(q)$. Note also that, as $J^+(q) \cap J^-(p)$ is closed, it will be a compact subset of Ω if and only if it is bounded, and contained in Ω.

Condition (ii) may be regarded as a causality postulate. It is a simplified version of Leray's condition of global hyperbolicity, suitable for domains that satisfy condition (i). It should be noted that the condition is time-reversible. First, if $p \notin J^+(q)$ then $J^+(q) \cap J^-(p)$ is empty, and so also $q \notin J^-(p)$. Again, $p \in J^+(q)$ means either that $p = q$, or that the unique geodesic $\widehat{pq} \subset \Omega$ is *causal* (time-like or null) and future-directed. As Ω is geodesically convex, the unique geodesic $\widehat{pq} \subset \Omega$ is $\widehat{qp}$, with reversed orientation. Hence $p \in J^+(q)$ if and only if $q \in J^-(p)$, and so condition (ii) is equivalent to: (ii') if p and q are in Ω_0, then $J^-(q) \cap J^+(p)$ is a compact subset of Ω_0.

It is easy to prove that every space–time is locally causal.

Theorem 4.4.1. *Every point of a space–time has a neighbourhood that is a causal domain.*

PROOF. Let p_0 be a point in a space–time M, and choose a coordinate chart that is Minkowskian at p_0, so that p_0 is at the origin of the local coordinates, and $g_{ij}(0) = \eta_{ij}$. By Theorem 1.2.2, there is a ball $B_0 = \{x; |x| < \epsilon_0\}$ which is geodesically convex. Causal vectors at p_0 are characterized by

$$(\xi^1)^2 \geqslant (\xi^2)^2 + (\xi^3)^2 + (\xi^4)^2.$$

By the continuity of the metric tensor, there is an ϵ, $0 < \epsilon \leqslant \tfrac{1}{2}\epsilon_0$, such that

$$(\xi^1)^2 \geqslant \tfrac{1}{4}((\xi^2)^2 + (\xi^3)^2 + (\xi^4)^2)$$

for all causal vectors ξ at every point of $B = \{x; |x| < \epsilon\}$. On a causal geodesic in B, the x^i, $i = 2, 3, 4$ are therefore monotonic functions of x^1 that satisfy the inequality

$$\left(\frac{dx^2}{dx^1}\right)^2 + \left(\frac{dx^3}{dx^1}\right)^2 + \left(\frac{dx^4}{dx^1}\right)^2 \leqslant 4.$$

Hence if y and z are points on a causal geodesic in B, then

$$(z^2 - y^2)^2 + (z^2 - y^3)^2 + (z^4 - y^4)^2 \leqslant 4(z' - y')^2. \qquad (4.4.2)$$

Now put $\Omega = \{x; |x^1| < \tfrac{1}{2}(\epsilon - r), \ r < \epsilon\},$

where $r = ((x^2)^2 + (x^3)^2 + (x^4)^2)^{\frac{1}{2}}$ (figure 4.4.1). Then $\Omega \subset \bar{B} \subset B_0$, and B_0 is a geodesically convex domain; also, it follows from (4.4.2) that, if y and z are points in B and $J^+(y) \cap J^-(z)$ is not void, then

$$J^+(y) \cap J^-(z) \subset \Omega.$$

So both conditions of Definition 4.4.1 hold, and Ω is a causal domain. $\square$

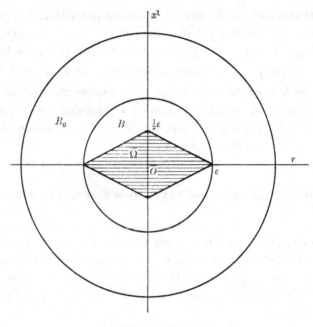

FIGURE 4.4.1

The principal result of this section is Theorem 4.4.2, which shows that the integral equation (4.4.1) can be solved by successive approximation in a causal domain. But we first prove two lemmas on causal curves in geodesically convex domains, which will frequently be needed from now on.

Lemma 4.4.1. *Let Ω be a geodesically convex domain; let q be a point in Ω, and let $[0, 1] \ni t \to f(t)$ be a parametrized C^1 curve in Ω. Suppose that $f(t) \in D^+(q)$ for $0 < t \leqslant 1$, and that $f'(t)$ is causal and future-directed. Then $t \to \Gamma(q, f(t))$ is strictly increasing on $0 < t \leqslant 1$.*

PROOF. As $f(t) \in C^1$, we have

$$(d/dt)\, \Gamma(q, f(t)) = \langle \mathrm{grad}_p\, \Gamma(q, p)|_{p=f(t)}, f'(t)\rangle.$$

Now $\mathrm{grad}_p\, \Gamma(q, p)$ is tangent to the geodesic from q to p at $p = f(t)$; hence it is time-like and future-directed. By hypothesis, $f'(t)$ is causal and future-directed. Choose a local coordinate system that is Minkowskian at $f(t)$, and in which the vector $(1, 0, 0, 0)$ is future-directed; let ξ^i and η^i denote the contravariant components of

$\operatorname{grad}_p \Gamma(q,p)|_{p=f(t)}$ and of $f'(t)$ respectively, in this coordinate system. Then

$$\xi^1 > ((\xi^2)^2 + (\xi^3)^2 + (\xi^4)^2)^{\frac{1}{2}} = \rho, \quad \eta^1 \geqslant ((\eta^2)^2 + (\eta^3)^2 + (\eta^4)^2)^{\frac{1}{2}} = \sigma,$$

say. But

$$\langle \xi, \eta \rangle = \xi^1\eta^1 - \xi^2\eta^2 - \xi^3\eta^3 - \xi^4\eta^4 \geqslant \xi^1\eta^1 - \rho\sigma \geqslant \sigma(\xi^1 - \rho) > 0,$$

and hence $(d/dt)\,\Gamma(q,f(t)) > 0$, as asserted. $\square$

Lemma 4.4.2. *Let Ω be a geodesically convex domain; let q be a point in Ω, and let $[0,1] \ni t \to f(t)$ be a parametrized C^1 curve in Ω. Suppose that $f(0) \in J^+(q)$, and that $f'(t)$ is causal and future directed for $0 \leqslant t \leqslant 1$. Then either $f(1) \in D^+(q)$, or $t \to f(t)$ is a bicharacteristic arc on $C^+(q)$.*

PROOF. If $f(0) \in J^+(q)$, then either $f(0) \in D^+(q)$, or $f(0) \in C^+(q)$. Suppose first that $f(0) \in D^+(q)$. If $f(1) \notin D^+(q)$, then the curve $t \to f(t)$ must meet $C^+(q)$, by continuity. So there is a T, such that $0 < \leqslant 1$, and

$$\text{(i) } f(t) \in D^+(q), \ 0 \leqslant t < T; \qquad \text{(ii) } f(T) \in C^+(q).$$

It follows from (i) and Lemma 4.4.1 that

$$\Gamma(q,f(T)) > \Gamma(q,f(0)) > 0.$$

But this contradicts (ii), as (ii) implies that $\Gamma(g,f(T)) = 0$. Hence $f(1) \in D^+(q)$; the argument also shows that $f(t) \in D^+(q)$ for $0 \leqslant t \leqslant 1$.

Next, suppose that $f(0) \in C^+(q)$. Let q' be an auxiliary point in $D^+(q)$; then $q \in D^+(q')$. As the geodesic from q to $f(0) \in C^+(0)$ is null and future-directed, it follows from the first part of the proof that $f(0) \in D^+(q')$. (This is trivial if $f(0) = q$.) The first part also implies that $f(t) \in D^+(q')$ for $0 \leqslant t \leqslant 1$, whence

$$\Gamma(q',f(t)) > 0, \ 0 \leqslant t \leqslant 1.$$

One can now let $q' \to q$, to conclude, by the continuity of Γ, that

$$\Gamma(q,f(t)) \geqslant 0, \ 0 \leqslant t \leqslant 1.$$

Now if $\Gamma(q,f(T)) > 0$ for some T, $0 < T < 1$, then $f(T) \in D^+(q)$; the first part of the proof shows that then $f(1) \in D^+(q)$. So $f(1) \in D^+(q)$ unless

$$\Gamma(q,f(t)) = 0, \ 0 \leqslant t \leqslant 1.$$

This means that the curve $t \to f(t)$ is on $C^+(q)$. By hypothesis, the curve is causal and future-directed. As $C^+(q)$ is a characteristic, the only causal future-directed curves on $C^+(q)$ are the null geodesics issuing from q,

which are the bicharacteristics of $C^+(q)$. Hence $t \to f(t)$ is a bicharacteristic arc in this case. So the proof is complete. $\square$

It is useful to note two simple consequences of this lemma concerning
causal domains. Suppose that Ω_0 is a causal domain, and that Ω_1 and
Ω_2 are two geodesically convex domains such that both $\Omega_0 \subset \Omega_1$ and
$\Omega_0 \subset \Omega_2$. If p and q are points in Ω_0 and $p \in J^+(q)$, then $J^+(q) \cap J^-(p)$ is
a non-empty compact subset of $\Omega_0 \subset \Omega_1 \cap \Omega_2$. It follows that the
unique geodesic $\widehat{qp} \subset \Omega_1$ coincides with the unique geodesic $\widehat{qp} \subset \Omega_2$.
This fact will be used constantly, and without explicit mention, in the
sequel.

Again, the lemma shows that if Ω is a geodesically convex domain,
if q_0 and p_0 are points in Ω such that $p_0 \in D^+(q_0)$, and

$$D^+(q_0) \cap D^-(p_0) \subset \Omega,$$

then $D^+(q_0) \cap D^-(p_0)$ is a causal domain (figure 4.4.2(a)). In fact, this is
the basic example of a causal domain. Another example is a 'lens-
shaped' domain, contained in a geodesically convex domain, bounded
by two space-like hypersurfaces each of which is met at most once by
every future-directed time-like curve in Ω (figure 4.4.2(b)).

We now return to the integral equation (4.4.1). It will be convenient
to introduce the following subset of $\Omega_0 \times \Omega_0$:

$$\Delta^+ = \{(p, q); (p, q) \in \Omega_0 \times \Omega_0, p \in J^+(q)\}. \qquad (4.4.3)$$

As Δ^+ is the closure of one of the components of the open set

$$\{(p, q); (p, q) \in \Omega_0, \Gamma(p, q) > 0\},$$

it is a closed set.

Theorem 4.4.2. *Let Ω_0 be a causal domain, and let $K(p, q)$ be a function
defined on $\Omega_0 \times \Omega_0$ whose support is contained in Δ^+, and which is
continuous on Δ^+. Then there exists a function $L(p, q)$, also supported in
Δ^+ and continuous on Δ^+, such that, for every $\phi \in C_0^\infty(\Omega_0)$ the equation*

$$\phi(q) + \int K(p, q)\,\phi(p)\,\mu(p) = \psi(q) \qquad (4.4.4)$$

implies that $\qquad \psi(q) + \int L(p, q)\,\psi(p)\,\mu(p) = \phi(q). \qquad (4.4.5)$

Moreover, if $K \in C^\infty(\Omega_0 \times \Omega_0)$, then $L \in C^\infty(\Omega_0 \times \Omega_0)$.

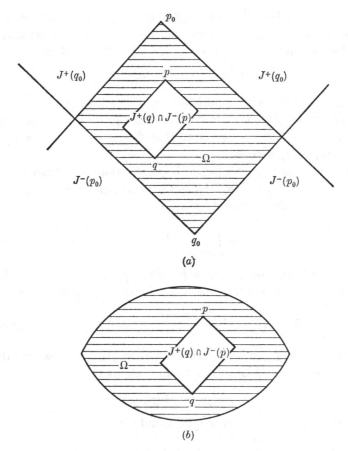

FIGURE 4.4.2

PROOF. The function $L(p, q)$, which is the resolving kernel of the integral equation (4.4.4), is defined by

$$L(p, q) = \sum_{\nu=0}^{\infty} (-1)^{\nu+1} K_\nu(p, q), \qquad (4.4.6)$$

where

$$K_0 = K, \quad K_{\nu+1}(p, q) = \int K(p, p') K_\nu(p', q) \mu(p') \quad (\nu = 0, 1, \ldots). \quad (4.4.7)$$

If it is assumed that supp $K_\nu \subset \Delta^+$, then the integrand of the integral defining $K_{\nu+1}$ vanishes unless $p' \in J^+(q)$ and $p \in J^+(p')$, that is to say unless $p' \in J^+(q) \cap J^-(p)$. As Ω_0 is causal, this is a compact subset of $\Omega_0 \times \Omega_0$ which is empty unless $(p, q) \in \Delta^+$. Hence the K_ν are well defined by (4.4.7), and vanish if $(p, q) \notin \Delta^+$.

6

Next, we prove that the series (4.4.6) converges uniformly on every compact subset of $\Omega_0 \times \Omega_0$. It can be shown that such a set can be covered by a finite number of sets of the form $\Phi \times \Phi$, where

$$\Phi = J^+(q_0) \cap J^-(p_0),$$

p_0 and q_0 being points in Ω_0 such that $p_0 \in D^+(q_0)$; the simple proof is left to the reader. It is therefore sufficient to prove the uniform convergence of (4.4.7) on $\Phi \times \Phi$. Note that Φ is a non-empty compact subset of Ω_0 as Ω_0 is causal.

By Lemma 4.4.1, $p' \to \Gamma(p', q)$ decreases along backward causal geodesics from p, if $p \in J^+(q)$. Hence the maximum of $p' \to \Gamma(p', q)$ on $J^+(q) \cap J^-(p)$ is $\Gamma(p, q)$, and so we can write (4.4.7) as

$$K_{\nu+1}(p, q) = \int_0^{\Gamma(p, q)} dt \int_{\Gamma(p', q)=t} K(p, p')\, K_\nu(p', q)\, \mu_\Gamma(p'), \quad (4.4.8)$$

where μ_Γ is a Leray form such that $d_{p'}\Gamma(p', q) \wedge \mu_\Gamma = \mu(p')$. We shall now prove by induction that there are constants M and N such that

$$|K_\nu(p, q)| \leqslant M \frac{(N\Gamma(p, q))^\nu}{\nu!} \quad (\nu = 0, 1, \ldots), \quad (4.4.9)$$

for $(p, q) \in (\Phi \times \Phi) \cap \Delta^+$. By Lemma 4.4.2, $(p, q) \in \Phi \times \Phi$ implies that $J^+(q) \cap J^-(p) \subset \Phi$, and Lemma 4.4.1 then shows that

$$0 \leqslant \Gamma(p, q) \leqslant \Gamma(p_0, q_0)$$

on $(\Phi \times \Phi) \cap \Delta^+$. Let N be an upper bound for

$$\int_{\Gamma(p', q)=t} |K(p, p')|\mu_\Gamma(p'),$$

when $(p, q) \in (\Phi \times \Phi) \cap \Delta^+$ and $0 \leqslant t \leqslant \Gamma(p_0, q_0)$. Then, if (4.4.9) holds for ν, (4.4.8) implies that

$$|K_{\nu+1}| \leqslant \int_0^{\Gamma(p, q)} MN \frac{(Nt)^\nu}{\nu!}\, dt = M \frac{(N\Gamma(p, q))^{\nu+1}}{(\nu+1)!}.$$

If we take M so that $|K| < M$ on $\Phi \times \Phi$, then (4.4.9) holds for $\nu = 0$, and so it follows by induction that it holds for all ν. As $0 \leqslant \Gamma \leqslant \Gamma(p_0, q_0)$ on $(\Phi \times \Phi) \cap \Delta^+$, and the K_ν vanish outside Δ^+, we therefore have the estimates

$$|K_\nu(p, q)| \leqslant M \frac{(N\Gamma(p_0, q_0))^\nu}{\nu!}, \quad (p, q) \in \Phi \times \Phi \quad (\nu = 0, 1, \ldots), \quad (4.4.10)$$

and so the series (4.4.7) converges uniformly on $\Phi \times \Phi$ to a function which is continuous on $\Delta^+ \cap (\Phi \times \Phi)$.

To prove (4.4.5), we replace q by p and p by p' in (4.4.4), multiply by $L(p,q)\,\mu(p)$, and integrate, to obtain, formally,

$$\int L(p,q)\,\psi(p)\,\mu(p)$$
$$= \int L(p,q)\,\phi(p)\,\mu(p) + \int L(p',q)\,\mu(p') \int K(p,p')\,\phi(p)\,\mu(p). \quad (4.4.11)$$

The function ψ must be considered as defined by (4.4.4), and its support is not compact, We must therefore show that the double integral on the right-hand side exists. The inner integral is a function of p' whose support is contained in

$$J^-(\operatorname{supp}\phi) = \bigcup_{p\in\operatorname{supp}\phi} J^-(p). \quad (4.4.12)$$

The support of $p' \to L(p',q)$ is contained in $J^+(q)$, and so the existence of the double integral is assured provided that $J^+(q) \cap J^-(\operatorname{supp}\phi)$ is relatively compact. This follows from a basic result that will be established in the next chapter (Theorem 5.1.1). For convenience, we repeat the relevant part of the argument here.

If $p \in \Omega_0$ and $p_1 \in D^+(p)$, then $p \in D^-(p_1)$, so that $D^-(p_1)$ is an open neighbourhood of p. So the support of ϕ, which is compact, can be covered by a finite number of such sets, $D^-(p_1), \ldots, D^-(p_N)$, say. By Lemma 4.4.2, $p \in D^-(p_j)$ implies that $J^-(p) \subset D^-(p_j) \subset J^-(p_j)$, for $j = 1, \ldots, N$. Hence

$$J^-(\operatorname{supp}\phi) \subset \bigcup_{j=1}^{j=N} J^-(p_j).$$

As Ω_0 is a causal domain, $J^+(q) \cap J^-(p_j)$ is compact, and so

$$J^+(q) \cap J^-(\operatorname{supp}\phi),$$

being contained in the union of a finite number of compact sets, is relatively compact. (It is actually a compact set.)

We conclude that (4.4.11) holds, and that the order of integration in the double integral on the right-hand side can be inverted. The integral then becomes

$$\int \phi(p)\,\mu(p) \int K(p,p')\,L(p',G)\,\mu(p').$$

But by (4.4.6) and (4.4.7), and the uniform convergence of the series (4.4.7),

$$\int K(p,p')\,L(p',q)\,\mu(p')$$
$$= \sum_{\nu=0}^{\infty} \int K(p,p')\,K_\nu(p',q)\,\mu(p') = -L(p,q) - K(p,q).$$

Hence (4.4.11) implies (4.4.5).

It remains to establish the last part of the theorem. As supp $K \subset \Delta^+$, the statement that $K \in C^\infty(\Omega_0 \times \Omega_0)$ means that K is C^∞ in the interior of Δ^+, and that all its derivatives (in some fixed coordinate chart covering Ω_0) vanish on $\partial\Delta^+$. Note that this is equivalent to

$$\Gamma^{-\mu} K(p,q) \to 0 \quad (\mu = 0, 1, \ldots) \tag{4.4.13}$$

as $\Gamma \downarrow 0$ for $(p,q) \in \Delta^+$, Now it follows from (4.4.7) that, for $\nu \geqslant 2$,

$$
\begin{aligned}
K_\nu(p,q) &= \int \cdots \int K(p, p^{(1)}) K(p^{(1)}, p^{(2)}) \ldots K(p^{(\nu-1)}, p^{(\nu)}) K(p^{(\nu)}, q) \mu(p^{(1)}) \\
&\qquad\qquad\qquad\qquad\qquad\qquad\qquad\qquad\qquad\qquad \ldots \mu(p^{(\nu)}) \\
&= \iint K(p, p^{(1)}) K_{\nu-2}(p^{(1)}, p^{(\nu)}) K(p^{(\nu)}, q) \mu(p^{(1)}) \mu(p^{(\nu)}).
\end{aligned}
$$

Writing this in local coordinates, and differentiating with respect to both sets of arguments, one obtains

$$
\begin{aligned}
&\partial_x^\alpha \partial_y^\beta K_\nu(x, y) \\
&= \iint \partial_x^\alpha K(x, x') K_{\nu-2}(x', x'') \partial_y^\beta K(x'', y) |g(x') g(x'')|^{\frac{1}{2}} dx' dx''.
\end{aligned}
$$

There are no additional terms from the variable boundaries of the integration domain, as K vanishes to all orders on $\partial\Delta^+$, by hypothesis. (For $\nu = 1$, the term $K_{\nu-2}$ must be omitted.) So it follows by induction that $K_\nu \in C^\infty(\Omega_0 \times \Omega_0)$ for all ν. Again, (4.4.9) shows that the formal series obtained from (4.4.6) by term-by-term differentiation all converge uniformly on $\Phi \times \Phi$, and hence on every compact subset of $\Omega_0 \times \Omega_0$. It therefore follows that L is C^∞ in the interior of Δ^+, and that its derivatives of all orders vanish on $\partial\Delta^+$. (In fact, we have shown that the series (4.4.6) converges in $C^\infty(\Omega_0 \times \Omega_0)$.) This completes the proof of Theorem 4.4.2. $\square$

4.5   The fundamental solutions

We are now ready to establish the principal result of this chapter. It will be recalled that a fundamental solution of P is a distribution $G_q(p)$, acting on test functions $\phi(p) \in C_0^\infty(\Omega_0)$ and functions of a parametric point $q \in \Omega_0$, such that (4.2.2) holds,

$$(G_q(p), {}^t P \phi(p)) = \phi(q), \quad \phi \in C_0^\infty(\Omega_0). \tag{4.5.1}$$

Theorem 4.5.1. Let Ω_0 be a causal domain, and let $P = \square + \langle a, \nabla \rangle + b$, where a and b are in $C^\infty(\Omega_0)$, be a scalar differential operator defined on

Ω_0. Then P has a fundamental solution G_q^+ in Ω_0, such that $PG_q^+ = \delta_q$ and $\operatorname{supp} G_q^+ \subset J^+(q)$. It is of the form

$$G_q^+ = \frac{1}{2\pi}(U\delta_+(\Gamma) + V^+), \qquad (4.5.2)$$

where $U = U(p,q)$ is the function of Theorem 4.2.1, and $V^+ = V^+(p,q)$ is such that $\operatorname{supp} V^+ \subset \Delta^+$ and $V^+ \in C^\infty(\Delta^+)$. When $\Gamma \downarrow 0$ for $(p,q) \in \Delta^+$, the Hadamard series is an asymptotic expansion for V^+,

$$V^+ \sim \sum_{\nu=0}^{\infty} V_\nu \frac{\Gamma^\nu}{\nu!}. \qquad (4.5.3)$$

Likewise, P has a fundamental solution G_q^- in Ω_0 such that $PG_q^- = \delta_q$ and $\operatorname{supp} G_q^- \subset J^-(q)$, which is of the form

$$G_q^- = \frac{1}{2\pi}(U\delta_-(\Gamma) + V^-), \qquad (4.5.4)$$

where V^- has properties similar to those of V^+, with past and future interchanged.

PROOF. It was pointed out in the comments on Definition 4.4.1 that condition (ii) is time-reversible. The theorem therefore only has to be proved for G_q^+; the corresponding results for G_q^- follow by reversal of the time-orientation.

It is obvious that the second member of (4.5.2) is a distribution if $p \to V^+(p,q)$ is locally integrable, and this is implied by $V^+ \in C^\infty(\Delta^+)$. By (4.5.1), it must be shown that, for all $\phi \in C_0^\infty(\Omega_0)$,

$$\phi(q) = \frac{1}{2\pi}\int_{C^+(q)} U(p,q)\,{}^tP\phi(p)\,\mu_\Gamma(p) + \frac{1}{2\pi}\int V^+(p,q)\,{}^tP\phi(p)\,\mu(p). \quad (4.5.5)$$

Let
$$\tilde{G}_q^+ = \frac{1}{2\pi}(U\delta_+(\Gamma) + \tilde{V}H_+(\Gamma))$$

be a C^∞ parametrix of P, as in Theorem 4.3.2. Then, taking $W = \tilde{V}$ in (4.4.1), one finds that

$$\phi(q) + \int K(p,q)\,\phi(p)\,\mu(p) = \psi(q), \qquad (4.5.6)$$

where

$$\psi(q) = \frac{1}{2\pi}\int_{C^+(q)} U(p,q)\,{}^tP\phi(p)\,\mu_\Gamma(p) + \frac{1}{2\pi}\int_{J^+(q)} \tilde{V}(p,q)\,{}^tP\phi(p)\,\mu(p),$$

$$\qquad (4.5.7)$$

and $K = (1/2\pi)P(\tilde{V})H_+(\Gamma)$. It follows from Theorem 4.3.2 that K

satisfies the hypotheses of Theorem 4.4.2, including $K \in C^\infty(\Omega_0 \times \Omega_0)$. One can therefore solve (4.5.6) by means of (4.4.5),

$$\phi(q) = \psi(q) + \int L(p,q)\,\psi(p)\,\mu(p), \qquad (4.5.8)$$

where L is the resolving kernel of K. Also, by (4.5.7),

$$2\pi \int L(p,q)\,\psi(p)\,\mu(p) = \int_{J^+(q)} L(p',q)\,\mu(p') \int_{J^+(p')} \tilde{V}(p,p')\,{}^tP\phi(p)\,\mu(p)$$

$$+ \int_{J^+(q)} L(p',q)\,\mu(p') \int_{C^+(p')} U(p,p')\,{}^tP\phi(p)\,\mu_{\Gamma(p,p')}(p). \quad (4.5.9)$$

To show that (4.5.8) is of the form (4.5.5), we must invert the orders of integration. For the first integral on the right-hand side, this is straightforward, as this is just a double integral over

$$\{(p,p');\, p' \in J^+(q),\, p \in J^+(p') \cap \operatorname{supp}\phi\}$$

$$= \{(p,p');\, p \in \operatorname{supp}\phi,\, p' \in J^+(q) \cap J^-(p)\}.$$

It was shown in the proof of Theorem 4.4.2 that this is a relatively compact subset of $\Omega_0 \times \Omega_0$, and so one can invert the order of integration, to obtain

$$\int_{J^+(q)} L(p',q)\,\mu(p') \int_{J^+(p')} \tilde{V}(p,p')\,{}^tP\phi(p)\,\mu(p)$$

$$= \int {}^tP\phi(p)\,\mu(p) \int_{J^+(q) \cap J^-(p)} \tilde{V}(p,p')\,L(p',q)\,\mu(p'). \quad (4.5.10)$$

The inversion of the order of integration in the second integral requires more work; we state the result as a lemma, and postpone the proof.

Lemma 4.5.1. *If $\phi \in C_0^\infty(\Omega_0)$, then*

$$\int_{J^+(q)} L(p',q)\,\mu(p') \int_{C^+(p')} U(p,p')\,{}^tP\phi(p)\,\mu_{\Gamma(p,p')}(p)$$

$$= \int {}^tP\phi(p)\,\mu(p) \int_{J^+(q) \cap C^-(p)} U(p,p')\,L(p',q)\,\mu_{r(p,p')}(p'), \quad (4.5.11)$$

where $\mu_{\Gamma(p,p')}(p')$ is a Leray form such that $d_{p'}\,\Gamma(p,p') \wedge \mu_\Gamma(p') = \mu(p')$.
Now when (4.5.10) and (4.5.11) are substituted in (4.5.9), and the

resulting expression is carried into (4.5.8), one obtains (seeing that ψ is given by (4.5.7)) an identity that is precisely of the form (4.5.5), with

$$V^+(p,q) = \tilde{V}(p,q) + \int_{J^+(q) \cap J^-(p)} \tilde{V}(p,p') L(p',q) \mu(p')$$
$$+ \int_{J^+(q) \cap C^-(p)} U(p,p') L(p',q) \mu_{\Gamma(p,p')}(p'), \quad (4.5.12)$$

when $(p,q) \in \Delta^+$, and of course $V^+ = 0$ when $(p,q) \notin \Delta^+$. Hence we have obtained a fundamental solution of P which is of the form (4.5.2).

To show that $V^+ \in C^\infty(\Delta^+)$, we choose a fixed coordinate chart (Ω, π) covering Ω_0; this can be done by taking Ω to be a geodesically convex domain containing Ω_0, and the local coordinate system π to be normal at some point of Ω. Put $\pi p = x$, $\pi q = y$, $\pi p' = x'$, and denote, for short, the functions $U(p,q)$, $\tilde{V}(p,q)$ and $L(p,q)$ in terms of the local coordinates by $U(x,y)$, $\tilde{V}(x,y)$ and $L(x,y)$ respectively. Now there is a diffeomorphism $\zeta \to x' = h(x,\zeta)$, such that the ζ^i are local coordinates normal and Minkowskian at p, and the tangent vector $(1, 0, 0, 0)$ at p is past-directed; also, the ζ^i are admissible local coordinates in Ω_0. Then

$$C^-(p) = \{\zeta; \zeta^1 = \rho\}, \quad J^-(p) = \{\zeta; \zeta^1 \geqslant \rho\}, \quad \rho = ((\zeta^2)^2 + (\zeta^3)^2 + (\zeta^4)^2)^{\frac{1}{2}}.$$

One then has $\mu(p') = k(x,\zeta)\,d\zeta$, where k is C^∞ (it is just $|g|^{\frac{1}{2}}$, in the local coordinates ζ^i), and so one can take

$$\mu_{\Gamma(p,q')}(p') = k(x,\zeta) \frac{d\zeta^2 \wedge d\zeta^3 \wedge d\zeta^4}{2\rho}. \quad (4.5.13)$$

Hence it follows from (4.5.12) that

$$V^+ - \tilde{V} = \int_{\zeta^1 \geqslant \rho} \tilde{V}(x, h(x,\zeta)) L(h(x,\zeta), y) k(x,\zeta)\,d\zeta$$
$$+ \frac{1}{2} \int_{\zeta^1 = \rho} U(x, h(x,\zeta)) L(h(x,\zeta), y) k(x,\zeta) \frac{d\zeta^2 d\zeta^3 d\zeta^4}{\rho}. \quad (4.5.14)$$

As $L = 0$ for $p' \notin J^+(q)$, the integration domains are compact; the integrands are C^∞ in x and y, and vanish to all orders on the variable boundaries of the integration domains. For these arise from L only, and $L \in C^\infty(\Omega_0 \times \Omega_0)$. It follows that all derivatives of $V^+ - \tilde{V}$, with respect to either x or y, can be computed by differentiation under the integral signs, that $V^+ - \tilde{V}$ is C^∞ in the interior of Δ^+, and that it vanishes to all orders on $\partial \Delta^+$. In view of Theorem 4.3.2, this implies that $V^+ \in C^\infty(\Delta^+)$. Also, it follows from (4.5.14) and (4.3.26) that the Hadamard series is an asymptotic expansion of V^+, as $\Gamma \downarrow 0$ with $(p,q) \in \Delta^+$. So the proof of the theorem is complete. $\square$

PROOF OF LEMMA 4.5.1. The first member of (4.5.11) can be considered as the integral of the exterior form

$$U(p,p')\,L(p',q)\,{}^{t}P\phi(p)\,\mu(p') \wedge \mu_{\Gamma(p,p')}(p)$$

over $C = \{(p,p');\, p' \in J^{+}(q),\, p \in C^{+}(p') \cap \operatorname{supp} \phi\}$
        $= \{(p,p');\, p \in J^{+}(q) \cap \operatorname{supp} \phi,\, p' \in J^{+}(q) \cap C^{-}(p)\}.$

As this is contained in $J^{+}(q) \cap J^{-}(\operatorname{supp} \phi)$, it is relatively compact; it is not a sub-manifold of $\Omega_0 \times \Omega_0$, because of the singularity on the diagonal $\{(p,p');\, p' = p\}$. However, the integral can be defined, say by taking it first over C with a neighbourhood of the diagonal removed, and then shrinking this neighbourhood to zero. It is clear that the integrand is then locally integrable; one can see this by writing it out in coordinates that are normal and Minkowskian at p' (compare (4.5.13)). So one can invert the order of integration. Now

$$\mu(p') \wedge \mu_{\Gamma(p,\,p')}(p) = d_{p'}\Gamma(p,p') \wedge {}_{\Gamma(p,\,p')}(p') \wedge \mu_{\Gamma(p,\,p')}(p),$$

where the second factor on the right-hand side is a Leray form such that $d_{p'}\Gamma(p,p') \wedge \mu_{\Gamma}(p') = \mu(p')$ But as $\Gamma(p,p') = 0$ on C, one has $d_{p'}\Gamma = -d_{p}\Gamma$; taking the parity of the Leray forms into account, one therefore finds that

$$\mu(p') \wedge \mu_{\Gamma(p,\,p')}(p) = d_{p}\Gamma(p,p') \wedge \mu_{\Gamma(p,\,p')}(p) \wedge \mu_{\Gamma(p,\,p')}(p')$$
$$= \mu(p) \wedge \mu_{\Gamma(p,\,p')}(p'),$$

and so (4.5.11) follows. The lemma is proved. $\square$

It was pointed out at the end of section 4.3 that there are infinitely many C^{∞} parametrices of P supported in $J^{+}(q)$. The proof of Theorem 4.5.1 only shows that one can construct a fundamental solution G_{q}^{+} from any one of these. However, it will be shown in the next chapter that there is, in fact, only one forward fundamental solution (Corollary 5.1.1). So (4.5.2) is the definitive form of the forward fundamental solution of P in Ω_0; it may be considered as the field of a point source at q, acting on a previously undisturbed background. It consists of a singular part $U\delta_{+}(\Gamma)$ which is a measure, supported on $C^{+}(q)$, and of a regular part V^{+}, which is a function; the function $p \to V^{+}(p,q)$ has its support contained in $J^{+}(q)$, and is in $C^{\infty}(J^{+}(q))$ (it is C^{∞} at the boundary of its support). This regular part, which is sometimes called the *tail term* of the fundamental solution, distinguishes the general case from the Minkowskian case, where G_{q}^{+} is 'sharp', its support being $C^{+}(q)$.

(We shall return to this point at the end of section 5.6.) It follows from Lemma 4.3.1 and $PG_q^+ = \delta_q$ that $p \to V^+(p,q)$ is a solution of the characteristic initial value problem (4.3.7),

$$PV^+ = 0, \quad p \in D^+(q), \quad V^+ = V_0, \quad p \in C^+(q). \qquad (4.5.15)$$

As a matter of fact, one can prove directly, by an energy estimate, that this problem has at most one C^2 solution.

The representation (4.5.13) of V^+ was useful for the purpose of establishing that $V^+ \in C^\infty(\Delta^+)$, and that it is asymptotically given by the Hadamard series when p is near $C^+(q)$. There are other, simpler, ways of determining V^+ by solving integral equations; such integral equations will be derived in section 5.4.

4.6   Trivial transformations

We conclude the chapter with some remarks about differential operators of the type

$$Pu = \square\, u + \langle a, \nabla u \rangle + bu,$$

that are trivially related to each other. The uniqueness of the forward and backward fundamental solutions of P in a causal domain will be assumed.

There are two types of so-called *trivial transformations*. The first consists in multiplying the unknown u by a non-vanishing C^∞ function σ, and then multiplying $P(\sigma u)$ by σ^{-1}, to obtain a new differential operator

$$P'u = \sigma^{-1} P(\sigma u) = \square\, u + \langle a', \nabla u \rangle + b'u, \qquad (4.6.1)$$

where

$$a' = a + 2\sigma^{-1}\nabla\sigma, \quad b' = \sigma^{-1} P\sigma = b + \sigma^{-1}\langle a, \nabla\sigma \rangle + \sigma^{-1}\square\,\sigma. \qquad (4.6.2)$$

A simple computation shows that

$$\left. \begin{array}{l} \nabla_i a'_j - \nabla_j a'_i = \nabla_i a_j - \nabla_j a_i \quad (i,j = 1, \ldots, 4), \\[2mm] b' - \tfrac{1}{4}\langle a', a' \rangle - \tfrac{1}{2}\operatorname{div} a' = b - \tfrac{1}{4}\langle a, a \rangle - \tfrac{1}{2}\operatorname{div} a. \end{array} \right\} \qquad (4.6.3)$$

The first condition can also be written in the invariant form

$$d\langle a', dx \rangle = d\langle a, dx \rangle.$$

These conditions are necessary, and locally also sufficient, for two differential operators P and P' to be related by a trivial transformation of this type.

For the forward fundamental solution of P' in a causal domain, one has

$$P'G_q'^+ = \sigma^{-1}P(\sigma G_q'^+) = \delta_q, \text{ supp } G_q'^+ \subset J^+(q).$$

As

$$\sigma(p)\,\delta_q(p) = \sigma(q)\,\delta_q(p),$$

it follows that

$$G_q'^+(p) = \frac{\sigma(q)}{\sigma(p)}\,G_q^+(p), \qquad (4.6.4)$$

and there is a similar relation between $G_q'^-$ and G_q^-.

For a differential operator P defined on an open set in $\mathbf{R}^n$, the second trivial transformation consists simply in the multiplication of P by a non-vanishing C^∞ function. In the present context, this is equivalent to the passage from a given space–time M to a conformal space–time $\tilde{M}$, with the same underlying C^∞ manifold, but a new metric tensor, defined in local coordinates by $d\tilde{s}^2 = \rho^2 ds^2$, where $\rho \in C^\infty(M)$. In the four-dimensional case, this implies

$$\tilde{g}_{ij} = \rho^2 g_{ij}, \quad \tilde{g}^{ij} = \rho^{-2}g^{ij}, \quad |\tilde{g}|^{\frac{1}{2}} = \rho^4 |g|^{\frac{1}{2}} \quad (i,j = 1, ..., 4). \quad (4.6.5)$$

A differential operator $\tilde{P}$ on $\tilde{M}$ can then be associated with P, as follows. By (4.6.5), the d'Alembertian in $\tilde{M}$ is

$$\tilde{\square}\,u = \rho^{-4}|g|^{-\frac{1}{2}}\,\partial_i(|g|^{\frac{1}{2}}g^{ij}\rho^2\,\partial_j u).$$

This can be rearranged as

$$\tilde{\square}\,u = \rho^{-4}|g|^{\frac{1}{2}}\,\partial_i(|g|^{\frac{1}{2}}g^{ij}(\rho\,\partial_j\,(\rho u) - \rho u\,\partial_j\rho))$$
$$= \rho^{-3}\,\square\,(\rho u) - \rho^{-3}u\,\square\,\rho.$$

Now it is known that the respective curvature scalars

$$\tilde{R} = \tilde{g}^{il}\tilde{g}^{jk}\tilde{R}_{ijkl}, \quad R = g^{il}g^{jk}R_{ijkl},$$

are related by

$$\tfrac{1}{6}\tilde{R} = \tfrac{1}{6}\rho^{-2}R + \rho^{-3}\,\square\,\rho.$$

Hence

$$(\tilde{\square} + \tfrac{1}{6}\tilde{R})\,u = \rho^{-3}\,(\square + \tfrac{1}{6}R)\,\rho u, \qquad (4.6.6)$$

so that $\square + \tfrac{1}{6}R$ is a self-adjoint conformally invariant scalar differential operator. Write $c + \tfrac{1}{6}R$ for b, so that

$$Pu = \square u + \langle a, \nabla u \rangle + (c + \tfrac{1}{6}R)\,u. \qquad (4.6.7)$$

Then it follows from (4.6.6) that

$$\rho^{-3}P(\rho u) = \tilde{P}u = \tilde{\square}\,u + \langle \tilde{a}, \nabla u \rangle + (\tilde{c} + \tfrac{1}{6}\tilde{R})\,u, \qquad (4.6.8)$$

where

$$\langle \tilde{a}, \nabla u \rangle = \rho^{-2}a^i\,\partial_i u, \quad \tilde{c} = \rho^{-2}c + \rho^{-3}\langle a, \nabla\rho \rangle. \qquad (4.6.9)$$

It is now easy to show that, if $G_q(p)$ is a fundamental solution of P, then

$$\tilde{G}_q(p) = \frac{1}{\rho(p)\rho(q)} G_q(p) \qquad (4.6.10)$$

is a fundamental solution of $\tilde{P}$. Conversely, if $\tilde{G}_q(p)$ is a fundamental solution of $\tilde{P}$, and $G_q(p)$ is defined by (4.6.10), then it is a fundamental solution of P. For in local coordinates, the Dirac measures in $\mathscr{D}'(M)$ and $\mathscr{D}'(\tilde{M})$ are respectively

$$\delta_q(p) = \frac{\delta(x-y)}{|g|^{\frac{1}{2}}}, \quad \tilde{\delta}_q(p) = \frac{\delta(x-y)}{|\tilde{g}|^{\frac{1}{2}}} = \frac{\delta(y-x)}{\rho^4|g|^{\frac{1}{2}}}.$$

Hence $\tilde{\delta}_q = \rho^{-4}\delta_q$, and so it follows from (4.6.8) that, if $\tilde{P}\tilde{G}_q = \tilde{\delta}_q$, then

$$\rho^{-3}P(\rho\tilde{G}_q) = \rho^{-4}\delta_q,$$

whence $\qquad P(\rho(p)\,\tilde{G}_q(p)) = \rho^{-1}\delta_q(p) = \dfrac{1}{\rho(q)}\delta_q(p),$

and this implies (4.6.10).

Locally, it follows from (4.6.5) that M and $\tilde{M}$ have the same null vectors, and that vectors that are time-like in M are also time-like in $\tilde{M}$; one can also associate the time-orientations of M and $\tilde{M}$. The two sets of null geodesics also coincide. But there is in general no simple relation between the exponential maps in M and $\tilde{M}$. In particular, a connected open set that is geodesically convex in M need not be geodesically convex in $\tilde{M}$, and the same is true for causal domains. Some care is therefore needed when (4.6.10) is used to relate the fundamental solutions of P and $\tilde{P}$. However, if a connected open set in the underlying manifold is a causal domain in both space–times, then (4.6.10) holds for both the forward and the backward fundamental solutions of P and $\tilde{P}$ respectively.

An example of a space–time that is conformally flat is a space–time with constant curvature. By definition, this means that the curvature tensor and the metric tensor are related by

$$R_{ijkl} = K(g_{ik}g_{jl} - g_{il}g_{jk}), \qquad (4.6.11)$$

where K, the curvature, is a (real) constant. It is known that at every point q of such a space–time there is a coordinate chart in which q is $x = 0$, and

$$\langle dx, dx \rangle = \frac{\eta_{ij}dx^idx^j}{(1+\frac{1}{4}K\eta_{ij}x^ix^j)^2}.$$

Let us take $\qquad\qquad P = \Box + b, \qquad (4.6.12)$

where b is a constant, which may be complex.

There is a simple method for deriving the two fundamental solutions of P. (By Theorem 4.4.1, these are always well defined locally.) It is known that a space–time with constant curvature is harmonic: this means that $\square \Gamma$ is a function of Γ only. (Conversely, a harmonic space–time has constant curvature.) In fact,

$$\square \Gamma = \begin{cases} 2 + 6(K\Gamma)^{\frac{1}{2}} \cot(K\Gamma)^{\frac{1}{2}} & \text{if} \quad K > 0, \\ 2 + 6(-K\Gamma)^{\frac{1}{2}} \coth(-K\Gamma)^{\frac{1}{2}} & \text{if} \quad K < 0. \end{cases} \qquad (4.6.13)$$

If a scalar field v is a function of Γ only, then

$$\langle \nabla \Gamma, \nabla v \rangle = 4\Gamma \frac{dv}{d\Gamma}, \quad \square v = 4\Gamma \frac{d^2 v}{d\Gamma^2} + \square \Gamma \frac{dv}{d\Gamma}. \qquad (4.6.14)$$

Hence the transport equations (4.2.12) and (4.3.5) become, if it is assumed that U and V_0 depend only on Γ,

$$8\Gamma \frac{dU}{d\Gamma} + (\square \Gamma - 8) U = 0, \quad U(0) = 1,$$

$$8\Gamma \frac{dV_0}{d\Gamma} + (\square \Gamma - 4) V_0 = -4\Gamma \frac{d^2 U}{d\Gamma^2} - \square \Gamma \frac{dU}{d\Gamma} - bU,$$

and V_0 must remain bounded as $\Gamma \to 0$. By means of (4.6.13), one can easily deduce from these equations that

$$V_0(0) = -\tfrac{1}{4}(b + 2K).$$

The other coefficients of the Hadamard series are difficult to compute. But it is evident that one can determine the solution of the boundary value problem

$$\square V + bV = 0, \quad V = V_0 \quad \text{when} \quad \Gamma = 0,$$

directly, by assuming that V is a function of Γ only. By (4.6.14), this gives

$$4\Gamma \frac{d^2 V}{dr^2} + \square \Gamma \frac{dV}{dr} + bV = 0, \quad V(0) = V_0(0) = -\tfrac{1}{4}(b + 2K), \quad (4.6.15)$$

and as $\Gamma = 0$ is a singular point of this ordinary differential equation, the single boundary condition determines V uniquely. One then has $V^+ = VH_+(\Gamma)$ and $V^- = VH_-(\Gamma)$. Note that $U(\Gamma)$ need not be determined, as only its restriction to $C(q)$, which is just $U = U(0) = 1$, intervenes in the fundamental solutions.

We omit the detailed computations, which are straightforward but somewhat tedious. It turns out that

$$G_q^+ = \frac{1}{2\pi}\delta_+(\Gamma) - \frac{K}{4\pi}P_\nu'(\cos{(K\Gamma)^{\frac{1}{2}}})\,H_+(\Gamma), \qquad (4.6.16)$$

for $K > 0$; if $K < 0$, then $\cos{(K\Gamma)^{\frac{1}{2}}}$ has to be replaced by $\cosh{(-K\Gamma)^{\frac{1}{2}}}$. The constant ν is a root of the equation

$$\nu(\nu+1) = 2+\frac{b}{K}, \qquad (4.6.17)$$

and $P_\nu'(z)$ is the derivative of the Legendre function $P_\nu(z)$ which is, by definition,
$$P_\nu(z) = {}_2F_1(-\nu,\nu+1;\,1;\,\tfrac{1}{2}(1-z)), \qquad (4.6.18)$$

and is the same for either root of (4.6.17). This function has a logarithmic singularity when $z = 3$. For $z < -1$, one can use the continuation formula

$$P_\nu(z) = \left(\frac{1+z}{2}\right)^\nu {}_2F_1\!\left(-\nu,\,-\nu;\,1;\frac{z-1}{z+1}\right).$$

There are two simple cases. If $b = 0$, then one can satisfy (4.6.17) by taking $\nu = 1$, whence $P_\nu(z) = P_1(z) = z$, so that $P_\nu'(z) = 1$, and hence

$$G_q^+ = \frac{1}{2\pi}\delta_+(\Gamma) - \frac{K}{4\pi}H_+(\Gamma), \quad (b = 0). \qquad (4.6.19)$$

Note, however, that the Hadamard series does not terminate.

If $b = -2K$, then $V_0(0) = 0$ and so V vanishes identically. Hence

$$G_q^+ = \frac{1}{2\pi}\delta_+(\Gamma) \quad (b = -2K). \qquad (4.6.20)$$

By (4.6.11), one finds that $R = -12K$. So this result is, in fact, also an immediate consequence of (4.6.6), as P is conformally equivalent to the d'Alembertian in a flat space-time.

NOTES

This chapter is based on Hadamard (1923, 1932) and M. Riesz (1949). Both Hadamard's book and Riesz' paper are still very well worth reading.

Section 4.1. The discussion of properly and improperly posed problems follows Hadamard (1923), who introduced these ideas and emphasized their importance. It is perhaps worth pointing out that, if P is a linear differential operator that can be extended to a closed operator $P\colon X \to Y$, where X and Y are, for example, Banach spaces, and it can be shown that P is bijective, then the continuity of the inverse of P is assured by the closed graph theorem.

The forms (4.1.32) and (4.1.33) for the fundamental solution of the four-dimensional d'Alembertian are due to Leray (1952). The identity (4.1.34) is related to the results of Méthée (1954, 1957).

Section 4.2. The expression (4.2.19) for the transport scalar is due to deWitt and Brehme (1960).

Section 4.3. The concept of a parametrix is due to Hilbert and E. E. Levi; it was introduced into the theory of linear hyperbolic equations by Hadamard. Lemma 4.3.2 is a classic extension lemma due to E. Borel; see for example Guillemin and Golubitsky (1973) p. 98, Lemma 2.5.

Section 4.4. Causal domains have been introduced here in order to ensure that fundamental solutions can be derived from a C^∞ parametrix by the integral equation method, and as an appropriate setting for the basic existence and uniqueness theorems in section 5.1. As these theorems show, it is equivalent to local causality in the sense of Hawking and Ellis (1973). Some hypothesis of this kind is needed; it is not enough to require only that the domain in question is to be geodesically convex. See, for instance, figure 13, and the remarks preceding it, on p. 194 of M. Riesz (1949). For global hyperbolicity, see Leray (1952) and Hawking and Ellis (1973), where further references will also be found.

Section 4.5. The integral equation method was used by Hadamard (1923) pp. 296–308. The argument in the text follows, in effect, M. Riesz (1949) pp. 198–206. The perspicuous designation 'tail term' for the regular part of the fundamental solution is due to deWitt and Brehme (1960). There is a different approach, due essentially to Sobolev (1936); (this paper contains an important anticipation of the theory of distributions). This has been developed by Y. Choquet-Bruhat (Fourès-Bruhat (1952, 1956), Bruhat (1962, 1964)). The distribution $(1/2\pi)\, U\delta_+(\Gamma)$ is already a parametrix. It follows from Theorem 4.2.1 that, for all $\phi \in C_0^\infty(\Omega)$,

$$\phi(q) + \frac{1}{2\pi} \int_{C^+(q)} (PU(p,q))\, \phi(p)\, \mu_\Gamma(p) = \frac{1}{2\pi} \int_{C^+(q)} U(p,q)\, {}^t\!P\phi(p)\, \mu_\Gamma(p).$$

This is an integral equation for ϕ, which can be solved by successive approximation. Replacing ${}^t\!P\phi$ in the second member by $\psi \in C_0^\infty(\Omega)$, one obtains a map $\psi \to \phi$ which turns out to be a distribution (for q fixed), and so is the fundamental solution $(G_q^+(p), \psi(p))$. The differentiability properties of this map (as a function of q) can then be determined separately. An advantage of this method is that it can be applied to space–times of finite differentiability class m; one can take $m = 4$, provided that the fourth-order derivatives of the co-ordinate transformations, and the third-order derivatives of the metric tensor, are locally Lipschitz (and appropriate restrictions are placed on the other coefficients of P).

Section 4.6. For harmonic spaces, see Ruse, Walker and Willmore (1961) and Lichnérowicz and Walker (1945).

5

Representation theorems

The two fundamental solutions derived in the last chapter will now be used to build up the theory of wave equations defined on a causal domain in a four-dimensional space–time. The basic results on the existence and uniqueness of solutions are proved in section 5.1. The important concepts of past-compact and future-compact subsets of a causal domain must be introduced for this purpose. Also reciprocity relations are obtained between the forward and backward fundamental solutions of a differential operator and of those of its adjoint operator.

These existence and uniqueness theorems are deduced from a representation formula for the solutions of an equation $Pu = f$, where f is a distribution whose support is either past-compact or future-compact. In section 5.3 a more explicit representation is derived for the solution of the classical Cauchy problem, with the data prescribed on a space-like hypersurface. (The existence and uniqueness of this solution are also proved.) The result in question (Theorem 5.3.3) is an extension of M. Riesz' elegant representation of the solution of the Cauchy problem in a flat space–time. In section 5.4, a similar representation theory is developed for the characteristic initial value problem, mainly when the data are given on a future null semi-cone.

The theory is extended to tensor wave equations in section 5.5. The only new elements needed here are tensor-valued distributions and bitensors. It turns out that, once these have been introduced, the results obtained in the scalar case generalize easily to the tensor case.

The last two sections deal with two questions that are of some interest in applications. A monopole of variable strength, travelling along a future-directed time-like curve (a world line) may be thought of as a line source. The field due to such a line source, in a causal domain, is computed in section 5.6; it generalizes the well-known Liénard–Wiechert potential. Finally, a wave equation is said to satisfy Huygens' principle if the solution of the Cauchy problem relative to a hypersurface S at a point p depends only on the data on the intersection of the null cone $C(p)$ and of S. It is shown, in section 5.7, that an equivalent condition is the vanishing of the 'tail terms' of the forward and backward fundamental solutions. Some progress has been made with the difficult problem of determining scalar wave equations with this property explicitly.

5.1 Existence and uniqueness theorems

Consider a scalar wave equation

$$Pu \equiv \square u + \langle a, \nabla u \rangle + bu = f, \qquad (5.1.1)$$

defined on a causal domain Ω_0 of a four-dimensional space–time. Such a domain is a connected open subset of a geodesically convex domain Ω, such that, for all points p and q in Ω_0, $J^+(q) \cap J^-(p)$ is a compact subset of Ω_0 which is empty if $p \notin J^+(q)$ (Definition 4.4.1). By Theorem 4.5.1, the differential operator P has two fundamental solutions in Ω_0. There is a *forward fundamental solution*,

$$G_q^+(p) = \frac{1}{2\pi} \left(U(p,q)\,\delta_+(\delta(p,q)) + V^+(p,q) \right), \qquad (5.1.2)$$

which is a distribution that acts on test functions $\phi(p) \in C_0^\infty(\Omega_0)$, and also depends on the source point $q \in \Omega_0$. The function U is in

$$C^\infty(\Omega_0 \times \Omega_0),$$

the support of V^+ is $\Delta^+ = \{(p,q)\,;\,(p,q) \in \Omega_0,\,p \in J^+(q)\}$, and the restriction of V^+ to this closed set is a C^∞ function of p and q. The *backward fundamental solution* is

$$G_q^+(p) = U(p,q)\,\delta_-(\Gamma(p,q)) + V^-(p,q), \qquad (5.1.3)$$

where U is the same function as before, while

$$\operatorname{supp} V^- \subset \Delta^- = \{(p,q)\,;\,(p,q) \in \Omega_0 \times \Omega_0,\,P \in J^-(a)\}$$

and $V^-|_{\Delta^-} \in C^\infty(\Delta^-)$. It will be proved later in this section that each of these two fundamental solutions is uniquely determined by a causal condition. To simplify the notation, we shall write $G_q^+(\phi)$ and $G_q^-(\phi)$ for $(G_q^+(p), \phi(p))$ and $(G_q^-(p), \phi(p))$ respectively, where $\phi(p) \in C_0^\infty(\Omega_0)$.

We first consider the equation (5.1.1) when f is a distribution with compact support.

Lemma 5.5.1. *Suppose that $f \in \mathscr{E}'(\Omega_0)$. Then both*

$$(u, \phi) = (f(q), G_q^+(\phi)), \quad \phi \in C_0^\infty(\Omega_0) \qquad (5.1.4)$$

and $(u, \phi) = (f(q), G_q^-(\phi))\ \phi \in C_0^\infty(\Omega_0),$ (5.1.5)

are distributions which satisfy the equation (5.1.1).

PROOF. It will be sufficient to prove this for (5.1.4), as the other case can then be deduced by reversing the time-orientation of Ω_0. By (5.1.2),

$$2\pi G_q^+(\phi) = \int_{C^+(q)} U(p, q)\, \phi(p)\, \mu_\Gamma(p) + \int V^+(p, q)\, \phi(p)\, \mu(p).$$

Let (Ω, π) be a fixed coordinate chart covering Ω_0; for instance, one can take Ω to be a geodesically convex domain containing Ω_0, and use local coordinates that are normal at a point of Ω. Denote πp by x and πq by y. Then there exists a diffeomorphism $h\colon \zeta \to x = h(y, \zeta)$, between $h^{-1}\Omega$ and Ω, such that the ζ^i, as local coordinates, are normal and Minkowskian at q, and the vector $(1, 0, 0, 0) \in TM_q$ is future-directed. Thus

$$J^+(q) = \{\zeta;\ \zeta^1 \geqslant \rho\}, \quad C^+(q) = \{\zeta;\ \zeta^1 = \rho\}, \quad \rho = ((\zeta^2)^2 + (\zeta^3)^2 + (\zeta^4)^2)^{\frac12},$$

and $\mu(p) = k(y, \zeta)\, d\zeta, \quad \mu_\Gamma(p)|_{C^+(q)} = k(y, \zeta)|_{\zeta^1 = \rho}\, \dfrac{d\zeta^2 \wedge d\zeta^3 \wedge d\zeta^4}{2\rho},$

where $k(y, \zeta)$ is a C^∞ function. Hence, for $q = \pi^{-1}y,$

$$2\pi G_q(\phi) = \int_{\zeta^1 \geqslant \rho} V^+(h(y, \zeta), y)\, \phi(h(y, \zeta))\, k(y, \zeta)\, d\zeta$$

$$+ \int_{\zeta^1 = \rho} U(h(y, \zeta), y)\, \phi(h(y, \zeta))\, k(y, \zeta)\, \frac{d\zeta^2\, d\zeta^3\, d\zeta^4}{2\rho}. \quad (5.1.6)$$

It follows from the differentiability properties of U and V^+ that $G_q^+(\phi) \in C^\infty(\Omega_0)$, and that the derivatives of $G_q^+(\phi)$ with respect to the y^i can be computed by differentiation under the integral signs. Also, if K and K' are compact subsets of Ω_0, and we have both $\operatorname{supp}\phi \subset K$ and $q \in K'$, then U, V^+ and k, and their derivatives of all orders, are bounded. So one obtains, from (5.1.6), estimates of the form

$$\sum_{|\alpha| \leqslant N} \sup_{K'} |\partial_y^\alpha G_q^+(\phi) \circ \pi^{-1}(y)| \leqslant C_N \sum_{|\alpha| \leqslant N} \sup |\partial_x^\alpha \phi \circ \pi^{-1}(x)| \quad (N = 0, 1, \ldots),$$

(5.1.7)

where the C_N are constants which depend on K and K', but not on ϕ.

As $f \in \mathscr{E}'(\Omega_0)$, there exists a compact set $K' \subset \Omega_0$ and numbers C and N such that

$$|(f, \psi)| \leqslant C \sum_{|\alpha| \leqslant N} \sup_{K'} |\partial_y^\alpha \psi \circ \pi^{-1}(y)|$$

for all $\psi \in C^\infty(\Omega_0)$. (This follows from (2.3.2), extended in the obvious way to distributions on Ω_0.) Setting $\psi(q) = G_q^+(\phi)$, and taking K' and N to be the same set and the same integer respectively, as in the semi-norm estimate for f, one obtains

$$|(u, \phi)| \leqslant C C_N \sum_{|\alpha| \leqslant N} \sup |\partial_x^\alpha \phi \circ \pi^{-1}(x)|, \qquad (5.1.8)$$

for all $\phi \in C_0^\infty(K)$. As K can be any subset of Ω_0, it follows that $u \in \mathscr{D}'(\Omega_0)$. It is now obvious that $Pu = f$, as

$$(Pu, \phi) = (u, {}^t P \phi) = (f(q), G_q^+({}^t P \phi)) = (f, \phi).$$

So the lemma is proved. It may be noted that u is of finite order, and that its order does not exceed the order of f. $\square$

This argument remains valid if $f \in \mathscr{D}'(\Omega_0)$ is such that $(f(q), G_q^+(\phi))$ can be defined; the validity of (5.1.5) can be extended in the same manner. Let us therefore consider the supports of $G_q^+(\phi)$ and $G_q^-(\phi)$. The future and past emissions of a subset Φ of Ω_0 are defined as

$$J^+(\Phi) = \bigcup_{q \in \Phi} J^+(q), \quad J^-(\Phi) = \bigcup_{q \in \Phi} J^-(q). \qquad (5.1.9)$$

As $p \in J^+(q)$ if and only if $q \in J^-(p)$, one also has

$$J^+(\Phi) = \{p ; J^-(p) \cap \Phi \neq \varnothing\}, \quad J^-(\Phi) = \{p ; J^+(p) \cap \Phi \neq \varnothing\}. \qquad (5.1.10)$$

Now $\operatorname{supp} G_q^+ \subset J^+(q)$; hence $G_q^+(\phi) = 0$ if $J^+(q)$ does not meet the support of ϕ. By (5.1.10), this means that $G_q^+(\phi) = 0$ if $q \notin J^-(\operatorname{supp} \phi)$; similarly, $G_q^-(q) = 0$ if $q \notin J^+(\operatorname{supp} \phi)$. It will follow from Theorem 5.1.1 below that the future and past emissions of a compact set are closed. Hence

$$\operatorname{supp} G_q^+(\phi) \subset J^-(\operatorname{supp} \phi), \quad \operatorname{supp} G_q^-(p) \subset J^+(\operatorname{supp} \phi). \qquad (5.1.11)$$

To exploit this fully, we introduce two classes of subsets of Ω_0 which are characterized by causality conditions.

Definition 5.1.1. *A set $\Phi \subset \Omega_0$ is called past-compact if $J^-(p) \cap \Phi$ is compact (or empty) for all $p \in \Omega_0$; it is called future-compact if $J^+(p) \cap \Phi$ is compact (or empty) for all $p \in \Omega_0$.*

Note that past-compact and future-compact sets are closed in Ω_0. Suppose, for example, that Φ is past-compact, and that $\{p_\nu\}$ is a sequence of points of Φ which converges to a point $p \in \Omega_0$. Let p' be an auxiliary point in $D^+(p)$. Then $p \in D^-(p')$, an open set, and so all the

p_ν are in $D^-(p')$ from some ν_0 on. But then, also, $p_\nu \in J^-(p') \cap \Phi$, which is a compact set, for $\nu \geqslant \nu_0$, and hence $p = \lim p_\nu \in J^-(p') \cap \Phi \subset \Phi$. So Φ is closed in Ω_0. The argument in the future-compact case is similar.

A compact subset of Ω_0 is both past-compact and future-compact. The causality condition can now be expressed by saying that the future emission of any point of Ω_0 is past-compact, and the past emission of any point of Ω_0 is future-compact. If the time-orientation of Ω_0 is reversed, then past-compact and future-compact sets are interchanged.

The intersection of a past-compact and a future-compact set need not be compact. For example, in a Minkowskian space–time, a half-space $\{x; x^1 \geqslant A\}$, where A is a constant, is past-compact, and a half-space $\{x; x^1 \leqslant B\}$ is future-compact. If $A < B$, then the intersection of these sets is the slab $\{x; A \leqslant x^1 \leqslant B\}$ which is not compact. However, it will turn out that if K is a compact set, then $J^-(K)$ meets every past-compact set in a compact set, and $J^+(K)$ meets every future-compact set in a compact set. In view of (5.1.11), this will be sufficient to ensure that (5.1.4) can be extended to distributions f with past-compact support, and (5.1.5) to distributions with future-compact support. However, we must first show that $J^+(\mathrm{supp}\,\phi)$ and $J^-(\mathrm{supp}\,\phi)$ are past-compact and future-compact respectively. This will be a consequence of the following result, which is of intrinsic importance.

Theorem 5.1.1. *If $\Phi \subset \Omega_0$ is past-compact, then $J^+ \subset \Phi)$ is past-compact; if $\Phi \subset \Omega_0$ is future-compact, then $J^-(\Phi)$ is future-compact.*

PROOF. It is sufficient to prove the first assertion, as the second one then follows by reversing the time-orientation of Ω_0. The proof is in two steps.

(i) $J^+(\Phi)$ is closed. Let $\{p_\nu\}_{1 \leqslant \nu < \infty}$ be a sequence of points in $J^+(\Phi)$ that converges to a point $p \in \Omega_0$; it has to be shown that $p \in J^+(\Phi)$. For each ν, there is a point $q_\nu \in \Phi$ such that the geodesic $\widehat{q_\nu p_\nu} \subset \Omega$ (it is actually in Ω_0, because Ω_0 is causal) is causal and future-directed. (A causal curve is time-like or null.) As Ω_0 is open, $D^+(p)$ is not empty. If $p' \in D^+(p)$, then $p \in D^-(p')$, and so $D^-(p')$ is an open neighbourhood of p. Hence all but a finite number of the p_ν are in $D^-(p')$. Deleting the others, we obtain a new sequence $(p_\nu)_{1 \leqslant \nu < \infty}$ converging to p, with $p_\nu \in D^-(p')$ for all ν. By Lemma 4.4.2, applied in time-reversed form to the geodesic $\widehat{p_\nu q_\nu}$, it follows that $q_\nu \in D^-(p') \subset J^-(p')$. By hypothesis, $J^-(p') \cap \Phi$ is compact. Hence $(q_\nu)_{1 \leqslant \nu < \infty}$ has a convergent sub-sequence whose limit q is a point in Φ. Deleting all other points, one is left with two sequences (p_ν) and (q_ν) that have the following properties: $q_\nu \to q \in \Phi$, $p_\nu \to p \in \Omega_0$, and the geodesic $\widehat{q_\nu p_\nu}$ is causal and future-directed for each ν. By the continuity of the exponential map, the geodesics $\widehat{q_\nu p_\nu}$ tend to $\widehat{qp}$. Also $\Gamma(p, q) = \lim \Gamma(p_\nu, q_\nu) \geqslant 0$. Hence $\widehat{qp}$ is causal and future-directed, and so $p \in J^+(q) \subset J^+(\Phi)$; hence $J^+(\Phi)$ is closed.

(ii) $J^-(p) \cap J^+(\Phi)$ is compact for all $p \in \Phi_0$. If $p \in \Omega_0$ then

$$J^-(p) \cap J^+(\Phi)$$

is empty unless $p \in J^+(\Phi)$. For it is obvious that $p \in J^+(\Phi)$ implies that $J^-(p) \cap J^+(\Phi)$ is not empty. On the other hand, if $J^-(p) \cap J^+(\Phi)$ is not empty, then there are points $q \in \Phi$ and $q' \in J^+(q)$ such that $q' \in J^-(p)$. But then $q \in J^-(q') \subset J^-(p)$, by Lemma 4.4.2, and so $J^-(p) \cap \Phi$ contains the point q.

One may therefore assume that $p \in J^+(\Phi)$. Denote $J^-(p) \cap \Phi$ by Φ_1, and $\Phi \backslash \Phi_1$ by Φ_2. Then $J^-(p) \cap J^+(\Phi_2)$ is empty, and hence

$$J^-(p) \cap J^+(\Phi) = J^-(p) \cap (J^+(\Phi_1) \cup J^+(\Phi_1)) = J^-(p) \cap J^+(\Phi_1).$$

We therefore only have to show that $J^-(p) \cap J^+(\Phi_1)$ is compact. Now $\Phi_1 = J^-(p) \cap \Phi_1$ is compact by hypothesis. So $J^+(\Phi_1)$ is closed, by part (i) of the proof, and hence $J^-(p) \cap J^+(\Phi_1)$ is closed. Again, every point $q' \in \Omega_0$ has an open neighbourhood of the form $D^+(q)$; one has only to take $q \in D^-(q^1)$ for this to be the case. As Φ_1 is compact, it can be covered by a finite number of future dependence domains,

$$D^+(q_1), ..., D^+(q_N),$$

say. Obviously, $J^+(D^+(q)) \subset J^+(q)$, and so $J^+(\Phi_1)$ is contained in the union of $J^+(q_1), ..., J^+(q_N)$. But as Ω_0 is a causal domain, each $J^-(p) \cap J^+(q_\nu)$ is either compact or empty. Hence $J^-(p) \cap J^+(\Phi_1)$ is closed, and contained in a compact set. Hence it is compact, and the proof is complete. $\square$

Lemma 5.1.2. *Let K, Φ and Ψ be subsets of Ω_0, and suppose that K is compact, Φ past-compact, and Ψ future-compact. Then both $J^-(K) \cap \Phi$ and $J^+(K) \cap \Psi$ are compact.*

PROOF. By Theorem 5.1.1, $J^-(K)$ is closed; hence $J^-(K) \cap \Phi$ is closed. Also, K can be covered by a finite number of past dependence domains, $D^-(p_1), ..., D^-(p_N)$ say. It follows that $J^-(K)$ is contained in the union of $J^-(p_1), \ldots, J^-(p_N)$. As Φ is past-compact, each $J^-(p_\nu) \cap \Phi$ is compact (or empty), and so $J^-(K) \cap \Phi$ is contained in the union of a finite number of compact sets, and closed. Hence it is compact. The compactness of $J^+(K) \cap \Psi$ now follows by reversing the time-orientation of Ω_0. The lemma is proved. $\square$

We now return to the wave equation (5.1.1), and prove the basic existence and uniqueness theorem.

It will be convenient to introduce some new classes of distributions and test functions. We shall denote the class of distributions in $\mathscr{D}'(\Omega_0)$ with past-compact supports by $\mathscr{D}'^+(\Omega_0)$, and the class of functions in $C^\infty(\Omega_0)$ with past-compact supports by $\mathscr{D}^+(\Omega_0)$; likewise, $\mathscr{D}'^-(\Omega_0)$ and $\mathscr{D}^-(\Omega_0)$ will denote the distributions and C^∞ functions with future-compact supports, respectively. Note that $\mathscr{D}'^+(\Omega_0)$ is not the dual of $\mathscr{D}^-(\Omega_0)$, nor $\mathscr{D}'^-(\Omega_0)$ that of $\mathscr{D}^+(\Omega_0)$. Each of these classes is a vector space that is stable under differentiation and multiplication by C^∞ functions. So $u \to Pu$ maps $\mathscr{D}'^+(\Omega_0)$ into $\mathscr{D}'^+(\Omega_0)$, $\mathscr{D}^+(\Omega_0)$ into $\mathscr{D}^+(\Omega_0)$, $\mathscr{D}'^-(\Omega_0)$ into $\mathscr{D}'^-(\Omega_0)$, and $\mathscr{D}^-(\Omega_0)$ into $\mathscr{D}^-(\Omega_0)$. We now prove that P has a unique inverse in each of these classes of distributions or functions.

Theorem 5.1.2. *If* $f \in \mathscr{D}'^+(\Omega_0)$, *then the wave equation* (5.1.1) *has a unique solution* $u \in \mathscr{D}'^+(\Omega_0)$, *given by* (5.1.4), *and*

$$\operatorname{supp} u \subset J^+(\operatorname{supp} f). \qquad (5.1.12)$$

If also $f \in C^\infty(\Omega_0)$, *so that* $f \in \mathscr{D}^+(\Omega_0)$, *then*

$$u(p) = \frac{1}{2\pi} \int_{C^-(p)} U(p,q) f(q) \mu_\Gamma(q) + \frac{1}{2\pi} \int V^+(p,q) f(q) \mu(q), \quad (5.1.13)$$

where $\mu_\Gamma(q)$ *is a Leray form such that* $d_q \Gamma(p,q) \wedge \mu_\Gamma(q) = \mu(q)$, *and* $u(p)$ *is then a* C^∞ *solution of* (5.1.1).

If $f \in \mathscr{D}'^-(\Omega_0)$, *then* (5.1.1) *has a unique solution* $u \in \mathscr{D}'^-(\Omega_0)$ *given by* (5.1.5), *and*

$$\operatorname{supp} u \subset J^-(\operatorname{supp} f). \qquad (5.1.14)$$

If also $f \in C^\infty(\Omega_0)$, *so that* $f \in \mathscr{D}^-(\Omega_0)$, *then*

$$U(p) = \frac{1}{2\pi} \int_{C^+(p)} U(p,q) f(q) \mu_\Gamma(q) + \frac{1}{2\pi} \int V^-(p,q) f(q) \mu(q), \quad (5.1.15)$$

and $u(p)$ *is then a* C^∞ *solution of* (5.1.1).

PROOF. (i) *Existence in the distribution case.* Suppose that $f \in \mathscr{D}'^{+}(\Omega_0)$. If $K \subset \Omega_0$ is a fixed compact set, and $\operatorname{supp} \phi \subset K$, then

$$\operatorname{supp} G_q^{+}(\phi) \subset J^{-}(K),$$

by (5.1.11). By Lemma 5.1.2, $K_1 = J^{-}(K) \cap \operatorname{supp} f$ is then a compact set. Let $\sigma \in C_0^{\infty}(\Omega_0)$ be such that $\sigma = 1$ on a neighbourhood of K_1. Then the second member of (5.1.4) is unaltered, for all $\phi \in C_0^{\infty}(K)$, if f is replaced by σf. As σf has compact support, this implies a semi-norm estimate similar to (5.1.8) for u, so that $u \in \mathscr{D}'(\Omega_0)$; it follows as before that $Pu = f$.

By (5.1.11), one has $(u, \phi) = 0$ when the support of f does not meet $J^{-}(\operatorname{supp} \phi)$. This is equivalent to $J^{+}(\operatorname{supp} f) \cap \operatorname{supp} \phi = \phi$. Hence (5.1.12) follows, in view of Theorem 5.1.1, which implies that $J^{+}(\operatorname{supp} f)$ is closed; the theorem also shows that $J^{+}(\operatorname{supp} f)$ is past-compact, and so (as a closed subset of a past-compact set is obviously past-compact) it follows that $u \in \mathscr{D}'^{+}(\Omega_0)$.

There is a similar argument when $f \in \mathscr{D}'^{-}(\Omega_0)$; one can also deduce the result in this case by reversing the time-orientation of Ω_0.

(ii) *Existence in the C^{∞} case.* If $f \in \mathscr{D}^{+}(\Omega_0)$, then (5.1.4) becomes

$$(u, \phi) = \frac{1}{2\pi} \int f(q) \mu(q) \int V^{+}(p, q) \phi(p) \mu(p)$$
$$+ \frac{1}{2\pi} \int f(q) \mu(q) \int_{C^{+}(q)} U(p, q) \phi(p) \mu_{\Gamma}(p). \quad (5.1.16)$$

One can now invert the orders of integration. As this step is almost identical with the corresponding step in the proof of Theorem 4.5.1, it will be omitted. The identity (5.1.16) is then transformed into

$$(u, \phi) = \int \tilde{u}(p) \phi(p) \mu(p), \quad (5.1.17)$$

where $\tilde{u}(p)$ denotes the second member of (5.1.13). To show that $\tilde{u}$ is C^{∞}, we write it out in local coordinates, and treat it like (5.1.4). In a notation similar to that used in the proof of Lemma 5.1.1, one obtains then

$$\tilde{u} \circ \pi^{-1}(x) = \frac{1}{2\pi} \int_{\zeta^1 \geqslant \rho} V^{+}(x, h(x, \zeta)) f(h(x, \zeta)) k(x, \zeta) d\zeta$$
$$+ \frac{1}{4\pi} \int_{\zeta^1 = \rho} U(x, h(x, \zeta)) f(h(x, \zeta)) k(x, \zeta) \frac{d\zeta^2 d\zeta^3 d\zeta^4}{\rho}, \quad (5.1.18)$$

where $\zeta \to y = h(x, \zeta)$ is now a coordinate transformation such that the ζ^i are normal and Minkowskian at p, and $(1, 0, 0, 0) \in TM_p$ is past-directed. It is apparent that $\tilde{u}(p) \in C^\infty(\Omega_0)$, taking into account that $f(p) \in \mathscr{D}^+(\Omega_0)$ vanishes to all orders on the boundary of its support. One can therefore identify the distribution with the function $\tilde{u}(p)$. By an obvious extension of Theorem 2.6.1, $\tilde{u}(p)$ is a classical solution of (5.1.1). The analogous result for the future-compact case (5.1.15) can be deduced by reversing the time-orientation of Ω_0.

(iii) *Uniqueness.* The adjoint of P is

$$^tPu = \Box\, u - \langle a, \nabla u \rangle + (b - \operatorname{div} a)\, u,$$

and is a differential operator of the same type as P. Hence Theorem 4.5.1, on the existence of the two basic fundamental solutions, also applies to ^tP; there are important relations between the fundamental solutions of P and ^tP, which will be derived in the next section. For the present, it is enough to observe that parts (i) and (ii) of the proof also apply to ^tP. In particular, if $\phi \in C_0^\infty(\Omega_0)$, then there is a $\psi \in \mathscr{D}^-(\Omega_0)$ such that $^tP\psi = \phi$, and $\operatorname{supp}\psi \subset J^-(\operatorname{supp}\phi)$. Now if $u \in \mathscr{D}'^+(\Omega_0)$, then it follows from Lemma 5.1.2 that $\operatorname{supp} u \cap J^-(\operatorname{supp}\phi)$ is compact, and so (Pu, ψ) exists. But

$$(Pu, \psi) = (u, {}^tP\psi) = (u, \phi).$$

As this holds for all $\phi \in C_0^\infty(\Omega_0)$, it follows that if $u \in \mathscr{D}'^+(\Omega_0)$ and $Pu = 0$, then $u = 0$. Hence $Pu = f \in \mathscr{D}'^+(\Omega_0)$ has at most one solution in $\mathscr{D}'^+(\Omega_0)$. There is a similar argument when $f \in \mathscr{D}'^-(\Omega_0)$. The proof of Theorem 5.1.2 is complete. $\Box$

Remark. Part (ii) of the proof shows that if f is (say) a bounded measurable function with past-compact support, then (5.1.13) still holds, and the second member is then a weak solution of $Pu = f$.

An important consequence of the theorem is the following:

Corollary 5.1.1. *The fundamental solutions $G_q^+(p)$ and $G_q^-(p)$ are the only fundamental solutions of P with pole q and supports contained in $J^+(q)$ and in $J^-(q)$ respectively.*

PROOF. It has already been noted that a causal domain is one in which $J^+(q)$ is past-compact, and $J^-(q)$ is future-compact, for every point $q \in \Omega_0$. So the corollary follows by taking $f = \delta_q(p)$ in the theorem. $\Box$

We have seen that P has a unique inverse in $\mathscr{D}'^+(\Omega_0)$ which is given by (5.1.4). It is evident from this representation that if (f_ν) is a sequence of distributions in $\mathscr{D}'^+(\Omega_0)$ which converges to $f \in \mathscr{D}'^+(\Phi_0)$ as

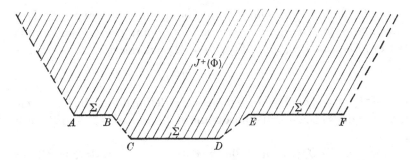

FIGURE 5.1.1

$\nu \to \infty$, and the supports of the f_ν are in a fixed past-compact set Φ, then the u_ν which are uniquely determined by $Pu_\nu = f_\nu$ and $u_\nu \in \mathscr{D}'^+(\Omega_0)$ also converge in a similar sense to the unique solution of $Pu = f$ that is in $\mathscr{D}'^+(\Omega_0)$. For the convergence follows from (5.1.4), and (5.1.12) gives $\operatorname{supp} u_\nu \subset J^+(\Phi)$, a fixed past-compact set. We can therefore say that, in this sense, the problem of solving $Pu = f$ in $\mathscr{D}'^+(\Omega_0)$ is properly posed. One can make a similar statement in the C^∞ case for $\mathscr{D}^+(\phi_0)$, if one defines convergence of a sequence (f_ν) of functions in $\mathscr{D}^+(\Omega_0)$ as convergence in $C^\infty(\Omega_0)$, together with the requirement that the supports of all the f_ν are to be in a fixed past-compact set.

We shall now consider the boundary of the future emission of a past-compact set. We first define the *spatial boundary* of a past-compact set Φ as the set Σ of all points $p \in \Phi$ such that $J^-(p) \cap \Phi = (p)$. This is illustrated by figure 5.1.1, where $n = 2$ and Φ is the broken line $ABCDEF$, BC and ED being characteristics. Then Σ consists of $AB\backslash(B)$, $EF\backslash(E)$, and CD.

It will be shown that $\partial J^+(\Phi)\backslash\Sigma$ is generated by null geodesics, so that it can be regarded as a generalized characteristic.

Theorem 5.1.3. *Let $\Phi \subset \Omega_0$ be a past-compact set, and let Σ denote its spatial boundary. Then Σ is not empty, and $J^+(\Sigma) = J^+(\Phi)$. If p is a point in $\partial J^+(\Phi)\backslash\Sigma$, then there is a point $q \in \Phi$ such that the geodesic $\widehat{qp}$ is a null geodesic, and every point of $\widehat{qp}$ other than q is in $\partial J^+(\Phi)\backslash\Sigma$.*

PROOF. If $p \in J^+(\Phi)$, then $J^-(p) \cap J^+(\Phi)$ is compact, by Theorem 5.1.1. Let p_1 be a point in $D^+(p)$. Then, by Lemma 4.4.2,

$$J^-(p) \cap J^+(\Phi) \subset D^-(p_1),$$

and $\Gamma(p_1, p') > 0$ for $p' \in J^-(p) \cap J^+(\Phi)$. So $\Gamma(p_1, p')$ has a positive maximum M on $J^-(p) \cap J^+(\Phi)$, which is attained at some point q of

$J^-(p) \cap J^+(\Phi)$. This point must be on the spatial boundary of Φ. For, in the first place, if $J^-(q) \cap J^+(\Phi)$ contains a point $q' \neq q$, then Lemma 4.4.1 implies that $\Gamma(p_1, q') > \Gamma(p_1, q) = M$; but clearly

$$q' \in J^-(p) \cap J^+(\Phi),$$

as $q' \in J^-(q) \subset J^-(p)$. This contradicts the definition of M. The same argument shows that q must be a point of Φ. Hence $q \in \Sigma$, and Σ is not empty.

Again, as $q \in J^-(p)$, it follows that $p \in J^+(q) \subset J^+(\Sigma)$. Hence $J^+(\Phi) \subset J^+(\Sigma)$. But $\Sigma \subset \Phi$, and so $J^+(\Sigma) \subset J^+(\Phi)$. Thus

$$J^+(\Sigma) = J^+(\Phi),$$

as asserted.

Suppose now that p is a point on $\partial J^+(\Phi) \backslash \Sigma$. As $J^+(\Phi)$ is past-compact, it is closed, and so $p \in J^+(\Phi) = J^+(\Sigma)$. Hence there is a point $q \in \Sigma$ such that $p \in J^+(q)$. If the geodesic $\widehat{qp}$ is time-like, then $p \in D^+(q)$. But $D^+(q)$ is an open subset of $J^+(\Phi)$, and so p would be an interior point of $J^+(\Phi)$, which contradicts the hypothesis that $p \in \partial J^+(\Phi)$. Hence $\widehat{qp}$ is a null geodesic.

To prove that every point of $\widehat{qp}$, other than q, is on $\partial J^+(\Phi) \backslash \Sigma$, we first show that $D^-(p)$ cannot meet $J^+(\Phi)$. For if it did, then there would be a point $q' \in \Phi$ and a point $q'' \in J^+(q')$ such that $q'' \in D^-(p)$. But Lemma 4.4.2, applied in time-reversed form to $D^-(p)$, shows that this implies that $q' \in D^-(p)$. But then $p \in D^+(q')$, which would mean that p is an interior point of $J^+(\Phi)$, contrary to the hypothesis.

As $\widehat{qp}$ is the common bicharacteristic along which $C^+(q)$ and $C^-(p)$ touch, every neighbourhood of a point $q' \in \widehat{qp}$ meets $D^+(q)$, and so q' must be on $\partial J^+(\Phi)$. If $q' \neq q$, then $J^-(q') \cap J^+(q)$ is the null geodesic arc $\widehat{q'q}$ and so q' cannot be on the spatial boundary. This completes the proof. $\square$

Remark. The argument can be carried a step further. Let (Ω, π) be a coordinate chart covering $\widehat{qp}$, and set $\pi p = x$, $\pi q = y$, $\pi q' = z$. If $q' \neq q$, $q' \neq p$, and $\pi^{-1}(z + \delta z)$ is a point on $\partial J^+(\Phi)$ near q', then $\pi^{-1}(z + \delta z) \notin D^+/q)$ and $\pi^{-1}(z + \delta z) \notin D^-(p)$, whence, in the local coordinates,

$$\Gamma(y, z + \delta z) \leqslant 0, \quad \Gamma(x, z + \delta z) \leqslant 0.$$

Now $\Gamma(y, z) = \Gamma(x, z) = 0$, as qp is a null geodesic. So

$$\Gamma(y, z + \delta z) - \Gamma(y, z) \leqslant 0, \quad \Gamma(x, z + \delta z) - \Gamma(x, z) \leqslant 0.$$

Let $\delta z \to 0$, in such a way that $\delta z / |\delta z|$, where $|\delta z|$ is the Euclidean norm, tends to a limit ζ, which is a vector in $\mathbf{R}^4$. Then one obtains

$$\langle \operatorname{grad}_z \Gamma(y, z), \zeta \rangle \leqslant 0, \quad \langle \operatorname{grad}_z \Gamma(x, z), \zeta \rangle \leqslant 0.$$

But $\operatorname{grad}_z \Gamma(y, z)$ and $\operatorname{grad}_z \Gamma(x, z)$ are both tangent to $\widehat{qp}$ at q', and have opposite orientations. Hence

$$\langle \operatorname{grad}_z \Gamma(y, z), \zeta \rangle = 0,$$

which means that ζ, considered as a tangent vector at q', is in the common tangent hyperplane of $C^+(q)$ and $C^-(p)$ at q', which is null. It follows that if q' has a neighbourhood Ω' such that $\Omega' \cap \partial J^+(\Phi)$ is a hypersurface, then this hypersurface is null.

The properties of the solutions of wave equations which have been derived are obviously related to the propagation of wave fronts, and to causality. Suppose that $f \in \mathscr{D}'^+(\Omega_0)$. The solution $u \in \mathscr{D}'^+(\Omega_0)$ of $Pu = f$ then evidently represents the disturbance produced by the source f in a previously undisturbed background. Theorem 5.1.2 shows that this disturbance is uniquely determined by f, and, by virtue of (5.1.12), that the front of the disturbance cannot propagate with a speed faster than the speed of light. It must be pointed out that the argument does not preclude the possibility that the front may propagate with a speed that is less than the speed of light. In the analytic case, this point can be settled by an appeal to Holmgren's uniqueness theorem.

Looking at the result in another way, one can infer from the fact that the support of $G_q^+(\phi)$ is contained in $J^-(\operatorname{supp} \phi)$ that the restriction of u to an open set $\omega \subset \Omega_0$ depends only on the restriction of f to $D^-(\omega)$, the union of all the $D^-(q)$ with $q \in \omega$. For if $\tilde{f}$ is another source field in $\mathscr{D}'^+(\Omega_0)$, and $\tilde{u}$ is the solution of $P\tilde{u} = \tilde{f}$ that is in $\mathscr{D}'^+(\Omega_0)$, then $f = \tilde{f}$ in $D^-(\omega)$ implies that $u = \tilde{u}$ in ω. This is why $D^-(\omega)$ is called the (past) dependence domain of ω. In the C^∞ case, one has the sharper result that $u(p)$ depends only on the restriction of f to $D^-(p)$.

These two basic properties of the solutions of wave equations in $\mathscr{D}'^+(\Omega_0)$ are a consequence of the existence of a fundamental solution with source point q and support contained in $J^+(q)$, and of the linearity of the equation. For one can look upon (5.1.4),

$$(u, \phi) = (f(q), G_q^+(\phi)),$$

as the superposition of point source fields. If $f \in \mathscr{D}^+(\Omega_0)$, then this representation becomes

$$(u, \phi) = \int f(q) \, G_q^+(\phi) \, \mu(q),$$

or, in the usual notation for the integrals of a distribution that depends on a parameter,

$$u(p) = \int f(q)\, G_q^+(p)\, \mu(q).$$

In general, one can see by regularization that any $f \in \mathscr{D}'^+(\Omega_0)$ is the limit, in the distribution topology, of a sequence of functions $f_\nu \in \mathscr{D}^+(\Omega_0)$ whose supports are contained in a pre-assigned neighbourhood of $\operatorname{supp} f$. It then follows that, in the sense of convergence in $\mathscr{D}'(\Omega_0)$,

$$u(p) = \lim_{\nu \to \infty} \int f_\nu(q)\, G_q^+(p)\, \mu(q). \qquad (5.1.19)$$

5.2   The reciprocity theorem

We have already used the observation that the adjoint of P is a differential operator of the same form as P. In fact, one can write it as

$$^tPu = \square\, u + \langle {}^t a, \nabla u \rangle + {}^t b u, \qquad (5.2.1)$$

where
$$^t a = -a, \quad {}^t b = b - \operatorname{div} a.$$

Note that the adjoint of ^tP is the original differential operator P. If $a \equiv 0$, then $^tP = P$, and P is self-adjoint. As the definition of a causal domain involves only the metric, and not the fields a and b, the adjoint operator also has a forward and a backward fundamental solution in a causal domain Ω_0, and these will be denoted by $^tG_q^+(p)$ and by $^tG_q^-(p)$ respectively. They are obtained in the same way as the fundamental solutions of P; one merely has to replace a by $^t a$ and b by $^t b$ in the construction described in sections 4.2–4.5. We have already used this fact in the proof of Theorem 5.1.2.

There is a simple relation between the coefficients of the singular parts of the fundamental solutions of P and ^tP. For a self-adjoint operator, this is the function $\kappa(p, q)$ defined by

$$2 \langle \nabla\Gamma, \nabla\kappa \rangle + (\square\, \Gamma - 8)\kappa = 0, \quad \kappa(q, q) = 1.$$

According to (4.2.20), $\kappa(p, q) = \kappa(q, p)$. Also, by (4.2.16),

$$U(p, q) = \kappa(p, q) \exp\left(-\frac{1}{2} \int_q^p \langle a(p'),\, dp' \rangle \right),$$

where the integral is taken along the geodesic $\widehat{qp} \subset \Omega_0$. As $^t a = -a$, it follows that the corresponding function for ^tP is

$$^tU(p, q) = \kappa(p, q) \exp\left(\frac{1}{2} \int_q^p \langle a(p'),\, dp' \rangle \right),$$

where the integral is again taken along the geodesic $\widehat{pq}$. But as $\widehat{pq}$ is just $\widehat{qp}$, with reversed orientation, we can conclude that

$$^tU(p,q) = U(q,p) \tag{5.2.2}$$

and this reciprocity relation holds, in fact, in any geodesically convex domain. A direct investigation of the coefficients of the two Hadamard series, along the same lines, would be difficult. It will however be shown in section 6.4 that they are related by similar reciprocity relations.

On the other hand, it is an easy consequence of Theorem 5.1.2, which holds for ^tP as well as for P, that there are reciprocity relations between the fundamental solutions of P and of ^tP.

Theorem 5.2.1. *For all ϕ and ψ in $C_0^\infty(\Omega_v)$,*

$$\int (G_q^+(p), \phi(p))\, \psi(q)\, \mu(q) = \int (^tG_q^-(p), \psi(p))\, \phi(q)\, \mu(q), \tag{5.2.3}$$

$$\int (G_q^-(p), \phi(p))\, \psi(q)\, \mu(q) = \int (^tG_q^+(p), \psi(p))\, \phi(q)\, \mu(q). \tag{5.2.4}$$

PROOF. As $^t(^tP) = P$, one has

$$\chi(q) = (^tG_q^-(p), P\chi(p)) \tag{5.2.5}$$

for all $\chi \in C_0^\infty(\Omega_0)$, by definition. This can be extended to $\chi \in \mathscr{D}^+(\Omega_0)$. For if $\chi \in \mathscr{D}^+(\Omega_0)$, then $J^-(q) \cap J^+(\operatorname{supp}\chi)$ is compact, by Theorem 5.1.1. Hence $J^-(q) \cap J^+(\operatorname{supp} P\chi)$ is also compact, as $\operatorname{supp} P\chi \subset \operatorname{supp}\chi$. Let $\sigma \in C_0^\infty(\Omega_0)$ be equal to unity on a neighbourhood of

$$J^-(q) \cap J^+(\operatorname{supp}\chi).$$

Then $\sigma\chi \in C_0^\infty(\Omega_0)$, and so (5.2.5) gives

$$\sigma(q)\,\chi(q) = (^tG_q^-(p), P(\chi(p)\,\sigma(p))).$$

Now $\operatorname{supp} {}^tG_q^- \subset J^-(q)$, so that $(^tG_q^-, P\chi\sigma) = (^tG_q^-, P\chi)$, and clearly $\sigma(q) = 0$; hence (5.2.5) still holds.

One can now appeal to Theorem 5.1.2. If $\psi \in C_0^\infty(\Omega_0)$ then, as a compact set is also future-compact, there is a unique χ such that $P\chi = \psi$ and $\chi \in \mathscr{D}^+(\Omega_0)$. Hence, by (5.2.5),

$$\chi(q) = (^tG_q^-(p), \psi(p)), \tag{5.2.6}$$

which is in effect an alternative representation of the inverse of P in $\mathscr{D}^+(\Omega_0)$. On the other hand, χ is given by (5.1.4), which becomes

$$\int \chi(q)\,\phi(q)\,\mu(q) = \int (G_q^+(p),\phi(p))\,\psi(q)\,\mu(q),$$

as χ is a locally integrable function. By the uniqueness theorem, one can substitute (5.2.6) in the first member, and one then obtains (5.2.3). The second identity (5.2.4) is proved similarly; it can also be deduced by reversing the time-orientation. The proof is complete. $\square$

Remark. For a self-adjoint operator ($a \equiv 0$), one has ${}^tG_q^+ = G_q^+$ and ${}^tG_q^- = G_q^-$, and (5.2.3), (5.2.4) then reduce to the single reciprocity relation

$$\int (G_q^+(p),\phi(p))\,\psi(q)\,\mu(q) = \int ({}^tG_q^-(p),\psi(p))\,\phi(q)\,\mu(q). \qquad (5.2.7)$$

When p and q are interchanged in the second member of (5.2.3), then the reciprocity relation assumes the form

$$\int (G_q^+(p),\phi(p))\,\psi(q)\,\mu(q) = \int ({}^tG_q^-(q),\psi(g))\,\phi(p)\,\mu(p),$$

which suggests the abbreviated formulation $G_q^+(p) = {}^tG_p^-(q)$. One can give this heuristic relation a precise meaning by introducing kernel distributions; how this is done is explained in section 6.3.

When Theorem 4.5.1 is applied to ${}^tP$, one obtains the two basic fundamental solutions in the explicit form

$$ {}^tG_q^+(p) = {}^tU\delta_+(\Gamma) + {}^tV^+, \quad {}^tG_q^-(p) = {}^tU\delta_-(\Gamma) + {}^tV^-. \qquad (5.2.8)$$

It has already been proved that ${}^tU(p,q) = U(q,p)$. We shall now prove similar reciprocity relations between the tail terms of the fundamental solutions of P and ${}^tP$.

Theorem 5.2.2. *The tail terms of the fundamental solutions of P and ${}^tP$ are related by*

$$V^+(p,q) = {}^tV^-(q,p), \quad V^-(p,q) = {}^tV^+(q,p). \qquad (5.2.9)$$

PROOF. Let us compare the two representations of the unique solution χ of $P\chi = \phi \in C_0^\infty(\Omega_0)$ that is in $\mathscr{D}^+(\Omega_0)$. By (5.2.6) and (5.2.9), one has, on interchanging p and q,

$$\chi(p) = \frac{1}{2\pi}\int_{C^-(p)} {}^tU(q,p)\,\phi(q)\,\mu_\Gamma(q) + \frac{1}{2\pi}\int {}^tV^-(q,p)\,\phi(q)\,\mu(q).$$

But χ is also given by (5.1.13), with f replaced by ϕ,

$$\chi(p) = \frac{1}{2\pi}\int_{C^-(p)} U(p,q)\,\phi(q)\,\mu_\Gamma(q) + \frac{1}{2\pi}\int V^+(p,q)\,\phi(q)\,\mu(q).$$

It follows from (5.2.2) that the integrals over $C^-(p)$ are identical. But we do not need to use this. Take p fixed, and suppose that

$$\mathrm{supp}\,\phi \subset D^-(p).$$

Then the integrals over $C^-(p)$ are absent, and so it follows that

$$\int ({}^t V^-(q,p) - V^+(p,q))\,\phi(q)\,\mu(q) = 0,$$

for all $\phi \in C_0^\infty(D^-(p))$. As V^+ and ${}^t V^-$ are continuous, this implies the first identity (5.2.9) when $q \in D^-(p)$, which is the same as $p \in D^+(q)$. It follows by continuity that (5.2.9) also holds when $p \in C^+(q)$, and of course both members vanish if $p \notin J^+(q)$. The second identity (5.2.9) follows by reversing the time-orientation of Ω_0; it can also be proved directly. $\square$

5.3 The Cauchy problem

The classical Cauchy problem for the wave equation (5.1.1) is to find a C^2 solution in a neighbourhood of a hypersurface S, given u and ∇u on S. The data must satisfy the compatibility relation

$$du = \langle \nabla u, dx \rangle$$

for all dx tangential to S, so that only one derivative of u, in a direction that is not tangent to S, can be prescribed. The Cauchy problem can evidently not be posed when S is a characteristic. We begin with a uniqueness theorem.

Theorem 5.3.1. *Let Φ be a past-compact set whose interior is not empty and whose boundary is a smooth hypersurface S. Let u be a function that is C^2 on a neighbourhood of Φ; suppose that $Pu = 0$ on Φ and that*

$$u = \nabla u = 0$$

on S. Then $u = 0$ on Φ.

PROOF. Let χ be the characteristic function of Φ. Then we have, for the distribution $u\chi$,

$$(P(u\chi), \phi) = (u\chi, {}^t P\phi) = \int_\Phi u\,{}^t P\phi\mu, \quad \phi \in C_0^\infty(\Omega_0).$$

Now $$\phi Pu - u\,^{t}P\phi = \nabla_i(\phi\nabla^i u - u\nabla^i\phi + a^i u\phi).$$

Hence it follows by integration over Φ that

$$(P(u\chi),\phi) = \int_\Phi \phi P(u)\,\mu + \int_S *(\phi\nabla u - u\nabla\phi + au\phi), \qquad (5.3.1)$$

where S has the orientation induced by the normal drawn into Φ. So the hypothesis implies that the distribution $u\chi$ satisfies $P(u\chi) = 0$, and as $u\chi$ has past-compact support, this implies, by Theorem 5.1.2, that $u\chi = 0$ as a distribution. But u is continuous, and so also $u = 0$ on Φ. $\square$

Remark. The hypothesis that S is a smooth hypersurface can obviously be weakened. For instance, S may consist of a finite number of hypersurfaces which meet transversally, or it may be a future null semi-cone. We shall return to this point later, in connection with characteristic initial value problems.

The theorem shows that the Cauchy problem for the equation $Pu = f$, with data given on S, cannot have more than one solution (of class C^2) in Φ. But this problem may not be properly posed. If v is a distribution with past-compact support, then

$$(v,\phi) = (Pv(q), G_q^+(\phi)), \quad \phi\in C_0^\infty(\Omega_0), \qquad (5.3.2)$$

where $G_q^+(\phi) = (G_q^+(p),\phi(p))$. For by Theorem 5.1.2, the second member is a solution of $P\tilde{v} = Pv$, and has past-compact support; but $\tilde{v} = v$ is a solution with past-compact support, and so $\tilde{v} = v$, as P has a unique inverse in $\mathscr{D}'^+(\Omega_0)$.

It therefore follows from (5.3.1) that, if $u\in C^2(\Omega_0)$, then

$$(u\chi,\phi) = (\chi(q)\,Pu(q),\,G_q^+(\phi)) + (g(q),\,G_q^+(\phi)), \qquad (5.3.3)$$

where g is the distribution

$$(g,\phi) = \int_S *(\phi\nabla u - u\nabla\phi + au\phi), \quad \phi\in C_0^\infty(\Omega_0).$$

This identity is in effect an integral representation of $u\chi$ in terms of the restrictions of Pu to $J^-(p)\cap\Phi$, and of u and ∇u to $J^-(p)\cap S$, valid for all $p\in\Omega_0\backslash S$. If $J^+(\Phi)\backslash\Phi$ is not empty, then one can take

$$p\in J^+(\Phi)\backslash\Phi$$

and thus obtain infinitely many relations between $Pu|_\Phi\, u|_S$ and $\nabla u|_S$; the Cauchy problem is then over-determined. This will always be the case if some portion of S is time-like. The situation is similar to that discussed in section 4.1 in connection with Kirchhoff's formula. Again, if $J^+(\Phi) = \Phi$ and S contains points that are not on the spatial boundary of $J^+(\Phi)$, one obtains 'internal' compatibility conditions that must be satisfied by f and the Cauchy data.

We shall therefore consider a Cauchy problem for a space-like hypersurface S that is past-compact, and is the boundary of its future emission $J^+(S)$. Then S is also the spatial boundary of $J^+(S)$. For, by Theorem 5.1.3, the non-spatial boundary of $J^+(S)$ is generated by null geodesics. So the hypothesis that the non-spatial boundary of $J^+(S)$ is not empty contradicts the assumption that $\partial J^+(S) = S$. Note also that if $p \in J^+(S)\backslash S$, then every past-directed causal geodesic from p meets $S = \partial J^+(S)$. Hence $J^+(S)\backslash S$ is a domain of determinacy for the Cauchy problem relative to S.

Theorem 5.3.2. *Let S be a past-compact space-like hypersurface, such that $\partial J^+(S) = S$. Suppose that f is C^∞ and that C^∞ Cauchy data are given on S. Then the Cauchy problem for $Pu = f$ has a unique solution u in $J^+(S)$, such that $u \in C^\infty(J^+(S))$.*

PROOF. We only have to prove the existence of such a solution, as its uniqueness then follows from Theorem 5.3.1. Consider first a coordinate neighbourhood $\omega \subset \Omega_0$ of a point of S, such that $\omega \cap S = \{x; x^1 = 0\}$. Formally, the Cauchy problem in ω can be solved by a power series in x^1 whose coefficients are determined as follows. Write $x = (x^1, x^*)$, where $x^* = (x^2, x^3, x^4)$. The Cauchy data give $u_0(x^*) = u(0, x^*)$ and $u_1(x^*) = \partial_1 u(0, x^*)$. As S is assumed to be space-like, we have $g^{11}(0, x^*) > 0$. So the $u_\nu(x^*) = \partial_1^\nu u(0, x^*)$ can be computed recursively, for $\nu \geq 2$, from u_0, u_1, and the equations $\partial_1^{\nu-2}(Pu-f) = 0$. To convert the formal Taylor series in x^1 into a convergent series, let $\sigma(t) \in C_0^\infty(\mathbf{R})$ be such that $0 \leq \sigma \leq 1$, and $\sigma = 1$ for $|t| \leq \frac{1}{2}$, $\sigma = 0$ for $|t| \geq 1$. Then one can show, as in the proof of Lemma 4.3.2, that if $(\lambda_\nu)_{0\leq\nu<\infty}$ is a sequence of positive numbers which increases sufficiently rapidly, then the series

$$\sum_{\nu=0}^\infty u_\nu(x^*)\frac{(x^1)^\nu}{\nu!}\sigma(\lambda_\nu x^1)$$

converges in $C^\infty(\omega)$ to a function $g(x)$. It is evident that g assumes the Cauchy data on $\omega \cap S$, and that $Pg-f$ and its derivatives of all orders vanish on $\omega \cap S$.

Now let (ω_μ) be a locally finite covering of S by coordinate neighbourhoods of this type, and denote the union of the ω_μ by Ω'. Let (ϕ_μ) be a partition of unity, defined on Ω', subordinated to this covering. By carrying out the construction just described in each ω_μ, one obtains a collection of functions v_μ, such that each v_μ assumes the Cauchy data on $\omega_\mu \cap S$, and $Pv_\mu - f$ vanishes to all orders on $\omega_\mu \cap S$. Put

$$v(p) = \sum_\mu v_\mu(p)\,\phi_\mu(p)$$

for $p \in \Omega_1$, and $v = 0$ on $\Omega_0 \backslash \Omega_1$. It is evident that $v \in C_0^\infty(\Omega_0)$, and that it assumes the Cauchy data on S. For this is true for each v_μ, and one also has, in terms of the local coordinates in each ω_μ,

$$\partial_i v = \sum_\mu \phi\,\partial_i v_\mu + \Sigma v\,\partial_i \phi,$$

$$\sum_\mu \phi_\mu = 1, \quad \sum_\mu \partial_i \phi_\mu = 0 \quad (i = 1, ..., 4).$$

A similar argument shows that $Pv - f$ vanishes to all orders on S.

Let χ denote the characteristic function of $J^+(S)$. As $J^+(S)$ is past-compact and $\partial J^+(S) = S$, the function $\chi(Pv - f)$ will therefore be in $\mathscr{D}^+(\Omega_0)$. Hence the equation $Pw = \chi(f - Pv)$ has a unique solution $w \in \mathscr{D}^+(\Omega_0)$ such that $\operatorname{supp}\omega \subset J^+(S)$. This implies that w, and its derivatives of all orders, vanish on S. It is then obvious that $u = v + w$ solves the Cauchy problem in $J^+(S)$, and that $u \in C^\infty(J^+(S))$. $\square$

Remark. It is assumed in Theorem 5.3.2 that S is contained in a causal domain. By Theorem 4.4.1, every point of a space–time has a neighbourhood that is a causal domain. It is also easy to verify that, in the neighbourhood constructed for the proof of this theorem, the coordinate hypersurface $S_1 = \{x; x^1 = 0\}$ is both past-compact and future-compact, and is the boundary of both $J^+(S_1)$ and $J^-(S_1)$. It follows from this that a space-like hypersurface S has a neighbourhood Σ in which the Cauchy problem has a unique C^∞ solution for C^∞ Cauchy data. It is characterized as follows. If $p \in \Sigma$, then either all the past-directed, or all the future-directed, causal geodesics from p meet Σ just once; also, the domain bounded by $C^-(p)$ and S, or by $C^+(p)$ and S, respectively, is relatively compact, and its closure is contained in a geodesically convex domain. One may call Σ the *local Cauchy development* of S.

We now turn to the problem of deriving a representation of the solution of the Cauchy problem whose existence follows from Theorem 5.3.2 in an explicit form. In principle, (5.3.3) is already such a representation, if one takes χ to be the characteristic function of $J^+(S)$, replaces Pu by f, and computes g from the Cauchy data. For zero Cauchy data, one has $g = 0$, and (5.3.3) reduces to

$$(u\chi, \phi) = (\chi(q)f(q), G_q^+(\phi)).$$

By the remark following the proof of Theorem 5.1.2, this implies that, for $p \in J^+(S)\backslash S$,

$$u(p) = \frac{1}{2\pi} \int_{C^-(p)\cap J^+(S)} U(p,q)f(q)\,\mu_\Gamma(q)$$
$$+ \frac{1}{2\pi} \int_{J^-(p)\cap J^+(S)} V^+(p,q)f(q)\,\mu(q). \quad (5.3.4)$$

It follows from the reciprocity relations (5.2.2) and (5.2.9) that the second member of (5.3.4) is $({}^tG_q^-(q), f(q)\chi(q))$. This suggests that the general representation theorem for the Cauchy problem can be obtained by evaluating $({}^tG_q^-(q), \chi(q)\,Pu(q))$ for a sufficiently smooth u. This is what we shall do now. We first introduce some geometrical quantities which will be needed.

Let p be a point in $J^+(S)\backslash S$. We set

$$D_p = J^-(p) \cap J^+(S), \quad C_p = C^-(p) \cap J^+(S),$$
$$S_p = J^-(p) \cap S, \quad \sigma_p = C^-(p) \cap S. \quad (5.3.5)$$

Note that $\partial D_p = C_p \cup S_p$ and $\sigma_p = \partial S_p$. The 2-surface σ_p is space-like, and inherits a non-degenerate metric from the metric of the space–time. The invariant 2-surface element on σ_p determined by this metric will be denoted by $d\sigma_p$, and σ_p will be oriented by $d\sigma_p > 0$.

As σ_p is space-like and compact, it has a neighbourhood in which there are two distinct null hypersurfaces that contain it (Theorem 3.3.3). One of these is $C^-(p)$; let us denote the other one by T_p (figure 5.3.1). At every point $q \in \sigma_p$ one can define a unique vector ξ tangent to the bicharacteristic of T_p by means of the condition

$$\langle \xi(q), \mathrm{grad}_q\,\Gamma(p,q)\rangle = -1. \quad (5.3.6)$$

We also need the dilatation of $d\sigma_p$ along the bicharacteristics of T_p. Let σ_p' denote the intersection of the pseudo-sphere $\{q; \Gamma(p,q) = \epsilon\}$ and T_p; let $\omega \subset \sigma_p$ be open in σ_p, and let ω' be the set of points in which σ_p' is met by bicharacteristics of T_p originating at points of ω. Then there is a scalar θ, defined on σ_p, such that

$$\lim_{\epsilon \to 0} \frac{1}{\epsilon}\left(\int_{\omega'} d\sigma_p' - \int_\omega d\sigma_p \right) = -\int_\omega \theta\,d\sigma_p,$$

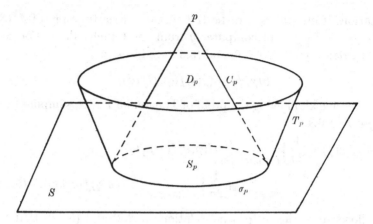

FIGURE 5.3.1

or, more briefly,

$$\theta \, d\sigma_p = \lim_{\epsilon \to 0} \frac{d\sigma_p - d\sigma'_p}{\epsilon}. \qquad (5.3.7)$$

Theorem 5.3.3. *Suppose that $u \in C^{\infty}(\Omega_0)$, and that p is a point in $J^+(S)\backslash S$. Then*

$$u(p) = u^{(1)}(p) + u^{(2)}(p) + u^{(3)}(p), \qquad (5.3.8)$$

where

$$u^{(1)}(\phi) = \frac{1}{2\pi} \int_{C_p} U P(u) \mu_\Gamma + \frac{1}{2\pi} \int_{D_p} V^+ P(u) \mu, \qquad (5.3.9)$$

$$u^{(2)}(p) = \frac{1}{2\pi} \int_{S_p} *(V^+ \nabla u - u \nabla V^+ + a V^+ u), \qquad (5.3.10)$$

with S_p oriented so that the positive normal points into D_p, and

$$u^{(3)}(p) = \frac{1}{2\pi} \int_{\sigma_p} (U(2\langle \xi, \nabla u \rangle + (\theta + \langle a, \xi \rangle) u) + V^+ u) \, d\sigma_p. \qquad (5.3.11)$$

Here $U = U(p,q)$, $V^+ = V^+(p,q)$, $u = u(q)$ in the integrands,

$$\mu_\Gamma = \mu_{\Gamma(p,q)}(q), \quad \mu = \mu(q),$$

and all differential operators act at q.

Remark. The identity (5.3.8) is a representation of u in $J^+(S)\backslash S$ in terms of the restriction of Pu to $J^+(S)$ and of u and ∇u to S. It is valid provided that every past-directed causal geodesic from p meets S, and that $\bar{D}_p$ is compact. If S is space-like and the boundary of its future emission, then it follows from Theorem 5.3.2 that (5.3.8) becomes the solution of the Cauchy problem in $J^+(S)\backslash S$ when Pu is replaced by $f \in C^\infty(\Omega_0)$ and both $u^{(2)}$ and $u^{(3)}$ are computed from the Cauchy data, which are also supposed to be C^∞. The result remains valid under much weaker differentiability hypotheses.

PROOF. The proof is a rather lengthy computation, which will require a number of lemmas.

Lemma 5.3.1. *If the support of u does not meet* σ_p, *then*

$$u(p) = u^{(1)}(p) + u^{(2)}(p). \qquad (5.3.12)$$

PROOF OF LEMMA 5.3.1. Let $\alpha(q) \in C_0^\infty(\Omega_0)$ be such that

$$\operatorname{supp}\alpha \subset J^+(S)\backslash S,$$

and that $\alpha = 1$ on a neighbourhood of $C_p \cap \operatorname{supp} u$. One can then put

$$u = \alpha u + (1-\alpha)u = u' + u'',$$

say. It is obvious that $u' \in \mathscr{D}^+(\Omega_0)$, and that the support of u'' does not meet C_p. It follows, first, that $u'(p) = ({}^tG_p^-(q), P'(q))$ whence, by the reciprocity theorems,

$$u'(p) = \frac{1}{2\pi}\int_{C_p} U(p,q)\,Pu'(q)\,\mu_\Gamma(q) + \frac{1}{2\pi}\int_{D_p} V^+(p,q)\,Pu'(q)\,\mu(q),$$
$$(5.3.13)$$

taking into account that $C^-(p) \cap \operatorname{supp} u' \subset C_p$ and that

$$\operatorname{supp} u' \subset J^+(S)\backslash S.$$

It follows from Theorem 5.2.2 that $V^+(p,q)$ and ${}^tV^-(q,p)$, as functions of q, satisfy

$${}^tPV^+(p,q) = {}^tP^t V^-(q,p) = 0,$$

for $q \in D^-(p)$. Hence

$$V^+(p,q)\,Pu''(q) = \operatorname{div}_q(V^+\nabla u'' - u''\nabla v^+ + aV^+u'').$$

As the support of u'' does not meet C_p, one obtains from this by integration over D_p

$$0 = \frac{1}{2\pi}\int_{D_p} V^+(p,q)\,Pu''(q)\,\mu(q) + \frac{1}{2\pi}\int_{S_p} *\,(V^+\nabla u'' - u''\nabla V^+ + aV^+u''),$$
$$(5.3.14)$$

where S_p has the orientation induced by the normal pointing into D_p. But it follows from the properties of α that $u'(p) = u(p)$, that

$$u'(q) = u(q)$$

on C_p, and that $u''(q) = u(q)$ on S_p. Hence (5.3.12) follows when (5.3.13) and (5.3.14) are added, and the lemma is proved. $\square$

Lemma 5.3.2. *There is a neighbourhood Ω' of σ_p, not containing p, in which there exists a space-like field $s(q)$ such that $s(q) > 0$ on $J^+(S)\backslash S$ and $\Omega' \cap S = \{q\,;\, s(q) = 0\}$, and a null field $t(q)$ such that $t(q) > 0$ on $J^+(S)\backslash S$ and $\Omega' \cap T_p = \{q\,;\, t(q) = 0\}$.*

Lemma 5.3.3. *Suppose that $u \in C_0^\infty(\Omega')$. Then*

$$u^{(1)}(p) + u^{(2)}(p) + \omega(p) = 0, \qquad (5.3.15)$$

where

$$w(p) = \frac{1}{2\pi}\int_{\sigma_p} (U(2\langle \nabla t, \nabla u\rangle + (\nabla t + \langle a, \nabla t\rangle u) - \langle \nabla t, \nabla r\rangle V^+ u)\mu_{t,\,\Gamma}$$
$$(5.3.16)$$

and $\mu_{\Gamma,\,t}$ is a Leray form such that $dt(q) \wedge dq\,\Gamma(p,q) \wedge \mu_{\Gamma,\,t} = \mu(q)$.
We postpone the proofs of these two lemmas.

END OF PROOF OF THEOREM 5.3.3. We first show that $w = u^{(3)}$. If $q \in \sigma_p$, then there is a coordinate chart at q such that $t(q) = x^1$, $\Gamma(p,q) = x^2$, and x^3, x^4 are constant on the bicharacteristics of the null hypersurfaces $t(q) = \text{const.}$ (See the remark following the proof of Theorem 3.3.2.) One then has

$$g^{11} = g^{13} = g^{14} = 0, \quad g^{12} = \langle \nabla\Gamma, \nabla t\rangle < 0, \qquad (5.3.17)$$

where the inequality follows from the fact that $\nabla\Gamma$ is past-directed and ∇t is future-directed. Hence also

$$g_{12} = 1/g^{12}, \quad g_{22} = g_{23} = g_{34} = 0,$$

and so $$-g(g^{12})^2 = g_{33}g_{44} - g_{34}^2.$$

But as, locally, $\sigma_p = \{x\,;\, x^1 = x^2 = 0\}$, this implies that

$$d\sigma_p = (g_{33}g_{44} - g_{34}^2)^{\frac{1}{2}}\, dx^2 \wedge dx^4 = -|g|^{\frac{1}{2}} g^{12}\, dx^3 \wedge dx^4. \quad (5.3.18)$$

As one can take $\mu_{t,\,\Gamma} = |g|^{\frac{1}{2}} dx^3 \wedge dx^4$, it follows that

$$\langle \nabla t, \nabla\Gamma\rangle \mu_{t,\,\Gamma} = -d\sigma_p. \qquad (5.3.19)$$

Again, (5.3.18) holds on all 2-surfaces $x^1 = 0$, $x^2 = $ const., and as $\Gamma = x^2$, it follows that

$$\theta = -\partial_2 \log(|g|^{\frac{1}{2}} |g^{12}|).$$

But $$\Box t = \Box x^1 = |g|^{-\frac{1}{2}} \partial_2(|g|^{\frac{1}{2}} g^{12}),$$

and so, by (5.3.19) and $g^{12} = \langle \nabla t, \nabla \Gamma \rangle$,

$$(\Box t)\mu_{t,\Gamma} = -\frac{1}{g^{12}|g|^{\frac{1}{2}}} \partial_2(|g|^{\frac{1}{2}} g^{12}) \, d\sigma_p = \theta \, d\sigma_p. \qquad (5.3.20)$$

Finally, ξ is tangential to the x^2-coordinate curves (which are the bicharacteristics of T_p when $x^1 = 0$) and is normalized by (5.3.6), $\langle \xi, \nabla \Gamma \rangle = -1$. As $\Gamma = x^2$ this means that $\xi^2 = -1$ and $\xi^i = 0$ for $i \ne 2$. As $\nabla^i t = g^{i1}$, it follows from (5.3.17) and (5.3.18) that

$$\langle \nabla t, \nabla u \rangle \mu_{t,\Gamma} = g^{12}(\partial_2 u) |g|^{\frac{1}{2}} dx^3 \wedge dx^4 = -(\partial_2 u) \, d\sigma_p,$$

and there is a similar identity for $\langle a, \nabla t \rangle \mu_{t,\Gamma}$. Hence

$$\langle \nabla t, \nabla u \rangle \mu_{t,\Gamma} = \langle \xi, \nabla u \rangle \, d\sigma_p, \quad \langle a, \nabla t \rangle \mu_{t,\Gamma} = \langle a, \xi \rangle \, d\sigma_p. \qquad (5.3.21)$$

When (5.3.19), (5.3.20) and (5.3.21) are substituted in (5.3.16), one finds that $w = u^{(3)}$, the second member of (5.3.10). This identity, which has been established locally, is obviously valid globally on σ_p.

Now Lemma 5.3.3 implies Theorem 5.3.3 when $u \in C_0^\infty(\Omega')$ and Lemma 5.3.1 implies it when the support of u does not meet σ_p. Let $\alpha(q) \in C_0^\infty(\Omega')$ be such that $\alpha = 1$ in a neighbourhood of σ_p. If $u \in C^\infty(\Omega_0)$, then $\alpha u \in C_0^\infty(\Omega')$ and $(1 - \alpha) u = 0$ in a neighbourhood of σ_p. But $u = \alpha u + (1 - \alpha) u$, and so the representation formula (5.3.8) follows. The theorem is proved.

It only remains to prove the two lemmas.

PROOF OF LEMMA 5.3.2. Choose a coordinate chart that is normal and Minkowskian at p, and such that $(1, 0, 0, 0) \in TM_p$ is past-directed. Then one can put, in a neighbourhood of $J^-(p)$,

$$x^1 = r(2 - \theta_2^2 - \theta_3^2 - \theta_4^2)^{\frac{1}{2}}, \; x^i = r\theta^i \quad (i = 2,3,4).$$

Then $r \geqslant 0$ and $\theta_2^2 + \theta_3^2 + \theta_4^2 \leqslant 1$ in $J^-(p)$, and $r \geqslant 0$, $\theta_2^2 + \theta_3^2 + \theta_4^2 = 1$ on $C^-(p)$; furthermore, the θ^i are constant on the null geodesics from p. As each past-directed causal geodesic from p meets S_p at just one point, the equation of S is of the form $r = \rho(\theta_1, \theta_2, \theta_3) \in C^\infty$. We can thus define $s(q)$ in a neighbourhood of S_p by setting $S(q) = \rho - r$. The hypersurfaces $s(q) = t$ meet $C^-(p)$ in a one-parameter family of space-like 2-surfaces. By part (ii) of Theorem 3.3.2, every point of σ_p therefore has a neighbourhood in which there exists a null field $t(q)$ such that $t(q) = s(q)$ on $C^-(p)$. This is actually all that is really needed in the proof of Theorem 5.3.3. But by arguing as in the proof of Theorem 3.3.3, one can show that $t(q)$ can be defined globally on a neighbourhood Ω' of σ_p. One can obviously choose Ω' so that $p \notin \Omega'$. $\square$

PROOF OF LEMMA 5.3.3. Let $\chi(q)$ denote the characteristic function of $J^+(S)$. We have already used the evident fact that the hypersurfaces $q \to \Gamma(p, q) = \text{const.}$ meet the characteristics $t(q) = \text{const.}$ transversally in Ω', by introducing t and Γ as local coordinates. This coordinate transformation also shows that

$$\chi(q)\,{}^tG_p^-(q) = \frac{1}{2\pi}\chi(q)\,{}^tU(q,p)\,\delta_-(\Gamma(q,p)) + \chi(q)\,{}^tV^-(q,p) \quad (5.3.22)$$

is a well defined distribution in Ω'. For $u \in C_0^\infty(\Omega')$ we therefore have

$$(\chi(q)\,{}^tG_p^-(q),\,Pu(q)) = ({}^tP(\chi(q)\,{}^tG_p^-(q)),\,u(q)). \quad (5.3.23)$$

By the reciprocity theorem, the first member of this identity is

$$\frac{1}{2\pi}(\chi(q)\,U(p,q)\,\delta_-(\Gamma(q,p)) + \chi(q)\,V^+(p,q),\,Pu(q)) = u^{(1)}(p). \quad (5.3.24)$$

We shall now compute the second member of (5.3.23). If H denotes the Heaviside function, then $H(s(q))$ is the restriction of χ to Ω', and $H(t(q))$ is, for $q \in C^-(p)$, the restriction of χ to $C^-(p) \cap \Omega'$. So

$${}^tP(\chi(q)\,{}^tU(p,q)\,\delta_-(\Gamma)) = {}^tP({}^tUH(t)\,\delta_-(\Gamma)).$$

As
$${}^tP = \square - \langle a, \nabla \rangle + b - \text{div}\, a,$$
one finds that

$${}^tP({}^tUH(t)\,\delta_-(r)) = {}^tP({}^tU\delta_-(\Gamma))\,H(t) + 2\nabla^i t \nabla_i({}^tU\delta_-(\Gamma))\,\delta(t)$$
$$+ {}^tU\delta_-(\Gamma)\,(\square - \langle a, \nabla \rangle)\,H(t).$$

But
$$\nabla^i t \nabla_i({}^tU\delta_-(\Gamma))\,\delta(t) = \nabla_i({}^tU V \nabla^i t \delta_-(\Gamma)\,\delta(t)) - {}^tU_-\delta(\Gamma)\,\square\,H(t)$$

and, as t is a null field,

$$\square\,H(t) = (\square\,t)\,\delta(t).$$

Hence
$${}^tP({}^tUH(t)\,\delta_-(\Gamma)) = {}^tP({}^tU\delta_-(\Gamma))\,H(t) + 2\nabla_i({}^tU V \nabla^i t \delta_-(\Gamma)\,\delta(t))$$

$$- {}^tU(\square t + \langle a, \nabla t \rangle)\,\delta_-(\Gamma)\,\delta(t). \quad (5.3.25)$$

Let us write, temporarily, ${}^t V^-(q,p) = \tilde{V}(q,p)\,H_-(\Gamma(q,p))$, where $\tilde{V}$ is a C^2 extension of ${}^t V^-$ to Ω'. Then one finds, after a little manipulation, that

$$
\begin{aligned}
{}^t P({}^t V^- &H(s)) \\
&= {}^t P({}^t V^-)\,H(s) + \nabla_i(\tilde{V}\nabla^i s H_-(\Gamma)\,\delta(s)) \\
&\quad + (\langle \nabla s, \nabla \tilde{V}\rangle - \langle a, \nabla s\rangle\,\tilde{V})\,H_-(\Gamma)\,\delta(s) + \tilde{V}\langle \nabla s, \nabla \Gamma\rangle\,\delta'_-(\Gamma)\,\delta(s).
\end{aligned}
\tag{5.3.26}
$$

The last term on the right-hand side can be treated as follows. Since $t(q) = s(q)$ on $C^-(p)$, one has

$$
\delta_-(\Gamma)\,H(s) = \delta_-(\Gamma)\,H(t).
$$

As $p \notin \Omega'$, one can differentiate this identity and form the scalar product with $\nabla\Gamma$ to obtain

$$
\begin{aligned}
\langle \Gamma, \nabla s\rangle\,\delta_-(\Gamma)\,\delta(s) &+ \langle \nabla\Gamma, \nabla\Gamma\rangle\,\delta'_-(\Gamma)\,H(s) \\
&= \langle \nabla\Gamma, \nabla t\rangle\,\delta_-(\Gamma)\,\delta(t) + \langle \nabla\Gamma, \nabla r\rangle\,\delta'_-(\Gamma)\,H(t).
\end{aligned}
$$

But
$$
\langle \nabla\Gamma, \nabla\Gamma\rangle\,\delta'_-(\Gamma) = 4\Gamma\delta'_-(\Gamma) = -4\delta_-(\Gamma),
$$

and so one is left with

$$
\langle \nabla\Gamma, \nabla s\rangle\,\delta_-(\Gamma)\,\delta(S) = \langle \nabla\Gamma, \nabla t\rangle\,\delta_-(\Gamma)\,\delta(t).
$$

Hence (5.3.26) becomes

$$
\begin{aligned}
{}^t P^t(V^- &H(s)) \\
&= {}^t P({}^t V^-)\,H(s) + \nabla_i(\tilde{V}\nabla^i s H_-(\Gamma)\,\delta(s)) \\
&\quad + (\langle \nabla s, \nabla \tilde{V}\rangle - \langle a, \nabla s\rangle\,\tilde{V})\,H_-(\Gamma)\,\delta(s) + \tilde{V}\langle \nabla t, \nabla \Gamma\rangle\,\delta_-(\Gamma)\,\delta(t).
\end{aligned}
\tag{5.3.27}
$$

We can now add (5.3.25) and (5.3.27); as $p \notin \Omega'$, we have

$$
{}^t P({}^t U\delta_-(\Gamma)) + {}^t P({}^t V^-) = 2\pi {}^t P({}^t G_p^-) = 2\pi \delta_p = 0.
$$

One therefore finds that

$$
2\pi {}^t P(\chi(q)\,{}^t G_p^-(q)) = W_2 + W_3,
\tag{5.3.28}
$$

where

$$
W_2 = \nabla_i(\tilde{V}\nabla^i s\,H_-(\Gamma)\,\delta(s)) + (\langle \nabla S, \nabla \tilde{V}\rangle - \langle a, \nabla s\rangle\,\tilde{V})\,H_-(\Gamma)\,\delta(s),
\tag{5.3.29}
$$

$$
\begin{aligned}
W_3 = 2\nabla_i({}^t U\nabla^i t\delta_-(\Gamma)\,\delta(t)) \\
- ({}^t U(\square t + \langle q, \nabla t\rangle) - \tilde{V}\langle \nabla\Gamma, \nabla t\rangle)\,\delta_-(\Gamma)\,\delta(t).
\end{aligned}
\tag{5.3.30}
$$

For $u \in C^\infty(\Omega')$,

$$(W_1, u) = -(H_-(\Gamma)\,\delta(s),\ \tilde{V}\langle\nabla s, \nabla u\rangle - u\langle\nabla s, \nabla\tilde{V}\rangle + \langle a, \nabla s\rangle\,\tilde{V}u)$$

$$= -\int_{S_p} (\tilde{V}\langle\nabla s, \nabla u\rangle - u\langle\nabla S, \nabla\tilde{V}\rangle + \langle a, \nabla s\rangle\,\tilde{V}u)\,\mu_s(q),$$

since the support of $H_-(\Gamma)\,\delta(s)$ is S_p. Now $\tilde{V} = {}^tV^-(q, p) = V^+(p, q)$ on S_p. Also, if ζ is a vector field, then $\langle\nabla s, \zeta\rangle\mu_s = *\zeta$ on S_p, where S_p is oriented by $\mu_s > 0$; this means that the positive normal to S_p points into D_p. Hence

$$(W_1, u) = -2\pi u^{(2)}(p),\ u \in C_0^\infty(\Omega'). \qquad (5.3.31)$$

Again, the distribution $\delta_-(\Gamma)\,\delta(t)$ is defined by

$$(\delta_-(\Gamma)\,\delta(t), \phi) = \int_{\sigma_p} \phi\mu_{t,\,\Gamma}(q),\ \phi \in C_0^\infty(\Omega'),$$

where σ_p is oriented by $\mu_{t,\,\Gamma} > 0$. (See (2.9.15).) Now (5.3.30) gives, for $u \in C_0^\infty(\Omega')$,

$$(W_2, u) = -(\delta_-(\Gamma)\,\delta(t),\ 2^t U\langle\nabla t, \nabla u\rangle + {}^tU(\Box t + \langle a, \nabla t\rangle) - \tilde{V}\langle\nabla\Gamma, \nabla t\rangle).$$

Replacing ${}^tU$ by $U(p, q)$ and $\tilde{V}$ by $V^+(p, q)$, it follows that

$$(W_2, u) = -2\pi w(p),\ u \in C_0^\infty(\Omega'), \qquad (5.3.32)$$

where w is defined by (5.3.16). In view of (5.3.24), (5.3.28), (5.3.31) and (5.3.32), the identity (5.3.23) therefore becomes

$$u^{(1)}(p) + u^{(2)}(p) + w(p) = 0,$$

which is (5.3.15), and so the lemma is proved. $\quad\Box$

We conclude the section with two observations on the representation theorem. It is evident that, as any function in $C^2(\Omega_0)$ can be approximated uniformly in every compact set by functions in $C^\infty(\Omega_0)$, that Theorem 5.3.3 holds if only $u \in C^2(\Omega_0)$. On the other hand, it is more difficult to answer the question whether the second member of (5.3.8), with Pu replaced by a function f, and $u^{(2)}$, $u^{(3)}$ computed from Cauchy data on S, is the solution of the Cauchy problem when f and the Cauchy data are in finite differentiability classes. One can in fact show that if $f \in C^2(\Omega_0)$, $u|_S = C^3(S)$, $\nabla u|_S \in C^2(S)$, then (5.3.8) furnishes a C^2 solution of the Cauchy problem in $J^+(S)$.

The second observation is that the hypothesis that S is space-like has not been used in the proof. It is, of course, essential, if the representation is to furnish the solution of the Cauchy problem when Pu,

$u|_S$ and $\nabla u|_S$ are replaced by given (smooth) functions. But (5.3.8) remains valid if every past-directed causal geodesic from p meets S once and only once, and $J^-(p) \cap J^+(S)$ is contained in a geodesically convex domain. One then obtains a generalization of Kirchhoff's formula for the ordinary wave equation. It should be added that the integrals over S_p and σ_p can also be put into different forms, by means of integration by parts. In particular, in the Minkowskian case, where $U = 1$ and $V^+ = 0$, one does not get Kirchhoff's formula (4.1.25) from (5.3.8), but a representation equivalent to it.

5.4 Characteristic initial value problems

The Cauchy problem for a space-like hypersurface is an initial value problem. One can also consider initial value problems for characteristics. Such problems have already arisen, in chapter 4, where it was shown that the tail terms of the two basic fundamental solutions are determined by characteristic initial value problems. We shall now discuss similar problems with more general data.

If Ω_0 is a causal domain and p_0 is a point of Ω_0, then $\Omega_0 \backslash C(p_0)$ consists of three components. They are $D^+(p_0)$, $D^-(p_0)$ and the set

$$\Omega_0 \backslash (J^+(p_0) \cup J^-(p_0)),$$

which consists of all points of Ω_0 which can be reached along space-like geodesics from p_0. One can pose a characteristic initial value problem in each of these sets. As the closure of $D^+(p_0)$ is $J^+(p_0)$, which is past-compact, it is to be expected that the characteristic initial value problem is properly posed in $D^+(p_0)$. We shall presently see that this is so. It then follows by reversal of the time-orientation that the characteristic initial value problem is also properly posed in $D^-(p_0)$. The third problem, to solve a wave equation $Pu = f$ in the exterior of $C(p_0)$, given u on $C(p_0)$, is not properly posed, and will not be discussed. In the analytic case, it can be solved in a neighbourhood of p_0 if the data are analytic, by a series similar to the Hadamard series.

It is obviously sufficient to consider the problem for $D^+(p_0)$, as the results obtained can be applied to $D^-(p_0)$ by reversing the time-orientation of Ω_0. It will be assumed that the second member f of the equation is in $C^\infty(\Omega_0)$. As regards the boundary data (the value of u on $C^+(p_0)$), the simplest plan, in view of the singularity of the boundary at p_0, is to require that they can be extended to a function in $C^\infty(\Omega_0)$. So the problem assumes the following form: given two functions f and

u_0 in $C^\infty(\Omega_0)$, to find u such that $Pu = f$ in $D^+(p_0)$ and $u = u_0$ on $C^+(p_0)$. We shall first prove uniqueness, and then the existence, of the solution, and finally derive a representation theorem.

Theorem 5.4.1. *Suppose that* $u \in C^\infty(\Omega_0)$, *that* $Pu = 0$ *in* $D^+(p_0)$, *and that* $u = 0$ *on* $C^+(p_0)$. *Then* $u = 0$ *on* $J^+(p_0)$.

PROOF. Let χ denote the characteristic function of $J^+(p_0)$. Then, by (5.3.3),

$$\left. \begin{aligned} (u\chi, \phi) &= (\chi(q)\, Pu(q), G_q^+(\phi)) + (g(q), G_q(\phi)), \\ (g, \psi) &= \int_{C^+(p_0)} * (\psi \nabla u - u \nabla \psi + au\psi), \end{aligned} \right\} \quad (5.4.1)$$

where $\phi \in C_0^\infty(\Omega_0)$ and $\psi \in C_0^\infty(\Omega_0)$. This is still valid because the divergence theorem holds for $D^+(p_0)$. Now

$$C^+(p_0) = \{p\,;\, p \in J^+(p_0),\, \Gamma(p_0, p) = 0\}.$$

Hence, by (2.9.5), one has, for any vector field ζ,

$$* \zeta = \langle \zeta, \mathrm{grad}_p\, \Gamma(p_0, p)\rangle \mu_{\Gamma(p_0, p)}(p),$$

where $C^+(p_0)$ is oriented by $\mu_\Gamma(p) > 0$. This is the required orientation, as $\Gamma(p_0, p) > 0$ in $D^+(p_0)$. Hence

$$(g, \psi) = \int_{C^+(p_0)} (\psi\langle\nabla\Gamma, \nabla u\rangle - u\langle\nabla\Gamma, \nabla\psi\rangle + \langle a, \nabla\Gamma\rangle u\psi)\, \mu_\Gamma. \quad (5.4.2)$$

But $\nabla\Gamma$ is tangent to the bicharacteristics if $p \in C^+(p_0)$. Hence $u = 0$ on $C^+(p_0)$ implies that $\langle\nabla\Gamma, \nabla u\rangle = 0$ on $C^+(p_0)$, and so it is sufficient to have $Pu = 0$ in $D^+(p_0)$ and $u = 0$ on $C^+(p_0)$ for the second member of (5.4.1) to vanish. As u is continuous, this implies that $u = 0$ on $J^+(p_0)$. $\square$

Theorem 5.4.2. *Let* f *and* u_0 *be given functions, both in* $C^\infty(\Omega_0)$. *Then there is one and only one* u *such that*

$$Pu = f \quad in \quad D^+(p_0), \qquad u = u_0 \quad on \quad C^+(p_0), \qquad (5.4.3)$$

and $u \in C^\infty(J^+(p_0))$.

PROOF. It will be sufficient to give this in outline only. As in the construction of the tail term of the parametrix in section 4.3, one can

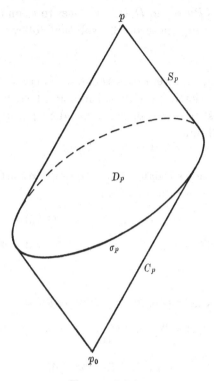

p

S_p

D_p

σ_p

C_p

p_0

FIGURE 5.4.1

determine a sequence of functions $u_\nu(p)$, defined on Ω_0, by solving the transport equations

$$2\langle \nabla\Gamma, \nabla u_\nu\rangle + (\Box\,\Gamma + \langle a, \nabla\Gamma\rangle + 4\nu - 4)\,u_\nu = -Pu_{\nu-1} \quad (\nu = 1, 2, \ldots),$$

subject to the condition that their solutions are to remain bounded when $p \to p_0$. Let $\sigma(t) \in C_0^\infty(\mathbf{R})$ be such that $0 \leqslant \sigma \leqslant 1$, that $\sigma = 1$ for $|t| \leqslant \frac{1}{2}$, and that $\sigma = 0$ for $|t| \geqslant 1$. The proof of Lemma 4.3.2 then shows that if (k_ν) is a sequence of positive numbers which increases sufficiently rapidly, then

$$\tilde{u} = u_0 + \sum_{\nu=1}^\infty u_\nu \frac{\Gamma^\nu}{\nu!}\sigma(k_\nu\Gamma), \quad \Gamma = \Gamma(p_0, p),$$

is a function in $C^\infty(\Omega_0)$ such that $P\tilde{u} - f$ vanishes to all orders on $C^+(p_0)$. Then $(P\tilde{u} - f)\chi$, where $\chi = H_+(\Gamma(p_0, p))$ is the characteristic function of $J^+(p_0)$, is in $\mathscr{D}^+(\Omega_0)$. Hence the equation $Pw = (f - P\tilde{u})\chi$ has a unique solution $w \in \mathscr{D}^+(\delta_0)$ whose support is contained in $J^+(p_0)$, and which therefore vanishes to all orders on $C^+(p_0)$. Clearly,

$u = \tilde{u} + w$ satisfies $Pu = f$ in $D^+(p_0)$, reduces to u_0 on $C^+(p_0)$, and is in $C^\infty(J^+(p_0))$. The uniqueness of the solution follows from Theorem 5.4.1.

We now turn to the representation theorem. This can be deduced from Theorem 5.3.3, by letting the initial hypersurface S tend to $C^+(p_0)$. But it is of interest to give a short independent proof. To simplify the notation, we shall put

$$\Gamma(p_0, p) = \gamma(p). \tag{5.4.4}$$

We also use a notation similar to that employed in the Cauchy problem for the various ranges of integration:

$$\left.\begin{array}{ll} J^-(p) \cap J^+(p_0) = D_p, & C^-(p) \cap J^+(p_0) = C_p, \\ J^-(p) \cap C^+(p_0) = S_p, & C^-(p) \cap C^+(p_0) = \sigma_p. \end{array}\right\} \tag{5.4.5}$$

(See figure 5.4.1.)

Theorem 5.4.3. *Suppose that* $u \in C^\infty(\Omega_0)$*, and put*

$$v(p) = 2\langle \nabla\gamma, \nabla u \rangle + (\Box\gamma + \langle a, \nabla\gamma \rangle - 4)\,u. \tag{5.4.6}$$

Then, for $p \in D^+(p_0)$*,*

$$u(p) = u^{(1)}(p) + u^{(2)}(p), \tag{5.4.7}$$

where

$$u^{(1)}(p) = \frac{1}{2\pi}\int_{C_p} U(p, q)\,Pu(q)\,\mu_\Gamma(q) + \frac{1}{2\pi}\int_{D_p} V^+(p, q)\,Pu(q)\,\mu(q) \tag{5.4.8}$$

and

$$u^{(2)}(p) = \frac{1}{2\pi}\int_{\sigma_p} U(p, q)\,v(q)\,\mu_{\Gamma,\gamma}(q) + \frac{1}{2\pi}\int_{S_p} V^+(p, q)\,v(q)\,\mu_\gamma(q). \tag{5.4.9}$$

PROOF. The characteristic function of $J^+(p_0)$ is $H_+(\Gamma(p, p_0)) = H_+(\gamma)$. By (4.3.4) and (5.4.6) we therefore have

$$P(uH_+(\gamma)) = P(u)\,H_+(\gamma) + v\delta_+(\gamma). \tag{5.4.10}$$

It follows that, for all $\phi \in C_0^\infty(\Omega_0)$,

$$(uH_+(\gamma), \phi) = \int_{J^+(p_0)} Pu(q)\,G_q^+(\phi)\,\mu(q) + \int_{C^+(p_0)} v(q)\,G_q^+(\phi)\,\mu_\gamma(q), \tag{5.4.11}$$

where

$$2\pi G_q^+(\phi) = \int_{C^+(q)} U(p, q)\,\phi(p)\,\mu_\Gamma(p) + \int_{J^+(q)} V^+(p, q)\,\phi(p)\,\mu(p). \tag{5.4.12}$$

One now obtains (5.4.7) by substituting (5.4.12) in (5.4.11) and inverting the orders of integration. As similar manipulations have been carried out before, we can omit the details, except for one term that is of a type that has not been met before.

The term in question arises when the second member of (5.4.12) is substituted in the last integral in the second member of (5.4.11). It is

$$\frac{1}{2\pi} \int_{C^+(p_0)} v(q)\,\mu_\gamma(q) \int_{C^+(q)} U(p,q)\,\phi(p)\,\mu_\Gamma(p),$$

and this is the integral of the 6-form

$$\frac{1}{2\pi} U(p,q)\,\phi(p)\,v(q)\,\mu_\gamma(q) \wedge \mu_\Gamma(p)$$

over

$$\{(p,q)\,;\, p\in C^+(q)\cap \operatorname{supp}\phi,\, q\in C^+(p_0)\}$$

$$= \{(p,q)\,;\, p\in J^+(p_0)\cap \operatorname{supp}\phi,\, q\in J^+(p_0)\cap C^-(p) = \sigma_p\}.$$

Formally, one can factorize $\mu_\gamma \wedge \mu_\Gamma$ as

$$\mu_\gamma(q) \wedge \mu_\Gamma(p) = d_q\Gamma(p,q) \wedge \mu_{\gamma,\,\Gamma}(q) \wedge \mu_\Gamma(p).$$

As $\Gamma(p,q) = 0$ on σ_p, and $\mu_{\gamma,\,\Gamma} = -\mu_{\Gamma,\,\gamma}$, it follows that

$$\mu_\gamma(q) \wedge \Gamma_\Gamma(p) = -d_p\Gamma(p,q) \wedge \mu_{\gamma,\,\Gamma}(q) \wedge \mu_\Gamma(p) = \mu(p) \wedge \mu_{\Gamma,\,\gamma}(q).$$

Hence $\displaystyle \frac{1}{2\pi} \int_{C^+(p_0)} v(q)\,\mu_\gamma(q) \int_{C^+(q)} U(p,q)\,\phi(p)\,\mu_\Gamma(p)$

$$= \frac{1}{2\pi} \int_{J^+(p_0)} \phi(p)\,\mu(p) \int_{\sigma_p} U(p,q)\,v(q)\,\mu_{\Gamma,\,\gamma}(q), \quad (5.4.13)$$

which formally gives the first term on the right-hand side of (5.4.9). However, one can show that

$$\langle \nabla\Gamma, \nabla\gamma \rangle \mu_{\Gamma,\,\gamma} = -d\sigma_p(q)$$

(compare (5.3.19)). If $p \to p_1 \in C^+(p_0)$, then σ_p shrinks to the bicharacteristic $\widehat{p_0 p_1}$, and $\langle \nabla\Gamma, \nabla\gamma \rangle \to 0$, as both vectors are, in the limit, tangent to this bicharacteristic. So the inner integral in the second member of (5.4.13) is not defined for $p \in C^+(p_0)$. But this difficulty does not arise when the support of ϕ is contained in $D^+(p_0)$. One can also show that the integral in question is C^∞ for $p \in D^+(p_0)$. So one obtains the representation (5.4.7) for $p \in D^+(p_0)$, and the theorem is proved. It is evident that (5.4.7) is also valid if only $u \in C^2(\Omega_0)$. $\square$

Remark. When $p \to p_1 \in C^+(p_0)$, in $D^+(p_0)$ then $u(p) \to u(p_1)$, and it can be shown that the integrals over D_p, S_p and C_p in (5.4.8) and (5.4.9) tend to zero. So the awkward integral over σ_p tends to $u(p_1)$. On the other hand, one can integrate (5.4.6), exactly like (4.3.5). One can therefore conclude that the following identity holds for all $u \in C^2(\Omega_0)$:

$$\lim_{p \to p_1} \frac{1}{2\pi} \int_{\sigma_p} U(p,q)\, v(q)\, \mu_{\Gamma,\,\gamma}(q) = -\tfrac{1}{4} U(p_1, p_0) \int_0^1 \frac{v(p_s)}{U(p_s, p_0)}\, ds = u(p_1),$$

$$(5.4.14)$$

where $s \to p_s$ is the equation of the geodesic $p_0 p_1$ in terms of an affine parameter; here, $p \in D^+(p_0)$, and $p \to p_1 \in C^+(p_0)$.

As an example, we shall consider a Minkowskian space–time and take $P \equiv \square$. Then $U = 1$ and $V^+ = 0$. As the term $u^{(1)}$ is just the retarded potential of $P(u) H_+(\gamma)$, we shall only compute $u^{(2)}$. Let us take p_0 to be the origin, and denote the coordinates of p by (t, X), where $X \in \mathbf{R}^3$, and those of q by (t', X'), where $X' \in \mathbf{R}^3$. Set

$$|X| = R, \quad |X'| = R', \quad X = R\xi, \quad X' = R'\xi', \quad \xi.\xi' = \cos\alpha, \quad (5.4.15)$$

so that ξ and ξ' are unit 3-vectors, and α is the angle between them. Then.

$$\gamma = t'^2 - R'^2, \quad \Gamma = (t-t')^2 - R^2 - R'^2 + 2RR'\cos\alpha,$$

and $C^+(p_0)$ is the null semi-cone $t' = R'$. On σ_p, since $\Gamma = \gamma = 0$, one has

$$R' = \rho(\alpha) = \frac{1}{2}\frac{t^2 - R^2}{t - R\cos\alpha}. \qquad (5.4.16)$$

Note that it is assumed that $t > R$, so that $\tfrac{1}{2}(t-R) < \rho < \tfrac{1}{2}(t+R)$. A simple calculation gives

$$v = 4 \frac{\partial}{\partial R'}\left(R'u(R', R'\xi')\right).$$

Again, if $d\omega'$ denotes the surface element of the unit 2-sphere at ξ', then

$$d\Gamma \wedge d\gamma \wedge d\omega' = 4((t'-t)\,dt' - (R' - R\cos\alpha)\,dR') \wedge (t'\,dt' - R'\,dR')$$

$$= 4(tR' - t'R\cos\alpha)\,dt' \wedge dR' \wedge d\omega'.$$

Now $t' = R'$ on σ_p, and $\mu(q) = R'^2 dt' \wedge dR' \wedge d\omega'$. Hence one can take, on σ_p

$$\mu_{\Gamma,\gamma} = \frac{1}{4} \frac{R' d\omega'}{t - R \cos \alpha}.$$

It then follows from (5.4.9) that, for $t > R$,

$$u^{(2)}(t, R\xi) = \frac{1}{2\pi} \int \left[R' \frac{\partial}{\partial R'} (R' u(R', R'\xi)) \right]_{R' = \rho(\alpha)} \frac{d\omega'}{t - R \cos \alpha}, \quad (5.4.17)$$

and this is the representation of the solution of the characteristic initial value problem for the ordinary wave equation, with the data given on $\{(t, X) ; t = R\}$.

It can be seen that the integrand in (5.4.17) becomes singular when $t \downarrow R$, as $t - R \cos \alpha$ then vanishes for $\xi' = \xi$. We shall now show that, if $u \in C^1(\mathbf{R}^4)$, then the second member of (5.4.17) tends to $u(R, R\xi)$ when $t \downarrow R$. Choose the coordinate axes in $\mathbf{R}^3$ so that $\xi = (0, 0, 1)$ and put $\xi' = (\sin \theta \cos \lambda, \sin \theta \sin \lambda, \cos \theta)$, where $0 \leqslant \theta \leqslant \pi$, $0 \leqslant \lambda \leqslant 2\pi$. Then $d\omega' = \sin \theta \, d\theta \, d\lambda$ and $\alpha = \theta$. Put, also

$$\bar{u}(R, \theta) = \frac{1}{2\pi} \int_0^{2\pi} u(R, R\xi') \, d\lambda.$$

Then $\bar{u}$ is a C^1 function of R and θ, and $\bar{u}(R, 0) = u(R, R\xi)$. The identity (5.4.17) becomes

$$u^{(2)}(t, R\xi) = \int_0^\pi (\rho^2 \bar{u}_R(\rho, \theta) + \rho \bar{u}(\rho, \theta)) \frac{\sin \theta \, d\theta}{t - R \cos \theta},$$

where $\bar{u}_R = \partial \bar{u} / \partial R$ and

$$\rho = \frac{1}{2} \frac{t^2 - R^2}{t - R \cos \theta}.$$

As

$$\frac{d\rho}{d\theta} = -\frac{R\rho \sin \theta}{t - R \cos \theta},$$

one therefore has

$$u^{(2)}(t, R\xi) = -\frac{1}{R} \int_0^\pi (\rho \bar{u}_R(\rho, \theta) + \bar{u}(\rho, \theta)) \frac{d\rho}{d\theta} d\theta$$

$$= -\frac{1}{R} \int_0^\pi \left(\frac{d}{d\theta} (\rho \bar{u}(\rho, \theta)) - \rho \bar{u}_\theta(\rho, \theta) \right) d\theta,$$

where $\bar{u}_\theta = \partial \bar{u} / \partial \theta$. Now $\rho(0) = \frac{1}{2}(t + R)$ and $\rho(\pi) = \frac{1}{2}(t - R)$, and so

$$u^{(2)}(t, R\xi) = \frac{t + R}{2R} \bar{u} \left(\frac{R + t}{2}, 0 \right) - \frac{t - R}{2R} \bar{u} \left(\frac{t - R}{2}, \pi \right) + \frac{1}{R} \int_0^\pi \rho \bar{u}_\theta(\rho, \theta) \, d\theta.$$

When $t \downarrow R$, the first term on the right-hand side tends to

$$\bar{u}(R, 0) = u(R, R\xi)$$

and the second term tends to zero. For the last term, one has the estimate

$$\frac{1}{R}\left|\int_0^\pi \rho\bar{u}_\theta(\rho,\theta)\,d\theta\right| \leqslant M\frac{t^2-R^2}{2R}\int_0^\pi \frac{\sin\theta}{t-R\cos\theta}\,d\theta = \frac{M\pi}{2R}(t^2-R^2)^{\frac{1}{2}},$$

where M is an upper bound for $\bar{u}_\theta(R',\theta)$ on $0 \leqslant R' \leqslant \frac{1}{2}(t+R), 0 \leqslant \theta \leqslant \pi$. Hence this term also tends to zero when $t \downarrow R$, and so

$$u^{(2)}(t,R\xi) \to u(R,R\xi) = u(|X|,X)$$

when $t \downarrow R = |X|$.

We can apply this to the characteristic initial value problem for the ordinary wave equation. Let $u_0(t,X) \in C^1(\mathbf{R}^4)$ be given, and put, for $t > R = |X|$,

$$u(t,X) = \frac{1}{2\pi}\int\left[R'\frac{\partial}{\partial R'}(R'u_0(R',R'\xi))\right]_{R'=\rho(\alpha)}\frac{d\omega'}{t-R\cos\alpha}, \quad (5.4.18)$$

where the notation is as in (5.4.17). Then

$$\square u = v_0\delta_+(\gamma), \quad \gamma = t^2-R^2, \quad v_0 = 4\frac{\partial}{\partial R}(Ru_0(R,R\xi))$$

and so in particular $\square u = 0$ in $D^+(0)$. Also, u is continuous, and $u \to u_0$ when $t \downarrow R$; hence u is a weak solution of the characteristic initial value problem in $D^+(0)$. If $u_0 \in C^3(\mathbf{R}^4)$, then it is a classical solution.

In section 4.5, the fundamental solution $G_q^+(p)$ of P was obtained by solving an integral equation. The regular part V^+ of $G_q^+(p)$ was expressed in terms of U, of the regular part $\tilde{V}$ of the parametrix, and of the resolving kernel of the integral equation. It is worth pointing out that one can derive other, simpler, integral equations for V^+ from Theorem 5.4.3.

First, one can take $u = V^+(p,p_0)$ in (5.4.7). Then, by (4.5.15) and (4.3.5), one has $Pu = 0$ in $D^+(p_0)$ and $v = -PU(p,p_0)$. Hence $u^{(1)} = 0$, and $u = u^{(2)}$ becomes

$$V^+(p,p_0) + \frac{1}{2\pi}\int_{C^+(p_0)\cap J^-(p)} V^+(p,q)\,PU(q,p_0)\,\mu_{\Gamma(p_0,q)}(q)$$

$$= -\frac{1}{2\pi}\int_{C^+(p_0)\cap C^-(p)} U(p,q)\,PU(q,p_0)\,\mu_{\Gamma(p,q),\,\Gamma(p_0,q)}(q). \quad (5.4.19)$$

This integral equation, in which the unknown is integrated over $C^+(p_0)$, is related to the integral equation (4.5.16) for the fundamental solution. The integrand on the right-hand side becomes singular if p tends to a point p_1 on $C^+(p_0)$. As we have proved that a function $V^+(p,p_0)$ satisfying the equation exists, it follows that the right-hand side of

(5.4.19) then tends to $V^+(p_1, p_0) = V_0(p_1, p_0)$. This also follows from (5.4.14) and (4.3.6), as it is in fact a special case of (5.4.14).

Again, one can take u in (5.4.7) to be the difference between V^+ and the first $N+1$ terms of the Hadamard series,

$$u = V^+(p, p_0) - \sum_{\nu=0}^{N} V_\nu(p, p_0) \frac{(\Gamma(p, p_0))^\nu}{\nu!}, \qquad p \in J^+(p_0).$$

Then $u = 0$ on $C^+(p_0)$, so that also $v = 0$ on $C^+(p_0)$, and, by (4.3.12),

$$PU = -P(V_N(p, p_0)) \frac{(\Gamma(p, p_0))^N}{N!}.$$

So one has $u^{(2)} = 0$, and (5.4.7) becomes

$$V^+(p, p_0) + \frac{1}{2\pi} \int_{J^+(p_0) \cap J^-(p)} V^+(p, q) \, PV_N(q, p_0) \frac{(\Gamma(q, p_0))^N}{N!} \mu(q)$$

$$= \sum_{\nu=0}^{N} V_\nu(p, p_0) \frac{(\Gamma(p, p_0))^\nu}{\nu!}$$

$$- \frac{1}{2\pi} \int_{J^+(p_0) \cap C^-(p)} U(p, q) \, PV_N(q, p_0) \frac{(\Gamma(q, p_0))^N}{N!} \mu_{\Gamma(p, q)}(q). \quad (5.4.20)$$

This is an integral equation for V^+ with a kernel that vanishes to the order $N-1$ when $\Gamma(p, p_0) \to 0$.

We conclude the section with a brief discussion of another characteristic initial value problem. Let T_1 and T_2 be two smooth characteristics in Ω_0 which intersect in a 2-surface $T_1 \cap T_2$ that is (necessarily) space-like. Assume that $T_1 \cap T_2$ is the spatial boundary of $J^+(T_1 \cap T_2)$, so that the boundary of $J^+(T_1 \cap T_2)$ consists of $T_1 \cap T_2$ and of the points on T_1 and T_2 that can be reached along future-directed null geodesics from $T_1 \cap T_2$. Then the characteristic initial value problem is properly posed in $J^+(T_1 \cap T_2)$. For one can prove both the existence and the uniqueness of the solution by arguments similar to those which were used to prove Theorems 5.3.2 and 5.4.2. We shall therefore only consider the representation of the solution. This can be obtained by imbedding T_1 and T_2 in two null fields $t_1(p)$ and $t_2(p)$ respectively, such that $t_1(p) = 0$ on T_1, $t_2(p) = 0$ on T_2, and

$$J^+(T_1 \cap T_2) = \{p; t_1(p) \geqslant 0, t_2(p) \geqslant 0\}.$$

These fields need only be defined in a neighbourhood of the boundary of $J^+(T_1 \cap T_2)$, which is the union of $T_1 \cap J^+(T_1 \cap T_2)$ and $T_2 \cap J^+(T_1 \cap T_2)$. For if $\Omega' \subset \Omega_0$ is such a neighbourhood, and $u \in C^\infty(\Omega_0)$, one can split u into two terms, $u = u' + u''$, such that supp $u' \subset \Omega'$ and the support of u'' does not meet $\partial J^+(T_1 \cap T_2)$. The contribution of u'' to the representation theorem is then just the retarded potential of the product of Pu''

and the characteristic function of $J^+(T_1 \cap T_2)$. We may therefore suppose that $\operatorname{supp} u \subset \Omega'$. One then finds, by a straightforward calculation, that

$$P(uH(t_1)\,H(t_2)) = P(u)\,H(t_1)\,H(t_2) + v_1\delta(t_1)\,H(t_2) + v_2 H(t_1)\,d(t_2)$$
$$+ 2\langle \nabla t_1, \nabla t_2\rangle\, u\,\delta(t_1)\,\delta(t_2), \qquad (5.4.21)$$

where
$$v_1 = 2\langle \nabla t_1, \nabla u\rangle + (\Box\,t_1 + \langle a, \nabla t_1\rangle)u, \left.\vphantom{\begin{matrix}1\\1\end{matrix}}\right\}$$
$$v_2 = 2\langle \nabla t_2, \nabla u\rangle + (\Box\,t_2 + \langle g, \nabla t_2\rangle)\, u. \qquad (5.4.22)$$

From this one can deduce that the restriction of u to $J^+(T_1 \cap T_2)$ is the sum of the following four groups of terms:

$$\frac{1}{2\pi}\int_{D_p} V^+(p.q)\,Pu(q)\,\mu(q) + \frac{1}{2\pi}\int_{C_p} U(p,q)\,Pu(q)\,\mu_{\Gamma(p,q)}(q), \quad (5.4.23)$$

where
$$D_p = \{q;\, t_1(q) \geqslant 0, t_2(q) \geqslant 0, q \in J^-(p)\},$$
$$C_p = \{q;\, t_1(q) \geqslant 0, t_2(q) \geqslant 0, q \in C^-(p)\};$$

$$\frac{1}{2\pi}\int_{\sigma'_p} U(p,q)\,v_1(q)\,\mu_{\Gamma,t_1}(q) + \frac{1}{2\pi}\int_{S'_p} V^+(p,q)\,v_1(q)\,\mu_{t_1}(q), \quad (5.4.24)$$

where
$$S'_p = \{q;\, t_1(q) = 0,\, t_2(q) \geqslant 0, q \in J^-(p)\},$$
$$\sigma'_p = \{q;\, t_1(q) = 0,\, t_2(q) \geqslant 0,\, q \in C^-(p)\},$$

and a similar sum with v_2 instead of v_1, and σ'_p, S'_p replaced by σ''_p, S''_p respectively, which are obtained by interchanging the sub-scripts 1 and 2; and, finally,

$$\frac{1}{\pi}\int_{C^-(p)\cap T_1\cap T_2} U(p,q)\,u(q)\,\langle \nabla t_1(a), \nabla t_2(q)\rangle\,\mu_{\Gamma,t_1,t_2}(q)$$
$$+\frac{1}{\pi}\int_{J^-(p)\cap T_1\cap T_2} V^+(p,q)\,u(q)\,\langle \nabla t_1(q), \nabla t_2(q)\rangle\,\mu_{t_1,t_2}(q). \quad (5.4.25)$$

Note that $\langle \nabla t_1, \nabla t_2\rangle\,\mu_{t_1,t_2}$ is the 2-surface element on $T_1 \cap T_2$.

As an example, let us suppose that the space-time is Minkowskian and that $P \equiv \Box$. As before, we write $x = (t, X)$, where $X \in \mathbf{R}^3$. Let us take $t_1 = t - X_1$, $t_2 = t + X_1$. Let $u_0(t, X)$ be a given function, say of class C^∞. Then the characteristic initial value problem

$$\Box u = 0 \quad \text{if} \quad t > |X_1|; \qquad u = u_0 \quad \text{if} \quad t = |X_1| \qquad (5.4.26)$$

has a unique C^∞ solution. One can deduce from (5.4.23), (5.4.24) and (5.4.25) that this solution is as follows. Define, for $t > |X_1|$,

$$\rho_1 = \tfrac{1}{2}(t + X_1) - \frac{r^2}{2(t - X_1)}, \quad \rho_2 = \tfrac{1}{2}(t - X_1) - \frac{r^2}{2(t + X_1)}, \quad \tau = (t^2 - X_1^2)^{\frac{1}{2}},$$

(5.4.27)

where r is a parameter. Then

$$u(t, X) = \frac{1}{2\pi} \int_0^\tau \int_0^{2\pi} \left[\frac{\partial}{\partial t'} u_0(t', t', X_2 + r\cos\lambda, X_3 + r\sin\lambda) \right]_{t' = \rho_1} \frac{r\,dr\,d\lambda}{t - \rho_1}$$

$$+ \frac{1}{2\pi} \int_0^\tau \int_0^{2\pi} \left[\frac{\partial}{\partial t'} u_0(t', -t, X_2 + r\cos\lambda, X_3 + r\sin\lambda) \right]_{t' = \rho_2} \frac{r\,dr\,d\lambda}{t - \rho_2}$$

$$+ \frac{1}{2\pi} \int_0^{2\pi} u_0(0, 0, X_2 + \tau\cos\lambda, X_3 + \tau\sin\lambda)\,d\lambda. \qquad (5.4.28)$$

5.5 Tensor wave equations

The theory which has been developed for scalar wave equations on a causal domain Ω_0 can easily be extended to tensor wave equations. We shall consider tensor fields of rank m, defined on Ω_0. Two fields of different types (r, s) and (r', s'), where $r + s = r' + s' = m$, will be identified if they can be obtained from each other, locally, by the raising or lowering of tensor sub- and superscripts. So the tensor analogue of our differential operator P can be written, for instance, in either of the forms

$$(Pu)_{i_1\dots i_m} = \nabla_j \nabla^j u_{i_1\dots i_m} + a_{i_1\dots i_m}{}^{j_1\dots j_m\,j} \nabla_j u_{j_1\dots j_m} + b_{i_1\dots i_m}{}^{j_1\dots j_m} u_{j_1\dots j_m},$$

(5.5.1)

or

$$(Pu)^{i_1\dots i_m} = \nabla_j \nabla^j u^{i_1\dots i_m} + a^{i_1\dots i_m}{}_{j_1\dots j_m}{}^j \nabla_j u^{j_1\dots j_m} + b^{i_1\dots i_m}{}_{j_1\dots j_m} u^{j_1\dots j_m},$$

(5.5.2)

where a and b are to be C^∞ fields. As in sections 3.5 and 3.6, the notation will be simplified by introducing tensor multi-indices $I = (i_1, \dots, i_m)$, $J = (j_1, \dots, j_m)$, ... Thus (5.5.1) and (5.5.2) can be replaced by

$$(Pu)_I = \nabla_j \nabla^j u_I + a_I{}^{Jj} \nabla_j u_J + b_I{}^J u_J \qquad (5.5.3)$$

and

$$(Pu)^I = \nabla_j \nabla^j u^I + a^I{}_J{}^j \nabla_j u^J + b^I{}_J u^J. \qquad (5.5.4)$$

The adjoint of P is then, in local coordinates,

$$({}^t Pu)^I = \nabla_j \nabla^j u^I - \nabla_j(a_J{}^{Ij} u^J) + b_J{}^I u^J. \qquad (5.5.5)$$

It follows that P is self-adjoint $(P = {}^tP)$ if, for all I, J and j,

$$a^{IJj} + a^{JIj} = 0, \quad b^{IJ} - b^{JI} = \nabla_j a^{IJj}. \tag{5.5.6}$$

The class of C^∞ tensor fields of rank m, defined on Ω_0, will be denoted by $\mathscr{E}^m(\Omega_0)$; it is a vector space over the complex numbers $\mathbf{C}$. The subspace of $\mathscr{E}^m(\Omega_0)$ consisting of fields with compact support will be denoted by $\mathscr{D}^m(\Omega_0)$. A tensor distribution $u \in \mathscr{D}'^m(\Omega_0)$ is then a continuous linear map $\mathscr{D}^m(\Omega_0) \to \mathbf{C}$, as explained in section 2.8. We must also introduce tensor-valued distributions. Let p' be a point of Ω_0; a tensor-valued distribution T is then a continuous linear map $\xi \to (T, \xi)$ of $\mathscr{D}^m(\Omega_0)$ into the (finite-dimensional) vector space of tensors of rank m at p'. (One can extend this to include continuous linear maps of $\mathscr{D}^m(\Omega_0)$ into the vector space of tensors of rank m' at p', but we shall not need this.) Evidently, T is also a function of p'; in the cases of interest here, $\xi \to (T, \xi)(p')$ will be a continuous linear map $\mathscr{D}^m(\Omega_0) \to \mathscr{E}^m(\Omega_0)$. Such a map is usually called a vector-valued distribution, but this term is ambiguous in the present context.

If (ω', π') is a coordinate chart at p', then each component of (T, ξ) is a (scalar-valued) tensor distribution. Let (ω, π) be another coordinate chart, and suppose that $\xi \in \mathscr{D}^m(\omega)$. Write $x = \pi p$, $x' = \pi' p'$, and $\xi_I(x)$ for $(\xi \circ \pi^{-1})_I$; then, by Definition 2.8.2,

$$(T, \xi)_{I'} = (T_{I'}^I(x, x'), \xi_I(x)\,|q(x)|^{\frac{1}{2}}),$$

where the $T_{I'}^I$ are distributions in $\mathscr{D}'(\pi\omega)$ that are functions of x'. It is obvious that the $T_{I'}^I(x, x')$ transform as the components of a bitensor, with respect to each set of arguments and tensor sub- and superscripts. (A *bitensor* of type $(0, m)$ at p and of type $(m', 0)$ at p' is a multilinear form on the product of m copies of T^*M_p and of m' copies of $TM_{p'}$; bitensors of other types are defined similarly. Bitensor sub- and superscripts can be raised and lowered in the usual way, by means of the appropriate components of the metric tensor at p and at p' respectively.)

The Dirac distribution of rank m, which will be denoted by $\delta_p^{(m)}(p)$, maps $\xi \in \mathscr{D}^m(\Omega_0)$ to $\xi(p')$,

$$(\delta_{p'}^{(m)}(p), \xi(p)) = \xi(p'). \tag{5.5.7}$$

For $m = 0$, it reduces to the Dirac measure $\delta_{p'}(p)$. Suppose that $\overset{m}{\tau}(p, p')$ is a continuous bitensor field such that

$$\overset{m}{\tau}{}_{I'}^I\big|_{p=p'} = \delta_{i_1'}^{i_1} \ldots \delta_{i_m'}^{i_m}, \tag{5.5.8}$$

where the same coordinate chart is used at p and p'. Then, obviously,

$$\delta_{p'}^{(m)}(p) = \overset{m}{\tau}(p,p')\,\delta_{p'}(p). \tag{5.5.9}$$

It is convenient to take $\overset{m}{\tau}$ to be C^∞.

We can now define a fundamental solution of the differential operator (5.5.1): it is a tensor-valued distribution $G_{p'}(p)$ such that

$$PG_{p'}(p) = \delta_{p'}^{(m)}(p).$$

As in the scalar case, there are two basic fundamental solutions $G_{p'}^+(p)$ and $G_{p'}^-(p)$ whose supports are contained in $J^+(p')$ and in $J^-(p')$ respectively. We shall only consider the first of these in detail, as the properties of the other one can then be inferred, as usual, by reversing the time-orientation of Ω_0. Most of the proofs can be carried over almost literally from the scalar to the tensor case; we shall not repeat them. There are some material simplifications when the tensor field a in (5.5.1) vanishes on Ω_0. As this is the most important case in the applications, we shall assume that

$$(Pu)_I = \nabla_j \nabla^j u_I + b_I{}^J u_J. \tag{5.5.10}$$

Note that P is not self-adjoint, unless $b_I{}^J = b^J{}_I$. The modifications that must be made when $a \neq 0$ will be indicated briefly at the end of the section.

By definition, a causal domain Ω_0 is contained in a geodesically convex domain Ω. We choose a fixed Ω, and introduce the *propagator* (also called the *transport bitensor*) in Ω. This is a bitensor field $\overset{m}{\tau}(p,p')$, of rank m at both p' and p, which satisfies, in local coordinates, the differential equations and initial conditions

$$\nabla^j \Gamma \nabla_j \overset{m}{\tau}{}_I^{I'} = 0, \quad \overset{m}{\tau}{}_I^{I'}\big|_{p=p'} = \delta_{i_1}^{i_1'}\dots\delta_{i_m}^{i_m'}, \tag{5.5.11}$$

where $\Gamma = \Gamma(p,p')$, and ∇_j, ∇^j act at p. (Note that the initial conditions are just (5.5.8).) If $\overset{1}{\tau}$ is a solution of (5.5.11) for $m = 1$, then

$$\overset{m}{\tau}(p,p') = \overset{1}{\tau}(p,p') \otimes \dots \otimes \overset{1}{\tau}(p,p'), \tag{5.5.12}$$

where there are m factors on the right-hand side, satisfies (5.5.11). This can be seen by writing out (5.5.12) in local coordinates,

$$\overset{m}{\tau}{}_I^{I'} = \overset{1}{\tau}{}_{i_1}^{i_1'}\dots\overset{1}{\tau}{}_{i_m}^{i_m'}.$$

We therefore only need to consider the case $m = 1$; we then write $\overset{1}{i'}$ for I' and i for I and τ for $\overset{1}{\tau}$. Let (Ω, π) be a coordinate chart covering Ω; write x' for $\pi p'$, x for πp, and $\Gamma(x, x')$, $\tau_i^{i'}(x, x')$ for the values of Γ and the components of τ in the local coordinates. Let $r \to z(r)$ be the equation of the geodesic $\overparen{p'p} \subset \Omega$ in the local coordinates, where r is an affine parameter such that $z(0) = x'$ and $z(1) = x$. Recall that

$$g^{jk}(z) \frac{\partial \Gamma(z, x')}{\partial z^k} \bigg|_{z=z(r)} = 2r \frac{dz^j}{dr} \quad (j = 1, \ldots, 4).$$

Hence the equations for τ become

$$\frac{d}{dr} \tau_i^{i'}(z(r), x') - \Gamma_{ik}^j(z(r)) \frac{dz^k}{dr} \tau_i^{i'}(z(r), x') = 0.$$

These are the differential equations of parallel transport along $\overparen{p'p}$. So τ — and likewise $\overset{m}{\tau}$ — is determined by parallel transport along $\overparen{p'p}$. The solutions of these differential equations are determined uniquely by the initial conditions

$$\tau_i^{i'}(z(0), x') = \tau_i^{i'}(x', x') = \delta_i^{i'}.$$

It is obvious that $\tau(p, p')$ is obtained by solving the differential equations of parallel transport subject to the initial conditions, and then putting $r = 1$. It follows that τ is uniquely determined by the differential equations and initial conditions, and a similar argument shows that $\overset{m}{\tau}$ is uniquely determined by (5.5.11); hence it is necessarily given by (5.5.12). It will be noted that $\overset{m}{\tau}$ depends on the choice of Ω. However, we shall only need $\overset{m}{\tau}$ when the geodesic $p'p$ is causal and in Ω_0, and it is then independent of the choice of Ω.

It follows from (5.5.11) that

$$\nabla^j \Gamma \nabla_j (\overset{m}{\tau}_{I'}^{I'} \overset{m}{\tau}_{J'}^I) = 0,$$

which becomes, on the geodesic $r \to z(r)$,

$$2r \frac{d}{dr} (\overset{m}{\tau}_I^{I'}(z(r), x') \overset{m}{\tau}_{J'}^I(z(r), x')) = 0,$$

as $\overset{m}{\tau}_I^{I'} \overset{m}{\tau}_{J'}^I$ is a scalar at p. Hence, in view of the initial conditions,

$$\overset{m}{\tau}_I^{I'}(p, p') \overset{m}{\tau}_{J'}^I(p, p') = \delta_{j_1'}^{i_1'} \ldots \delta_{j_m'}^{i_m'}. \tag{5.5.13}$$

Note also that $\overset{m}{\tau}$ is a symmetric function of its arguments,

$$\overset{m}{\tau}(p,p') = \overset{m}{\tau}(p',p). \tag{5.5.14}$$

This can be proved by considering the relation between $\overset{m}{\tau}(p,p')$, $\overset{m}{\tau}(p'',p)$ and $\overset{m}{\tau}(p'',p')$, where p, p' and p'' are points on the same geodesic, and letting $p'' \to p$. Finally, it should be observed that the construction of $\overset{m}{\tau}$ by integration along the geodesics from p' implies that $\overset{m}{\tau}$ is C^∞ on $\Omega_0 \times \Omega_0$.

We must also recall the definition of the biscalar $\kappa(p,p')$ which is the factor multiplying $\delta_+(\Gamma)$ in the fundamental solution of the scalar d'Alembertian. It is determined by

$$2\langle \nabla\Gamma, \nabla\kappa \rangle + (\Box\,\Gamma - 8)\kappa = 0, \quad \kappa(p',p') = 1, \tag{5.5.15}$$

and is given, in local coordinates, by (4.2.18) or (4.2.19).

We can now state the tensor analogue of Theorem 4.5.1, for the forward fundamental solution.

Theorem 5.5.1. *Let Ω_0 be a causal domain. Then the tensor differential operator (5.5.10) has a fundamental solution $G_{p'}^+(p)$ in Ω_0, such that*

$$PG_{p'}^+(p) = \delta_{p'}^{(m)}(p), \mathrm{supp}\, G_{p'}^+(p) \subset J^+(p'). \tag{5.5.16}$$

It is of the form

$$G_{p'}^+(p) = \kappa(p,p')\overset{m}{\tau}(p,p')\,\delta_+(\Gamma) + V^+(p,p'), \tag{5.5.17}$$

where $V^+(p,p')$ is a bitensor field that vanishes if $p \notin J^+(p')$ and which is C^∞ on the closed set $\{(p',p); (p',p)\in\Omega_0\times\Omega_0, p\in J^+(p')\}$.

PROOF. As the proof is in all respects similar to the proof of Theorem 4.5.1, we shall only discuss the analogue of Theorem 4.2.1 and the construction of the Hadamard series.

Let $U(p,p')$ be a C^∞ bitensor field, of rank m at both p and p', and let ϵ be a positive number. A straightforward calculation gives

$$P(U\delta_+(\Gamma-\epsilon))_I^{I'}$$
$$= (PU)_I^{I'}\delta_+(\Gamma-\epsilon) + (2\nabla^j\Gamma\nabla_j U_I^{I'} + (\Box\,\Gamma - 8)\, U_I^{I'})\delta_+'(\Gamma-\epsilon)$$
$$+ 4\epsilon U_I^{I'}\delta_+''(\Gamma-\epsilon). \tag{5.5.18}$$

It is clear from (5.5.8), (5.5.9), and the proof of Theorem 4.2.1, that the last term on the right-hand side will tend to $2\pi\delta_{p'}^{(m)}(p)$ as $\epsilon\downarrow 0$ provided that

$$U_I^{I'}|_{p=p'} = \delta_{i_1}^{i_1'}\ldots\delta_{i_m}^{i_m'},\tag{5.5.19}$$

where the same local coordinates are used at p and p'.

To obviate the difficulties caused by the term $\delta'_+(\Gamma-\epsilon)$ which does not converge as $\epsilon\downarrow 0$, one can impose the conditions

$$2\nabla^j\Gamma\nabla_j U_I^{I'}+(\square\,\Gamma-8)\,U_I^{I'} = 0.\tag{5.5.20}$$

As in the scalar case, it would be sufficient to require these to hold only on the null cone $C(p')$. But, as before, we shall suppose that (5.5.20) holds for all p and p' in Ω. It is now obvious from (5.5.8), (5.5.11) and (5.5.15) that

$$U(p,p') = \kappa(p,p')\overset{m}{\tau}(p,p'),\tag{5.5.21}$$

and it is not difficult to see, as in the scalar case, that this is the unique solution of (5.5.20) which satisfies (5.5.19).

With this choice of U, one can make $\epsilon\downarrow 0$ in (5.5.18), to obtain the tensor analogue of (4.2.9),

$$P(U\delta_+(\Gamma)) = P(U)\,\delta^+(\Gamma)+2\pi\delta_{p'}^{(m)}.\tag{5.5.22}$$

The next step is to generalize Lemma 4.3.1. For this, one has to determine a bitensor field $V_0(p,p')$ which satisfies the transport equation

$$2\nabla^j\Gamma\nabla_j V_{0,\,I}^{\ I'}+(\square\,\Gamma-4)\,V_{0,\,I}^{\ I'} = -(PU)_I^{I'},\tag{5.5.23}$$

and remains bounded as $p\to p'$. Let us put

$$V_{0,\,I}^{\ I'}(p,p') = U_I^{J'}(p,p')\,T_{J'}^{I'}(p,p'),\tag{5.5.24}$$

where T is a bitensor field that is of rank $2m$ at p' and of rank $0-$a scalar$-$at p. When this is substituted in (5.5.24) one finds, because of (5.5.20), that

$$U_I^{J'}(2\nabla^j\Gamma\nabla_j T_{J'}^{I'}+4T_{J'}^{I'}) = -(PU)_I^{I'}.$$

As $U = \kappa\overset{m}{\tau}$, one can use (5.5.13) to convert this into

$$2\nabla^j\Gamma\nabla_j T_{J'}^{I'}+4T_{J'}^{I'} = -\kappa^{-1}\overset{m}{\tau}{}_{J'}^{J}(PU)_J^{I'}.$$

As T is a scalar at p, this can be integrated for each I' and J' like the scalar equation (4.3.5); substituting in (5.5.24) one therefore finds that

$$V_{0,\,I}^{\ I'} = -\tfrac{1}{4}\kappa\overset{m}{\tau}{}_I^{J'}\int_0^1 \kappa^{-1}\overset{m}{\tau}{}_{J'}^{J}(PU)_J^{I'}|_{p=z(r)}\,dr,\tag{5.5.25}$$

where $r \to z(r)$ is, as before, the equation of the geodesic $\widehat{p'p}$ in local coordinates. One can then show, as in the proof of Lemma 4.3.1, that if $W(p, p')$ is a C^∞ bitensor field such that $W = V_0$ when $p \in C^+(p')$, then

$$P(U\delta_+(\Gamma) + WH_+(\Gamma)) = P(W)\,H_+(\Gamma) + 2\pi\delta_{p'}^{(m)}. \qquad (5.5.26)$$

Finally, the coefficients of the Hadamard series are defined by the transport equations

$$2\nabla^j\Gamma\nabla_j V_{\nu,I}^{\;\;I'} + (\square\,\Gamma + 4\nu - 4)\,V_{\nu,I}^{\;\;I'} = -(PV_{\nu-1})_I^{\;\;I'} \quad (\nu = 1, 2, \ldots) \quad (5.5.27)$$

and the conditions that the V_ν are to remain bounded as $p \to p'$. These equations can be treated like (5.5.23), by the method of variation of parameters; they are thus replaced by

$$V_{\nu,I}^{\;\;I'} = -\tfrac{1}{4}\kappa\tau_I^{\overset{m}{J'}}\int_0^1 \kappa^{-1}\tau_{J'}^{\overset{m}{}}(PV_{\nu-1})|_{p=z(r)}\,r^\nu\,dr, \qquad (5.5.28)$$

and these equations give the V_ν recursively.

The remainder of the argument is now as in the scalar case. If the space–time and the tensor field b in (5.5.10) are analytic, then the Hadamard series converges in a neighbourhood of the diagonal of $\Omega_0 \times \Omega_0$, and one obtains a tensor analogue of Theorem 4.3.1. In general, an obvious modification of Lemma 4.3.2 is used to form a C^∞ parametrix, and $G_{p'}^+(p)$ is constructed by solving an integral equation similar to (4.5.6), which can be solved by successive approximation in a causal domain. It then follows that the fundamental solution is indeed of the form (5.5.17). $\square$

Remark. The other properties of the fundamental solution asserted in Theorem 4.5.1 for the scalar case also extend to the tensor case. It follows from (5.5.26), (5.5.16) and (5.5.17) that the 'tail' V^+ of the forward fundamental solution, considered as a function of p for fixed p', is determined by the characteristic initial value problem

$$PV^+ = 0, \quad \text{if} \quad p \in D^+(p'); \quad V^+ = V_0, \quad \text{if} \quad p \in C^+(p'). \quad (5.5.29)$$

Also, the Hadamard series is an asymptotic expansion of V^+, valid as $\Gamma(p, p') \to 0$ for $p \in J^+(p')$.

We can now consider the material in the earlier sections of the present chapter. The forward fundamental solution $G_{p'}^+(p)$ of the scalar wave equation has two basic properties. First, the map $\phi \to G_{p'}^+(\phi)$ is a continuous linear map $C_0^\infty(\Omega_0) \to C^\infty(\Omega_0)$. This is the essential content of Lemma 5.1.1, and the meaning of the semi-norm estimates (5.1.7); we shall return to this point in section 6.3. The forward fundamental solution (5.5.17) of the tensor wave equation has a similar property; if $\xi \in \mathscr{D}^m(\Omega_0)$ then $\xi \to G_{p'}^+(\xi)$ is a continuous linear map

$$\mathscr{D}^m(\Omega_0) \to \mathscr{E}^m(\Omega_0).$$

This is proved as before; corresponding to (5.1.7), one obtains semi-norm estimates in which the sums run over the maxima of the moduli of the derivatives of all the components of ξ and $G_{p'}^+(\xi)$ respectively, in some fixed coordinate chart covering Ω_0. The second property of $G_{p'}^+(p)$ is that it is causal – its support is contained in $J^+(q')$. This also holds in the tensor case. It is therefore clear that the proof of Theorem 5.1.2 goes through as before, and leads to the following:

Theorem 5.5.2. Let $f \in \mathscr{D}'^m(\Omega_0)$ have past-compact support. Then the equation $Pu = f$, where P is given by (5.5.10), has one and only one solution $u \in \mathscr{D}'^m(\Omega_0)$ with past-compact support. It is given by

$$(u, \xi) = (f(p'), (G_{p'}^+(p, p'), \xi(p))), \quad \xi \in \mathscr{D}^m(\Omega_0); \qquad (5.5.30)$$

also, $$\operatorname{supp} u \subset J^+(\operatorname{supp} f). \qquad (5.5.31)$$

If, moreover, $f \in \mathscr{E}^m(\Omega_0)$, then $u \in \mathscr{E}^m(\Omega_0)$, and one has, in local coordinates,

$$u_I(x) = \frac{1}{2\pi} \int_{C^-(x)} U_I^{I'}(x, x') f_{I'}(x') \mu_\Gamma(x') + \frac{1}{2\pi} \int (V^+)_I^{I'}(x, x') f_{I'}(x') \mu(x'), \qquad (5.5.32)$$

where $U = \overset{m}{\kappa\tau}.$ $\square$

The reciprocity theorems also hold in the tensor case; in fact, all the material in section 5.2, can be taken over as it stands.

Let $S \subset \Omega_0$ be a space-like surface, such that $\partial J^+(S) = S$. One proves as before that the Cauchy problem for $Pu = f \in \mathscr{E}^m(\Omega_0)$, with C^∞ data on S, has a unique C^∞ solution in $J^+(S)$. The calculations which yield the representation theorem are similar to those in the scalar case, and are left to the reader. We state the result for the homogeneous wave equation $Pu = 0$, as the contribution from a non-zero second member $f \in \mathscr{E}^m(\Omega_0)$ is just (5.5.32), with f put equal to zero in $\Omega_0 \backslash J^+(S)$. The solution is the sum of an integral over $S_p = J^-(p) \cap S$ and an integral over $\sigma_p = C^-(p) \cap S$. The first of these is, in local coordinates,

$$\frac{1}{2\pi} \int * (V_I^{+I'}(x, x') \nabla^j u_{I'}(x') - u_{I'}(x') \nabla^j V_I^{+I'}(x, x'), \qquad (5.5.33)$$

where both ∇^j and the operator $*$ act at x'. The integral over σ_p is most easily written down in terms of the null field $t(p)$ introduced in Lemma 5.3.2, in a form similar to (5.3.16): it is

$$\frac{1}{2\pi}\int_{\sigma_p} (U_I^{I'}(x,x')\,(2\nabla^j t(x')\,\nabla_j u_{I'}(x') + u_{\Gamma'}(x')\,\Box\, t(x'))$$
$$- \langle\nabla t(x'),\nabla'\Gamma(x,x')\rangle\, V^{+I'}_I(x,x')\,u_{I'}(x'))\mu_{t,\,r}(x'). \quad (5.5.34)$$

There is a similar extension of the representation theorem for the characteristic initial value problem, with u given on a future null semi-cone $C^+(p_0)$.

It remains to indicate what modifications must be made when the differential operator P is given by (5.5.1). The transport equation (5.5.20) must be replaced by

$$2\nabla^j\Gamma\nabla_j U_I^{I'} + (\Box\,\Gamma - 8)\,U_I^{I'} + a_I{}^{Jj}(\nabla_j\Gamma)\,U_J^{I'} = 0.$$

If one regards p' as fixed, this is again a system of ordinary linear differential equations on each geodesic from p', which must be integrated subject to the initial conditions (5.5.19). On the other hand, the V_ν, which now satisfy the transport equations

$$2\nabla^j\Gamma\nabla_j V_{\nu,\,I}{}^{I'} + (\Box\,\Gamma + 4\nu - 4)\,V_{\nu,\,I}{}^{I'} + a_I{}^{Jj}(\nabla_j\Gamma)\,V_{\nu,\,J}{}^{I'}$$
$$= -(PV_{\nu-1})_I^{I'} \quad (\nu = 0, 1, \ldots), \quad (5.5.35)$$

where $V_{-1} = U$, can be calculated, successively, by quadratures. One can define a bitensor field $\bar{U}$ such that

$$\bar{U}_{I'}^I(p,p')\,U_I^{J'}(p,p') = \delta_{i_1'}^{j_1'}\ldots\delta_{i_m'}^{j_m'};$$

it is easily verified that

$$2\nabla^j\Gamma\nabla_j(\kappa\bar{U}_{I'}^I) - a_J{}^{Ij}(\nabla_j\Gamma)\,\kappa\bar{U}_{I'}^J = 0.$$

By making a substitution similar to (5.5.24), one can reduce the integration of the equations for the V_ν to the scalar case, and deduce that

$$V_{\nu,\,I}{}^{I'} = -\tfrac{1}{4}U_I^{J'}\int_0^1 (PV_{\nu-1})_J^{I'}\bar{U}_{J'}^I\big|_{x=z(r)}\,r^\nu dr \quad (\nu = 0, 1, \ldots). \quad (5.5.36)$$

The representation of the solution of the Cauchy problem must also be modified. One has to add the term

$$* (a_{I'}{}^{J'j}(x')\,(V^+)_I^{I'}(x,x')\,u_{J'}(x'))$$

in the integrand of (5.5.33), and the term

$$a_I{}^{J'j}\nabla_j t(x')\,u_{J'}(x')\,\mu_t\ _\Gamma(x')$$

in the integrand of (5.5.34).

5.6   The field of a time-like line source

If q is a point in space–time, and Ω_0 is a neighbourhood of q that is a causal domain, then the fundamental solution $G_q^+(p)$ in Ω_0 can be thought of as the field radiated by an instantaneous point source – a monopole – at q. In any causal neighbourhood of q it is determined uniquely by the condition that its support is to be past-compact.

In this section, we shall consider the field due to a monopole of varying strength which is travelling along a future-directed time-like curve Λ (a 'world line'). Such a curve can be given as a C^∞ imbedding $t \to \gamma(t)$ of an open interval $I \subset \mathbf{R}$ into the space–time M; the interval I need not be finite. We set $\gamma'(t) = v(t)$; this vector is assumed to be time-like and future-directed, and it is convenient to normalize the parametrization so that t is the proper time, which means that $\langle v, v\rangle = 1$.

We first consider the case of a scalar self-adjoint wave equation

$$Pu \equiv \square u + bu = f, \tag{5.6.1}$$

where $b \in C^\infty(M)$, and f is the distribution

$$f(p) = \int_I \delta_{\gamma(t)}(p)\,dm(t), \tag{5.6.2}$$

$t \to m(t)$ being a function on I that is locally of bounded variation. The equation (5.6.2) means that, for all $\phi \in C_0^\infty(M)$, such that the support of $t \to \phi(\gamma(t))$ is a compact subset of I,

$$(f, \phi) = \int_I \phi(\gamma(t))\,dm(t) = (m'(t)), \phi(\gamma(t))), \tag{5.6.3}$$

where $m'(t)$ is the derivative of the distribution $m(t) \in \mathscr{D}'(I)$ with which the function m must be identified.

We must begin with some geometrical considerations. Let $\Omega_0 \subset M$ be a causal domain, and suppose that Λ meets Ω_0. As Ω_0 is open, $\gamma^-(\Omega_0 \cap \Lambda)$ is an open set in $\mathbf{R}$, and so a union of disjoint open intervals. Let $I_0 = (t_1, t_2)$, where $t_1 < t_2$, be one of these. Note that I_0 need not be finite, as $\overline{\Omega}_0$ may not be compact. However, if either, or both, of t_1 and t_2 are finite, then we shall assume that $\gamma(t_1)$ and $\gamma(t_2)$ are respectively on $\partial\Omega_0$. We set $\gamma(I_0) = \Lambda_0$, consider this as fixed, and treat $m(t)$, restricted to I_0, as the data. The rest of I (and Λ) will be ignored. Obviously, $f(p)$ will have past-compact support if and only if $m(t)$ is constant for $t_1 < t < t_0$, for some $t_0 \in I$; one may suppose without loss of generality that $m(t) = 0$ on (t_1, t_0), and we shall suppose this to be the case. In the vector case, this requirement is incompatible with the law of charge conservation; we shall discuss this difficulty at the end of the section.

By definition, a causal domain Ω_0 is contained in some geodesically convex domain Ω. We choose a fixed Ω, and note that it follows from Lemma 4.4.2 that if p and q are in Ω_0, and the geodesic $\widehat{qp} \subset \Omega$ is causal, then it is independent of the choice of Ω. Let us now apply Lemma 4.4.2 to Λ_0. It implies that, if $t' \in I_0$, then $J^+(\gamma(t)) \subset D^+(\gamma(t'))$ for $t' < t < t_2$. It follows that

$$J^+(\Lambda_0) = \lim_{t \downarrow t_1} J^+(t),$$

and that $J^+(\Lambda_0)$ is open. For if $p \in J^+(\Lambda_0)$, then $p \in J^+(\gamma(t'))$ for some $t' \in I_0$, and so $p \in D^+(\gamma(t'')) \subset J^+(\Lambda_0)$ for $t_1 < t'' < t'$. Hence p has an open neighbourhood contained in $J^+(\Lambda_0)$.

Lemma 5.6.1. (i) *If p is a point of Ω_0, then $C^-(p)$ meets Λ_0 if and only if $p \in J^+(\Lambda_0)$.*

(ii) *If $p \in J^+(\Lambda_0)$, then $C^-(p)$ meets Λ_0 in only one point $q = \gamma(T)$, say.*

(iii) *On $J^+(\Lambda_0) \backslash \Lambda_0$, $p \to T(p)$ is a null field, with* grad $T(p) \neq 0$ *and*

$$\text{grad}_p \Gamma(p, \gamma(T)) + \langle \text{grad}_q \Gamma(p, \gamma(T)), v(T) \rangle \text{grad}\, T(p) = 0. \quad (5.6.4)$$

PROOF. (i) If $C^-(p)$ meets Λ_0 at q, then $p \in C^+(q) \subset J^+(q)$, and so $p \in J^+(\Lambda_0)$. Conversely, suppose that $p \in J^+(\Lambda_0)$, so that there is a t' such that $p \in J^+(\gamma(t'))$. If $p \in C^+(\gamma(t'))$, then $\gamma(t') \in C^-(p) \cap \Lambda_0$; for one would, otherwise, have $p \in D^+(\gamma(t'))$. By definition,

$$\Delta = J^+(\gamma(t')) \cap J^-(p),$$

which is not empty, is a compact subset of Ω_0. Hence $t \to \Gamma(p, \gamma(t))$ has a minimum on $\gamma^{-1}(\Delta \cap \Lambda_0)$ which is attained at some point t''. It follows from Lemma 4.4.2 that $\gamma(t) \in D^+(\gamma(t'))$ for $t' < t < t_2$; hence $\gamma(t'') \in D^+(\gamma(t'))$. If also $\Gamma(p, \gamma(t'')) > 0$, then $\gamma(t'')$ is an interior point of Δ, and so there is a $\delta > 0$ such that $\gamma(t) \in D^-(p)$ for $t'' < t < t'' + \delta$. But then Lemma 4.4.1 shows that $t \to \Gamma(p, \gamma(t))$ is strictly decreasing on $(t'', t'' + \delta)$, while $\gamma(t) \in \Delta$. This contradicts the definition of t'', and so $\Gamma(p, \gamma(t'')) = 0$: hence $\gamma(t'') \in C^-(p)$, and (i) is proved.

Again, if $p \in J^+(\Lambda_0)$, and $C^-(p)$ meets Λ_0 at $q = \gamma(T)$, say, then it follows from Lemma 4.4.2 that $y(t) \in D^-(\gamma(T))$ for $t_1 < t < T$. So $C^-(p)$ cannot meet Λ_0 in more than one point. This proves (ii).

Finally, we note that (i) and (ii) imply that if $p \in J^+(\Lambda_0)\backslash\Lambda_0$, then there is one and only one T such that $\Gamma(p, \gamma(T)) = 0$ and the null geodesic from $\gamma(T)$ to p is future-directed. Also,

$$-(\partial/\partial t)\,\Gamma(p, \gamma(t))|_{t=T} = \langle -\mathrm{grad}_q\,\Gamma(p, \gamma(T)), v(T)\rangle > 0,$$

since $v(T)$ is future-directed and time-like, and $\mathrm{grad}_q\,\Gamma(p, \gamma(T))$, which is tangent to the oriented geodesic from $\gamma(T)$ to p at $\gamma(T)$, is null and future-directed. (See the proof of Lemma 4.4.1.) Hence

$$p \to T(p) \in C^\infty(J^+(\Lambda_0)\backslash\Lambda_0),$$

and differentiation of $\Gamma(p, \gamma(T)) = 0$ gives (5.6.4). As $\mathrm{grad}_p\,\Gamma(p, \gamma(T))$ is a (non-zero) null vector, it follows that $\nabla T \neq 0$ and $\langle \nabla T, \nabla T\rangle = 0$. This proves (iii); obviously, the null hypersurfaces $T(p) = $ const. are just the null semi-cones $C^+(\gamma(t))$, $t \in I_0$. $\square$

We now consider the equation (5.6.1) restricted to Ω_0, with $f(p)$ defined by (5.6.2), but I replaced by I_0:

$$Pu \equiv \square u + bu = \int_{I_0} \delta_{\gamma(t)}(p)\,dm(t). \tag{5.6.5}$$

We recall that, as P is self-adjoint, its forward fundamental solution in Ω_0 is

$$G_q^+(p) = \frac{1}{2\pi}(\kappa(p, q)\,\delta_+(\Gamma(p, q)) + V^+(p, q)), \tag{5.6.6}$$

where κ is the biscalar discussed in detail in section 4.2.

Theorem 5.6.1. *Suppose that there is a $t_0 \in I_0$ such that $m'(t) = 0$ for $t_1 < t < t_0$. Then the unique solution of (5.6.5) with past-compact support is*

$$u(p) = -\frac{1}{2\pi}\frac{\kappa(p, \gamma(T))}{\langle \mathrm{grad}_q\,\Gamma(p, \gamma(T)), v(T)\rangle}m'(T) + \frac{1}{2\pi}\int_{I_0} V^+(p, \gamma(t))\,dm(t), \tag{5.6.7}$$

where $T(p)$ is defined by $\Gamma(p, \gamma(T)) = 0$, $\gamma(T) \in C^-(p)$ when $p \in J^+(\Lambda_0)$, and $T = t_1$ when $p \in \Omega_0\backslash J^+(\Lambda_0)$.

PROOF. The general representation of a solution of the wave equation with past-compact support is (5.1.4),

$$(u, \phi) = (f(q), (G_q^+(p), \phi(p))), \quad \phi \in C_0^\infty(\Omega_0).$$

By (5.6.5) and (5.6.3) one therefore has

$$(u, \phi) = (u^{(1)}, \phi) + (u^{(2)}, \phi), \qquad (5.6.8)$$

where $\quad (u^{(1)}, \phi) = \dfrac{1}{2\pi}\left((m'(t), \displaystyle\int_{C^+(\gamma(t))} \kappa(p, \gamma(t))\, \phi(p)\, \mu_\Gamma(p)\right) \qquad (5.6.9)$

and $\quad (u^{(2)}, \phi) = \dfrac{1}{2\pi}\displaystyle\int_{I_0}\left(\int V^+(p, \gamma(t))\, \phi(p)\, \mu(p)\right) dm(t); \qquad (5.6.10)$

$\mu_\Gamma(p)$ is our usual Leray form, such that $d_p \Gamma(p, q) \wedge \mu_\Gamma(p) = \mu(p)$.

We first consider $u^{(1)}$. By (iii) of Lemma 5.6.1, one can define a Leray form μ_T, such that $dT(p) \wedge \mu_T(p) = \mu(p)$ on $J^+(\Lambda_0)\backslash\Lambda_0$. On the null semi-cone $C^+(\gamma(t))\backslash(\gamma(t))$ one has $T(p) = t$, and it follows from (5.6.4) that one can take μ_T such that the following relation then holds:

$$\mu_\Gamma(p) = -\frac{\mu_T(p)}{\langle \mathrm{grad}_q\, \Gamma(p, \gamma(t)), v(t)\rangle}.$$

Hence (5.6.9) becomes

$$(u^{(1)}, \phi) = -\frac{1}{2\pi}\left(m'(t), \int_{T(p)=t} \frac{\kappa(p, \gamma(t))\, \phi(p)}{\langle \mathrm{grad}_q\, \Gamma(p, \gamma(t)), v(t)\rangle}\, \mu_T(p)\right),$$

which is just the first term in the second member of (5.6.7), applied to $\phi \in C_0^\infty(J^+(\Lambda_0)\backslash\Lambda_0)$. It can be shown that the singularity at Λ_0 does not invalidate this, and the extension to $C_0^\infty(\Omega_0)$ is immediate because of $m'(t) = 0$ for $t_1 < t < t_0$. Note that the value of $T(p)$ on $\Omega_0\backslash J^+(\Lambda_0)$ is immaterial; the convention of the theorem makes $T(p)$ continuous on Ω_0.

The identification of the second term on the right-hand side of (5.6.7) with $u^{(2)}$ follows from (5.6.10) and Fubini's theorem for Stieltjes integrals, as the inner integral is a continuous function of t whose support is a compact subset of I_0. So the theorem is proved. $\square$

Remark. Fubini's theorem also implies that

$$p \to \int_{I_0} V^+(p, \gamma(t))\, dm(t) = \int_{t_r}^{T(p)} V^+(p, \gamma(t))\, dm(t)$$

is locally integrable. (Note that the upper limit of integration is $T(p)$ because $V^+(p, \gamma(t)) = 0$ when $p \notin J^+(\gamma(t))$.) But the integral does not exist if $m(t)$ is discontinuous when $t = T(p)$, as $t \to V^+(p, \gamma(t))$ is then discontinuous. As $m(t)$ is locally of bounded variation, the set of its points of discontinuity is countable, and so $u^{(2)}$ is an essentially bounded function which may fail to exist on the union of a countable set of null semi-cones $C^+(\gamma(t))$.

The first term in the second member of (5.6.7) can be put into a different form by means of Theorem 4.4.2. One must first choose a C^∞ map

$$I_0 \times \mathbf{R}^+ \times S^2 \ni (t, r, \xi) \to p(t, r, \xi) \in J^+(\Lambda)\backslash\Lambda,$$

which is, for fixed t, a parametrization of $C^+(\gamma(t))\backslash\gamma(t))$ of the type used in Theorem 4.4.2. Thus ξ is constant, and r is an affine parameter, on each bicharacteristic of $C^+(\gamma(t))$. This map has a unique inverse for every point. $p \in J^+(\Lambda_0)\backslash\Lambda_0$; it gives the retarded time $t = T(p)$ which has already been introduced, and also an initial direction

$$\eta(p) = (\partial/\partial r)\, p(t, r, \xi)|_{r=0+}$$

of the null geodesic from $\eta(T)$ to p, and an affine distance $r = r(p)$ of p from Λ_0, measured along this null geodesic. Then, by (4.2.21),

$$\kappa(p, \gamma(T)) = r\left(\frac{d\sigma_0}{d\sigma}\right)^{\frac{1}{2}},$$

where $d\sigma$ is the 2-surface element at p of a cross-section of $C^+(\gamma(T))$, and $d\sigma_0 = \lim(d\sigma/r^2)$, as $r \downarrow 0$ (figure 5.6.1). Also, by (1.2.15),

$$\langle \operatorname{grad}_q \Gamma(p, \gamma(T)) \rangle = -2r(p)\, v(p).$$

Hence (5.6.7) becomes

$$u(p) = \frac{1}{4\pi} \frac{(d\sigma_0/d\sigma)^{\frac{1}{2}}}{\langle \eta(p), v(T) \rangle} m'(T) + \frac{1}{2\pi}\int V^+(p, \gamma(t))\, dm(t). \quad (5.6.11)$$

The first term on the right-hand side is similar to the Liénard–Wiechert potential. It is also, as comparison with (3.6.16) shows, a simple progressing wave. The amplitude, considered as a function on $C^+(\gamma(T))$, is the product of two factors. One, $(d\sigma_0/d\sigma)^{\frac{1}{2}}$, is the reciprocal of a quantity that has the character of a luminosity distance, measured along the null geodesic from $\gamma(T)$ to p. The other factor, $1/\langle \eta, v \rangle$, is a function of ξ only, giving an angular distribution of amplitude which is a maximum on the null geodesic which is tangent to Λ_0 at p; it would tend to infinity if $v(T)$ were to approach the velocity of light.

It follows from (5.6.7) or (5.6.11) that $u(p)$ can be expanded as a progressing wave of degree N, for any $N \geqslant 0$, with a remainder term. For $N = 0$, (5.6.7) is already of this form. For $N > 0$, we note first that the integral in the second member of (5.6.7) is over the open interval I_0. Hence one finds, integrating by parts, that

$$\int_{t_1}^{t_2} V^+(p, \gamma(t))\, dm(t) = -\int_{t_1}^{t_2} m(t)\, d_t\, V^+(p, \gamma(t)),$$

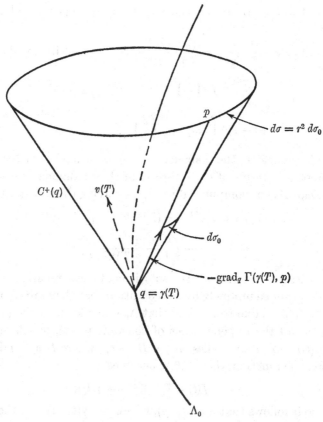

FIGURE 5.6.1

taking into account that $V^+(p, \gamma(t)) = 0$ when $t > T(p)$, and that (as we may suppose) $m(t) = 0$ for $t < t_0$. Now $t \to V^+(p, \gamma(t))$ is C^∞ for $t_1 < t \leqslant T(p)$, and

$$V^+(p, \gamma(T+)) - V^+(p, \gamma(T-1)) = V^+(p, \gamma(T+)) = V_0(p, \gamma(T)).$$

It follows that

$$\int V^+(p, \gamma(t)) \, dm(t) = V_0(p, \gamma(T)) \, m(T) - \int_{t_1}^{T} m(t) \frac{\partial}{\partial t} V^+(p, \gamma(t)) \, dt;$$

$$(5.6.12)$$

we may define $m(t)$ at points of discontinuity by setting $m(t) = m(t+)$. (As u is a distribution, the value of $m(t)$ at a discontinuity is, strictly speaking, irrelevant.) Put, for $\nu \geqslant 1$,

$$m_\nu(t) = \frac{1}{(\nu-1)!} \int_t^t (t-s)^{\nu-1} m(s) \, ds = \int_{t_1}^t m_{\nu-1}(s) \, ds. \quad (5.6.13)$$

Then it follows from (5.6.7) and (5.6.12), by repeated integration by parts, that

$$u(p) = -\frac{1}{2\pi} \frac{\kappa(p, \gamma(T))}{\langle \text{grad}_q \Gamma(\phi, \gamma(T)), v(T) \rangle} m'(T) + V_0(p, \gamma(T)) m(T)$$

$$+ \sum_{\nu=1}^{N-1} (-1)^\nu \left(\frac{\partial}{\partial t}\right)^\nu V^+(p, \gamma(t))|_{t=T} m_\nu(T)$$

$$+ (-1)^N \int_{t_1}^T m_{N-1}(t) \left(\frac{\partial}{\partial t}\right)^N V^+(p, \gamma(t)) \, dt. \tag{5.6.14}$$

In this expansion, the coefficients $u_\nu(p)$ of the $m_\nu(T)$ (with $m_0 = m$) are expressed in terms of the tail term of the fundamental solution. But they can also be computed directly by solving the transport equations

$$\left. \begin{aligned} 2\langle \nabla T, \nabla u_\nu \rangle + u_\nu \,\square\, T &= -(\square + b)\, U_{\nu-1} \quad (\nu = 0, 1, \ldots), \\ u_{-1} &= -\frac{1}{2\pi} \frac{\kappa(p, \gamma(T))}{\langle \text{grad}_q \Gamma(p, \gamma(T)), v(T) \rangle}. \end{aligned} \right\} \tag{5.6.15}$$

For it can be shown that the solutions of these recurrence equations are determined uniquely by the condition that they are to be bounded on Λ_0. This is due to the fact that Λ_0 is the locus of the (degenerate) caustics of the hypersurfaces of constant phase, which are just the $C^+(\gamma(t))$. Now if one takes $m(t) = H(t - t_0)$, where H again denotes the Heaviside function, then (5.6.2) becomes

$$f(t) = \delta_q(p), \quad q = \gamma(t_0),$$

and so it follows that $u = G_q^+(p)$, where $q = \gamma(t_0)$. Hence the transport equations (5.6.15) give an alternative algorithm for the derivation of an asymptotic expansion of $V^+(p, q)$, valid as p tends to a point of $C^+(q)$. In the analytic case, the series converges when p is sufficiently close to q. Note, however, that the leading term of the expansion must be determined by a separate argument; in fact, its form can be deduced from Theorem 4.2.1.

Another noteworthy consequence of Theorem 5.6.1 is obtained by taking

$$m'(t) = e^{i\omega t} H(t - t_0),$$

where ω is a fixed real number. The source is then a monopole which starts to radiate, with circular frequency ω, at (proper) time t_0. By (5.6.11) and (5.6.12) the resulting field is, for $T(p) \geqslant t_0$,

$$u(p) = \frac{1}{u\pi} \frac{1}{\langle \eta, v \rangle} \left(\frac{d\sigma_0}{d\sigma}\right)^{\frac{1}{2}} e^{i\omega T} + V_0(p, \gamma(T)) \frac{e^{i\omega T} - e^{i\omega t_0}}{i\omega}$$

$$- \int_{t_0}^T \frac{e^{i\omega T} - e^{i\omega t_0}}{i\omega} \frac{\partial}{\partial t} (V^+(p, \gamma(t))) \, dt,$$

while $u = 0$ for $T(p) < t_0$. It can be seen that in the high frequency limit $\omega \to \infty$, u is given by the leading term, which is the geometrical optics approximation.

It is not difficult to derive similar results for vector and tensor wave equations. For simplicity, we shall only consider the vector case, which is of particular interest. We take the equation to be, in local coordinates,

$$(Pu)_i \equiv \nabla_j \nabla^j u_i + b_i{}^j u_j = f_i \quad (i = 1, \ldots, 4). \tag{5.6.16}$$

The source will be defined by

$$(f, \xi) = \int \langle \xi(\gamma(t)), v(t) \rangle \, dm(t), \tag{5.6.17}$$

where $\xi \in \mathscr{D}^{(1)}(\Omega_0)$ is a test vector field, or

$$f_i(p) = \int g_{i'j'}(\eta(t)) \tau_i^{i'}(p, \gamma(t)) \, v^{j'}(t) \, \delta_{\gamma(t)}(p) \, dm(t). \tag{5.6.18}$$

The $\tau_i^{i'}(p, p')$ are the components of a bitensor of class C^∞, such that $\tau_i^{i'} = \delta_i^{i'}$ when $p = p'$, in the same coordinate chart at both points. For simplicity, we take τ to be the transport bitensor of rank unity.

The fundamental solution of P is given by (5.5.17), with $m = 1$,

$$(G_{p'}^+(p))_i^{i'} = \kappa(p, p') \tau_i^{i'}(p, p') \, \delta_+(\Gamma(p, p')) + (V^+)_i^{i'}(p, p'), \tag{5.6.19}$$

and the solution of (5.6.16) is, in general form, given by (5.5.30). Proceeding as in the scalar case, one finds that

$$u_i(p) = -\frac{1}{2\pi} \frac{\kappa(p, \gamma(T)) \, \tau_i^{i'}(p, \gamma(T)) \, v_{i'}(T)}{\langle \operatorname{grad}_q \Gamma(p, \gamma(T)), v(T) \rangle} m'(T)$$

$$+ \frac{1}{2\pi} \int_{I_0} (V^+)_i^{i'}(p, \gamma(t)) \, v_{i'}(t) \, dm(t) \quad (i = 1, \ldots, 4). \tag{5.6.20}$$

The first term on the right-hand side is again a simple progressing wave. It can be put into the geometrical form

$$\frac{1}{4\pi} \left(\frac{d\sigma_0}{d\sigma} \right)^{\frac{1}{2}} \frac{\tau_i^{i'}(p, \gamma(T)) \, v_{i'}(T)}{\langle \eta(p), v(T) \rangle} m'(T),$$

which reduces to the Liénard–Wiechert potential when the space–time is flat and $P \equiv \square$. By the definition of the transport bitensor, the terms $\tau_i^{i'}(p, \gamma(T)) v_{i'}(T)$ in the numerator are the components of the covector field at p which is obtained by parallel transport from the vector field $v(T)$ at $\gamma(T)$ along the null geodesic from $\gamma(T)$ to p. As in the scalar case, one can derive a progressing wave expansion by integration by parts.

We can now consider the situation that arises when the u_i are the components of the electromagnetic vector potential, so that $b_i{}^j = R_i{}^j$, the mixed components of the Ricci tensor, and the source f represents a charged particle moving along the world line Λ_0. Then the divergence of f must vanish. By (5.6.17) one has, for $\phi \in C_0^\infty(\Omega_0)$,

$$(\operatorname{div} f, \phi) = -(f, \operatorname{grad} \phi) = -(m'(t), \langle \nabla \phi(\gamma(t)), \gamma'(t) \rangle)$$

since $v(t) = \gamma'(t)$. But

$$-(m'(t), \langle \nabla \phi(\gamma(t)), (\gamma'(t)) \rangle)$$
$$= -(m'(t), (d/dt)\, \phi(\gamma(t))) = (m''(t), \phi(\gamma(t))).$$

Hence $$(\operatorname{div} f, \phi) = (m''(t), \phi(\gamma(t))). \qquad (5.6.21)$$

It follows that $\operatorname{div} f = 0$ if and only if $m'(t)$ is constant: this is the law of charge conservation.

Now if $m'(t) = k = \text{const.}$, then

$$(f, \xi) = k \int \langle \xi(\gamma(t)), v(t) \rangle \, dt, \quad \xi \in \mathscr{D}^1(\Omega_0),$$

and f does not have past-compact support. So the theory developed in this chapter cannot be applied to a charged particle. One can, however, make some comments on this problem.

The problem is obviously global, rather than local. In a space–time that is globally hyperbolic, in the sense of Leray, the fundamental solution $G_p^+(p)$ can be defined globally. The solution of $Pu = f$, when both u and f have past-compact support, is still unique, and given by

$$(u, \xi) = (f(p'), (G_{p'}^+(p), \xi(p))), \quad \xi \in \mathscr{D}'(M).$$

But the fundamental solution will not necessarily be of the simple form (5.6.19), as the extended backward null semi-cone $C^-(p)$ may have caustics, which give rise to further singularities. However, let us assume for the sake of the argument that (5.6.19) holds globally, as it certainly will if the space–time is a causal domain.

One can then set $m'(t) = kH(t - t_0)$, where k and t_0 are constants, and let $t_0 \to -\infty$. It may happen that the integral tends to a limit, and one then obtains, in the limit,

$$u_i(p) = -\frac{k}{2\pi} \frac{\kappa(p, \gamma(T)) \tau_i^{i'}(p, \gamma(T))\, v_{i'}(T)}{\langle \operatorname{grad}_q \Gamma(p, \gamma(T)), v(T) \rangle}$$
$$+ \frac{1}{2\pi} \int_{-\infty}^{T(p)} (V^+)_i^{i'}(p, \gamma(t))\, v_{i'}(t)\, dm(t). \qquad (5.6.22)$$

One must also consider the gauge condition $\operatorname{div} u = 0$. It is well known that

$$\nabla^i(\nabla_j\nabla^j u_i + R_i{}^j u_j) = \square\,(\operatorname{div} u), \tag{5.6.23}$$

so that $\qquad\qquad\qquad \square\,(\operatorname{div} u) = \operatorname{div} f.$

It follows from (5.6.21) that, for $m'(t) = kH(t-t_0)$,

$$\operatorname{div} f = k\delta_q(p), \quad q = \gamma(t_0).$$

Hence $\qquad\qquad \operatorname{div} u = kG_q^{(0)+}(p), \quad q = \gamma(t_0),$

where $G_q^{(0)+}$ denotes the forward fundamental solution of the scalar d'Alembertian. Suppose that this tends to zero when $t_0 \to -\infty$; then one may regard (5.6.22) as the vector potential of the field of a charged particle.

5.7 Huygens' principle

It was noted in section 5.3, in the Remark following the proof of Theorem 5.3.2, that a space-like hypersurface S has a neighbourhood in which the Cauchy problem relative to S is properly posed. For the homogeneous wave equation, $Pu = 0$, one then obtains, by Theorem 5.3.3 and its time-reversed counterpart, a representation of the solution u at a point p which is the sum of two integrals. One of these is over the hypersurface $(D^-(p) \cup D^+(p)) \cap S$. (Note that either $D^-(p) \cap S$ or $D^+(p) \cap S$ is empty, in the neighbourhood in question.) The other one is an integral over the 2-surface $C(p) \cap S$. For the ordinary wave equation, the first of these integrals is absent, and so $u(p)$ depends only on the Cauchy data restricted to $C(p) \cap S$. A differential operator P with this property is said to satisfy *Huygens' principle*. Strictly speaking, Huygens' principle, as originally enunciated, rests on an argument that was put into the form of a syllogism by Hadamard, who called the property just defined, the 'minor premise'. But it has become customary to restrict the term 'Huygens' principle' to Hadamard's minor premise.

It clearly follows from Theorem 5.3.3, and from the corresponding representation theorem in the past of S, that P will satisfy Huygens' principle if and only if the tail terms V^+ and V^- of the forward and backward fundamental solutions vanish. The supports of these fundamental solutions are then $C^+(p)$ and $C^-(p)$ respectively: they are often said to be 'sharp'. Suppose this is the case. If Ω_0 is a causal domain, and $f \in \mathscr{E}'(\Omega_0)$, then the support of the solution of $Pu = f$ that has past-compact support is contained in

$$C^+(\operatorname{supp} f) = \bigcup_{q \,\in\, \operatorname{supp} f} C^+(q).$$

(It can be shown that this set is past-compact.) This implies that the support of u meets every time-like curve (world line) in a compact set. Hence the disturbance represented by u, which is produced by the source f acting in a previously undisturbed background, which always has a sudden beginning, also has a sudden end. On the other hand, for a wave equation that does not satisfy Huygens' principle, there is no sudden cut-off, and the disturbance 'rings on'. One also speaks of the 'diffusion of waves', although the phenomenon is more akin to dispersion.

There is a simple, but not very informative, characterization of wave equations that satisfy Huygens' principle.

Theorem 5.7.1. *A scalar or tensor differential operator P, defined on a causal domain in a four-dimensional space–time, satisfies Huygens' principle if and only if $p \to U(p,q)$ satisfies $PU = 0$ when $p \in C(q)$.*

PROOF. It has already been noted that the vanishing of V^+ and V^- is a necessary and sufficient condition for the validity of Huygens' principle. These functions are solutions of the characteristic initial value problems

$$PV^+ = 0 \quad \text{if} \quad p \in D^+(q), \qquad V^+ = V_0 \quad \text{if} \quad p \in C^+(q),$$

$$PV^- = 0 \quad \text{if} \quad p \in D^-(q), \qquad V^- = V_0 \quad \text{if} \quad p \in C^-(q).$$

The uniqueness theorem for such problems (Theorem 5.4.1) extends to tensor wave equations. Hence an equivalent condition, both necessary and sufficient, is that $V_0(p,q) = 0$ when $p \in C(q)$. By (5.5.35) and (5.5.36), with $\nu = 0$ and $V_{-1} = U$, this will be the case if and only if $PU = 0$ on $C(q)$. So the theorem is proved. □

In principle, this theorem gives an algorithm that enables one to decide whether a given differential operator satisfies Huygens' principle. But in order to calculate U, one has first to integrate the differential equations of the geodesics. Even when this can be done, say by means of quadratures, the calculation of PU may prove difficult. (This remark also applies quite generally to the practical use of the Hadamard series.) The tensor case presents the additional difficulty that another, though linear, system of ordinary differential equations (the equations of parallel transport in the simplest case) has to be integrated. The theorem can therefore be used successfully only in a few isolated cases.

The scalar case has received a good deal of attention, and the literature is extensive. It has mainly been considered for linear second-order hyperbolic equations on $\mathbf{R}^n$. It follows from the representation theorems that will be derived in the next chapter that Huygens' principle is always invalid if n is odd. When n is even and $n \geqslant 4$, then the d'Alembertian on a Minkowskian space–time satisfies Huygens' principle (see (6.1.15)), and so do the differential operators that can be derived from it by trivial transformations. Hadamard posed the question whether there are any other linear second-order hyperbolic equations of Huygens type. It was thought for some time that there were no other equations with this property; this is sometimes called Hadamard's conjecture, although it does not seem to have been asserted by him. In fact, Matthison (1939) proved the conjecture for a scalar operator of the form $\square + \langle a(x), \nabla \rangle + b(x)$ on a four-dimensional Minkowskian space–time. But counter examples were found, for n even and $n > 4$, by Stellmacher (1953) and, finally, Günther (1965) showed that the d'Alembertian on a well-known relativistic space–time – the plane gravitational wave – satisfies Huygens' principle.

The metric in question can be given, in local coordinates, by the line element

$$\langle d\tilde{x}, d\tilde{x} \rangle = 2 d\tilde{x}^1 d\tilde{x}^2 + c_{\alpha\beta} \tilde{x}^\alpha \tilde{x}^\beta (d\tilde{x}^2)^2 - (d\tilde{x}^3)^2 - (d\tilde{x}^4)^2, \quad (5.7.1)$$

where $\tilde{x} \in \mathbf{R}^4$, α and β take only the values 3 and 4, and $(c_{\alpha\beta})$ is a symmetric matrix that is a function of x^1 only. The Ricci tensor of this metric vanishes if $c_{11} + c_{22} = 0$, and the metric is then a solution of the empty space Einstein gravitational equations. It can be shown that, locally, there is a coordinate transformation $\tilde{x} \to x$ which carries (5.7.1) into

$$\langle dx, dx \rangle = 2 dx^1 dx^2 - a_{\alpha\beta}(x^1) \, dx^\alpha dx^\beta, \quad (5.7.2)$$

with the same convention regarding α and β; the matrix $(a_{\alpha\beta})$ is symmetric and positive definite, and will be assumed to be a C^∞ function of x^1. This is easier to handle than (5.7.1). We shall now outline the construction of the fundamental solutions of the d'Alembertian

$$\square u \equiv 2 \partial_1 \partial_2 u - a^{\alpha\beta}(x^1) \partial_\alpha \partial_\beta u, \quad (5.7.3)$$

where the matrix $(a^{\alpha\beta})$ is the inverse of $(a_{\alpha\beta})$.

The differential equations of the geodesics are, in canonical form,

$$\frac{dx^i}{dr} = \frac{1}{2}\frac{\partial H}{\partial \xi_i}, \quad \frac{d\xi_i}{dr} = -\frac{1}{2}\frac{\partial H}{\partial x^i} \quad (i = 1, \ldots, 4),$$

where the Hamiltonian is, by (5.7.3),

$$H = H(x, \xi) = 2\xi_1\xi_2 - a^{\alpha\beta}(x^1)\,\xi_\alpha\xi_\beta. \tag{5.7.4}$$

Hence $$\frac{dx^1}{dr} = \xi_2, \quad \frac{dx^2}{dr} = \xi_1, \quad \frac{dx^\alpha}{dr} = -a^{\alpha\beta}(x^1)\,\xi_\alpha\xi_\beta,$$

$$\frac{d\xi_1}{dr} = \tfrac{1}{2}\xi_\alpha\xi_\beta\frac{da^{\alpha\beta}}{dx^1}, \quad \frac{d\xi_2}{dr} = \frac{d\xi_3}{dr} = \frac{d\xi_4}{dr} = 0,$$

and these equations are to be integrated subject to the initial conditions $x^i(0) = y^i$, $\xi_i(0) = \theta_i$, $i = 1, \ldots, 4$. Obviously, $\xi_i = \theta_i$ for $i = 2, 3, 4$. A simple calculation shows that, if $\theta_2 \neq 0$, then the solutions are

$$x^1 = y^1 + \theta_2 r,$$

$$x^2 = y^2 + \theta_1 r + \frac{1}{2\theta_2}(A^{\alpha\beta}(x^1, y^1) - a^{\alpha\beta}(y^1))\,\theta_\alpha\theta_\beta r,$$

$$x^\alpha = y^\alpha - A^{\alpha\beta}(x', y')\,\theta_\beta r \quad (\alpha = 3, 4),$$

where the $A^{\alpha\beta}$ are defined by

$$A^{\alpha\beta}(x^1, y^1) = \frac{1}{x^1 - y^1}\int_{y^1}^{x^1} a^{\alpha\beta}(t)\,dt \quad (\alpha, \beta = 3, 4). \tag{5.7.5}$$

These equations also hold when $\theta_2 = 0$, provided that the second members are replaced by the limits to which they tend when $\theta_2 \to 0$. The exponential map is obtained by putting $r = 1$, and is

$$x^1 = y^1 + \theta_2,$$

$$x^2 = y^2 + \theta_1 + \frac{1}{2\theta_2}(A^{\alpha\beta}(x^1, y^1) - a^{\alpha\beta}(y^1))\,\theta_\alpha\theta_\beta, \tag{5.7.6}$$

$$x^\alpha = y^\alpha - A^{\alpha\beta}(x^1, y^1)\,\theta_\beta \quad (\alpha = 3, 4),$$

where $x^1 = y^1 + \theta_2$ in the last two equations. Now $(a^{\alpha\beta}(x^1))$ is positive definite, and so there is, for every finite interval $I \subset \mathbf{R}$, a positive constant $M = M(I)$ such that

$$a^{\alpha\beta}(x^1)\,\lambda_\alpha\lambda_\beta \geqslant M(\lambda_3^2 + \lambda_4^2)$$

holds for all (λ_3, λ_4) and $x' \in I$. From this and (5.7.5) it follows that

$$A^{\alpha\beta}(x^1, y^1)\,\lambda_\alpha\lambda_\beta \geqslant M(\lambda_3^2 + \lambda_4^2)$$

for all (λ_3, λ_4) and $x^1 \in I$, $y^1 \in I$. Hence $(A^{\alpha\beta})$ is also symmetric and positive definite, and this implies that $\det A^{\alpha\beta}(x^1, y^1) > 0$. The equations (5.7.6) can therefore be solved for the θ_i, for all pairs of points (x, y), and this implies that the coordinate neighbourhood Ω defined by $x \in \mathbf{R}^4$ is a geodesically convex domain. As $\Gamma = H(y, \theta)$, a simple calculation gives, in view of (5.7.4) and (5.7.6), that

$$\Gamma = 2(x^1 - y^1)(x^2 - y^2) - A_{\alpha\beta}(x^1, y^1)(x^\alpha - y^\alpha)(x^\beta - y^\beta), \quad (5.7.7)$$

where $(A_{\alpha\beta}) = (A^{\alpha\beta})^{-1}$ is again a positive definite symmetric matrix. From this, one can easily conclude that Ω is also a causal domain. It should be pointed out that the coordinate neighbourhood in which (5.7.1) holds is not necessarily geodesically convex.

For the d'Alembertian, U reduces to the biscalar which can be computed by means of (4.2.18),

$$\kappa(x, y) = \left| \frac{g(y)}{g(x)} \right|^{\frac{1}{4}} |\det \partial x^i / \partial \theta^j|^{-\frac{1}{2}},$$

which can obviously be replaced by

$$\kappa(x, y) = \left| \frac{g(y)}{g(x)} \right|^{\frac{1}{4}} \left| \frac{\det \partial x^i / \partial \theta_j \big|_{x=y}}{\det \partial x^i / \partial \theta_j} \right|^{\frac{1}{2}}.$$

Now it can easily be seen from (5.7.6) that

$$\det (\partial x^i / \partial \theta_j) = \det (\partial x^\alpha / \partial \theta_\beta) = \det A^{\alpha\beta}(x^1, y^1)$$

Also, $A^{\alpha\beta}(y^1, y^1) = a^{\alpha\beta}(y^1)$ and $g(x^1) = -\det a_{\alpha\beta}(x^1)$. Hence

$$\kappa(x, y) = \frac{1}{|\det a_{\alpha\beta}(x^1) \det a_{\alpha\beta}(y^1)|^{\frac{1}{4}} |\det A^{\alpha\beta}(x^1, y^1)|^{\frac{1}{2}}}. \quad (5.7.8)$$

Thus $x \to \kappa(x, y)$ is a function of x^1 only, and it follows from (5.7.3) that $\Box \kappa = 0$. Hence the d'Alembertian of the metric (5.7.2) satisfies Huygens' principle.

It is worth mentioning that it follows from this and Theorem 6.4.2 that the forward fundamental solution of $\Box + \zeta^2$, where ζ is a complex constant, is

$$\frac{1}{2\pi} \kappa(x, y) \left(\delta_+(\Gamma) - \frac{\zeta}{2\Gamma_+^{\frac{1}{2}}} J_1(\zeta \Gamma_+^{\frac{1}{2}}) \right).$$

where J_1 denotes the Bessel function of order unity, of the first kind, and $\Gamma_+^{\frac{1}{2}} = \Gamma^{\frac{1}{2}}$ if $x \in D^+(y)$, $\Gamma_+^{\frac{1}{2}} = 0$ otherwise. The backward fundamental solution is obtained by changing $\delta_+(\Gamma)$ to $\delta_-(\Gamma)$ and $\Gamma_+^{\frac{1}{2}}$ to $\Gamma_-^{\frac{1}{2}}$.

The problem of finding an explicit characterization of the wave equations which satisfy Huygens' principle is a difficult one. Some progress has been made by the laborious method of expanding $PU(p, q)$ as a power series in the neighborhood of q. The first term is easy to compute; it is

$$PU|_q = (b - \tfrac{1}{4}\langle a, a \rangle - \tfrac{1}{2}\operatorname{div} a - \tfrac{1}{6}R)|_q,$$

where R is the curvature scalar. As $PU = 0$ for $p \in C(q)$ implies that $PU|_q = 0$ for all q, it follows that the first necessary condition for the validity of Huygens' principle is

$$b - \tfrac{1}{4}\langle a, a \rangle - \tfrac{1}{2}\operatorname{div} a = \tfrac{1}{6}R.$$

This result is – as it should be – invariant under trivial transformations. One can infer this from (4.6.4) and (4.6.10), or check it by direct calculation, using (4.6.3) and (4.6.9). By taking the calculation as far as the fifth condition, and using results obtained in general relativity on the classification of space–times, one can prove the following theorem, for space–times with vanishing Ricci tensor.

Theorem 5.7.2. *The wave equation $\square\, u = 0$ on an empty space–time satisfies Huygens' principle if and only if the space–time is flat or a plane wave space–time which admits a coordinate system in which the metric takes the form* (5.7.1), *with* $c_{11} + c_{22} = 0$.

NOTES

Section 5.1. Past-compact and future-compact sets were introduced by Leray (1952), in globally hyperbolic manifolds. Theorems 5.1.1 and 5.1.3 and Lemma 5.1.2 are adapted from Leray. Theorem 5.1.2 shows that the differential operator P, which is both a continuous map $C^\infty \to C^\infty$ and $\mathscr{D}' \to \mathscr{D}'$, has a unique continuous inverse if it is restricted to functions or distributions with past-compact supports respectively. For intermediate results, the appropriate spaces are Sobolev spaces of functions and distributions, as these are linked naturally with P by integral estimates of energy type. This approach, already implicit in Sobolev (1936), is also due to Leray (1952). It leads to the introduction of functional analysis, which yields 'abstract' direct proofs, based on the Hahn–Banach theorem or on orthogonal projection (Lax (1955), Gårding (1956), and Hörmander (1963), Theorem 9.3.1).

Section 5.2. For the reciprocity theorem in distribution form, see Lichnérowicz (1961).

Sections 5.3 and 5.4. The representation theorems given here are based on the work of M. Riesz (1960), who gave them for flat space–times of even dimension. Other representation theorems will be found in Hadamard (1923, 1932), M. Riesz (1949) and Lichnérowicz (1961).

Section 5.5. See Lichnérowicz (1961).

Section 5.6. Essentially, these results are due to deWitt and Brehme (1960). The difficulty caused by charge conservation was pointed out by Mme Choquet-Bruhat (Bruhat (1964)).

Section 5.7. As indicated in the text, the question of Huygens' principle was raised, and discussed, by Hadamard (1923, 1932) who proved Theorem 5.7.1; see also Courant and Hilbert (1962). The literature is extensive; the reader is referred to the papers by Günther (1952, 1965), McLenaghan (1969) and Wünsch (1970), where further references will be found. Theorem 5.7.2 is due to McLenaghan. The fundamental solution of a Huygens equation has a *lacuna*, as defined by Petrovsky (1945); for a recent treatment of Petrovsky's work, see Atiyah, Bott and Gårding (1970).

6

Wave equations on n-dimensional space-times

The theory developed in the last two chapters applies to four-dimensional space–times only. The corresponding theory for n-dimensional space–times is treated, more briefly, in this chapter. Two- and three-dimensional models are frequently used in applications; also, results valid for $n = 4$ can be obtained by the method of descent from results proved for $n > 4$. There is also the intrinsic advantage that the general theory allows one to separate those features which are peculiar to four-dimensional space–times from the basic properties of the solutions of wave equations on manifolds with a Lorentzian metric.

The general plan of this chapter is to retrace the argument as set out in chapters 4 and 5 for scalar wave equations on a four-dimensional space–time. The first section deals with the ordinary wave equation. There is an important distinction between odd and even n. For example, if $n = 4$ then the forward fundamental solution of the d'Alembertian is $(1/2\pi)\,\delta_+(\Gamma)$, while for $n = 3$ it is $(1/2\pi)\,\Gamma_+^{-\frac{1}{2}}$. To treat both odd and even n on the same footing, the method of analytic continuation, which was introduced by M. Riesz in the theory of hyperbolic equations, is used. It can be applied in general space–times by means of normal coordinates. It turns out that the proofs of the differentiability properties of fundamental solutions, and of the solutions of a wave equation with a C^∞ second member, can then be simplified materially by appealing to an elementary result in distribution theory.

The two basic fundamental solutions of a wave equation defined on a causal domain are derived in section 6.2. It is found that it is possible to take over most of the arguments developed in chapter 4. Section 6.3 corresponds to the material in sections 5.1–5.4. The last section stands apart, as it deals with a special topic, the method of descent.

6·1]   THE ORDINARY WAVE EQUATION   229

6.1   The ordinary wave equation in n dimensions

The n-dimensional wave equation is

$$\square u = \eta^{ij}\partial_i\partial_j u = (\partial_1^2 - \partial_2^2 - \ldots - \partial_n^2)\,u = f, \qquad (6.1.1)$$

where u and f are distributions. It will be assumed that $n > 2$, as the case $n = 2$ is elementary. As the d'Alembertian is invariant under translations, it is sufficient to consider fundamental solutions G such that $\square G = \delta(x)$.

Let
$$\gamma(x) = \Gamma(0,x) = \eta_{ij}x^ix^j \qquad (6.1.2)$$

denote the square of the geodesic distance of a point $x \in \mathbf{R}^n$ from the origin. Following M. Riesz, we first consider the functions

$$\left.\begin{aligned} G_\lambda^+(x) &= C_{n,\lambda}\gamma_+^{\lambda-\frac12 n} = C_{n,\lambda}\exp\left((\lambda - \tfrac12 n)\log\gamma\right) \quad \text{if} \quad x \in D^+(0), \\ &= 0 \quad \text{if} \quad x \notin D^+(0), \end{aligned}\right\} \quad (6.1.3)$$

where $\log\gamma$ is real, and λ is a complex parameter, such that

$$\operatorname{Re}\lambda > \tfrac12 n - 1;$$

then $x \to G_\lambda^+(x)$ is a locally integrable function. The constant $C_{n,\lambda}$ is to be chosen so that

$$C_{n,\lambda}\int_{D^+(0)}\gamma^{\lambda-\frac12 n}e^{-x^1}\,dx = 1. \qquad (6.1.4)$$

To compute it, put

$$x = (s, r\xi_1, \ldots, r\xi_{n-1}),\ \xi \in \mathbf{S}^{n-2},\ dx = r^{n-2}ds\,dr\,d\omega_{n-2}, \qquad (6.1.5)$$

where $d\omega_{n-2}$ is the $(n-2)$-surface element on $\mathbf{S}^{n-2}$. Then

$$\frac{1}{C_{n,\lambda}} = \omega_{n-1}\int_0^\infty\int_0^s e^{-s}(s^2-r^2)^{\lambda-\frac12 n}r^{n-2}\,ds\,dr,$$

where
$$\omega_{n-1} = 2\pi^{\frac12(n-1)}\Big/\left(\frac{n-3}{2}\right)!$$

is the area of $\mathbf{S}^{n-2}$. Now put $r = s\rho^{\frac12}$, and keep s as the second variable of integration; this gives

$$\frac{1}{C_{n,\lambda}} = \tfrac12\omega_{n-1}\int_0^\infty\int_0^1 e^{-s}s^{2\lambda-1}(1-\rho)^{\lambda-\frac12 n}\rho^{\frac12(n-3)}\,ds\,d\rho,$$

whence, substituting the value of ω_{n-1},

$$\frac{1}{C_{n,\lambda}} = \frac{\pi^{\frac12(n-1)}}{\left(\dfrac{n-3}{2}\right)!}(2\lambda-1)!\frac{(\lambda-\frac12 n)!\left(\dfrac{n-3}{2}\right)!}{(\lambda-\frac12)!}.$$

This can be simplified by means of the duplication formula for the gamma function,

$$(2\lambda - 1)! = \pi^{-\frac{1}{2}} 2^{2\lambda-1}(\lambda-1)!\,(\lambda-\tfrac{1}{2})!,$$

and one finds that

$$C_{n,\lambda} = \frac{1}{\pi^{\frac{1}{2}n-1}2^{2\lambda-1}(\lambda-1)!\,(\lambda-\tfrac{1}{2}n)!}. \tag{6.1.6}$$

Lemma 6.1.1. *The distribution $G_\lambda^+(x)\,\mathscr{D}'(\mathbf{R}^n)$, which is equal to the function (6.1.3) when $\mathrm{Re}\,\lambda > \tfrac{1}{2}n - 1$, can be defined for all $\lambda \in \mathbf{C}$ by analytic continuation.*

PROOF. For $\mathrm{Re}\,\lambda > \tfrac{1}{2}n - 1$,

$$(G_\lambda^+, \phi) = C_{n,\lambda} \int_{D^+(0)} \gamma^{\lambda-\frac{1}{2}n}\,\phi(x)\,dx, \quad \phi \in C_0^\infty(\mathbf{R}^n), \tag{6.1.7}$$

and as $\lambda \to C_{n,\lambda}$ is analytic in $\{\lambda;\ \lambda \in \mathbf{C},\ \mathrm{Re}\,\lambda > \tfrac{1}{2}n - 1\}$, G_λ^+ is an analytic function of λ in this half-plane. Now for $\mathrm{Re}\,\lambda > \tfrac{1}{2}n + 1$, one has

$$\Box\,\gamma^{\lambda-\frac{1}{2}n} = \partial_i(2(\lambda-\tfrac{1}{2}n)\,x^i\,\gamma^{\lambda-\frac{1}{2}n-1}) = 4(\lambda-1)\,(\lambda-\tfrac{1}{2}n)\,\gamma^{\lambda-\frac{1}{2}n-1}.$$

It therefore follows from (6.1.6) and (6.1.3) that

$$\Box\,G_\lambda^+ = G_{\lambda-1}^+ \tag{6.1.8}$$

holds, in the classical sense, for $\mathrm{Re}\,\lambda > \tfrac{1}{2}n + 1$. We now define G_λ^+ as $\Box^k G_{\lambda+k}^+$, with a suitable k. As $\Box$ is self-adjoint, this means that we put

$$(G_\lambda^+, \phi) = C_{n,\lambda+k} \int_{D^+(0)} \gamma^{\lambda+k-\frac{1}{2}n}\,\Box^k\phi(x)\,dx, \quad \phi \in C_0^\infty(\mathbf{R}^n), \tag{6.1.9}$$

where the integer k is chosen so that $\mathrm{Re}\,(\lambda + k - \tfrac{1}{2}n) > -1$, to ensure convergence of the integral.

For $\mathrm{Re}\,\mu > \tfrac{1}{2}n + 1$, we have from (6.1.8) that

$$\int (G_\mu^+ \Box \phi - G_{\mu-1}^+ \phi)\,dx = \int (G_\mu^+ \Box \phi - \phi \Box G_\mu^+)\,dx$$

$$= \int \partial_i(\eta^{ij}(G_\mu^+ \partial_j \phi - \phi \partial_j G_\mu^+))\,dx = 0,$$

by the divergence theorem, as $\phi \in C_0^\infty(\mathbf{R}^n)$ and $G_\mu^+ \in C^1(\mathbf{R}^n)$. Repeated application of this identity shows that (6.1.9) can be reduced to (6.1.7), if $\mathrm{Re}\,\lambda > \tfrac{1}{2}n$. As both the second members of (6.1.7) and (6.1.9) are analytic in λ, the second member of (6.1.9) is the analytic continuation of (6.1.7) to the half-plane $\{\lambda;\ \lambda \in \mathbf{C},\ \mathrm{Re}\,\lambda > \tfrac{1}{2}n - k - 1\}$. As k can be taken arbitrarily large, G_λ^+ can thus be extended to an analytic function on $\mathbf{C}$, and the lemma is proved. $\Box$

One can also write (6.1.7) as

$$(G_\lambda^+, \phi) = C_{n,\lambda} \int_0^\infty t^{\lambda - \frac{1}{2}n} \, dt \int_{\Sigma_t} \phi(x) \, \mu_\gamma(x),$$

where
$$\Sigma_t = \{x; \, \gamma(x) = t \geqslant 0, \, x \in J^+(0)\},$$

and μ_γ is a Leray form, such that $d\gamma \wedge \mu_\gamma = dx$. One can take

$$\mu_\gamma = (2x^1)^{-1} \, dx^2 \wedge \ldots \wedge dx^n;$$

introducing s, r, and ξ, as in (6.1.5), it follows that

$$\mu_\gamma|_{\Sigma_t} = \frac{r^{n-2} \, dr \wedge d\omega_{n-2}}{2(r^2 + t)^{\frac{1}{2}}}.$$

Put
$$\overline{\phi}(s, r) = \frac{1}{2} \int \phi(s, r\xi_1, \ldots, r\xi_{n-1}) \, d\omega_{n-1}$$

so that $\overline{\phi}$ is proportional to the spatial spherical mean of ϕ; obviously, $\overline{\phi} \in C_0^\infty(\mathbf{R} \times \overline{\mathbf{R}}^+)$. Then

$$(G_\lambda^+, \phi) = C_{n,\lambda} \int_0^\infty t^{\lambda - \frac{1}{2}n} \, \tilde{\phi}(t) \, dt,$$

where
$$\tilde{\phi}(t) = \int_0^\infty \overline{\phi}((r^2 + t)^{\frac{1}{2}}, r) \, \frac{r^{n-2} \, dr}{(r^2 + t)^{\frac{1}{2}}}. \tag{6.1.10}$$

In the notation introduced in section 2.7, we can therefore write

$$(G_\lambda^+, \phi) = C_{n,\lambda}'\left(\frac{t_+^{\lambda - \frac{1}{2}n}}{(\lambda - \frac{1}{2}n)!}, \, \tilde{\phi}(t)\right), \quad C_{n,\lambda}' = \frac{1}{\pi^{\frac{1}{2}n - 1}(\lambda - 1)! \, 2^{2\lambda - 1}} \tag{6.1.11}$$

if we consider $\tilde{\phi}(t)$ as the restriction to $\overline{\mathbf{R}}^+$ of a function defined on $\mathbf{R}$.

It was shown in section 2.7 that the function $t_+^\mu / \mu!$, which is locally integrable for $\mathrm{Re}\,\mu > -1$, can be extended to a distribution in $\mathscr{D}'(\mathbf{R})$ by analytic continuation. For $\mathrm{Re}\,\mu > -m - 1$, where m is a positive integer, this distribution is given by (2.7.5), which can be written as

$$\left(\frac{t_+^\mu}{\mu!}, \, \psi(t)\right) = \frac{1}{\mu!} \int_0^1 t^\mu \left(\psi(t) - \sum_{k=0}^{m-1} \psi^{(k)}(0) \frac{t^k}{k!}\right) dt$$

$$+ \frac{1}{\mu!} \sum_{k=0}^{m-1} \frac{\psi^{(k)}(0)}{(k + \mu + 1) \, k!} + \frac{1}{\mu!} \int_1^\infty t^\mu \psi(t) \, dt, \tag{6.1.12}$$

where $\psi \in C_0^\infty(\mathbf{R})$. When μ is a negative integer, say $\mu = -k - 1$, $k = 0, 1, \ldots$, then

$$\lim_{\mu \to -k-1} \left(\frac{t_+^\mu}{\mu!}, \, \psi(t)\right) = (-1)^k \psi^{(k)}(0) = (\delta^{(k)}, \psi).$$

Obviously, $(t_+^\mu / \mu!, \psi)$ is determined uniquely by the restriction of ψ to $\overline{\mathbf{R}}^+$.

It is natural to ask whether the same method can be used to continue (6.1.11) analytically into $\operatorname{Re}\lambda \leqslant \tfrac{1}{2}n-1$. When this is possible, then, by the uniqueness of the analytic continuation, one obtains an alternative form of (G_λ^+, ϕ). Some care is needed, however, as $\check{\phi}(t)$ is not C^∞ at $t = 0$. We shall prove the following result.

Lemma 6.1.2. *The distribution G_λ^+ can be computed by means of* (6.1.11)

  (i) *if $\phi \in C_0^\infty(\mathbf{R}^n\backslash(0))$;*
  (ii) *if $\phi \in C_0^\infty(\mathbf{R}^n)$ and $\operatorname{Re}\lambda > 0$.*

PROOF. It is obvious that $\check{\phi}(t) \in C^\infty(\mathbf{R}^+)$, and that its derivatives can be computed by differentiation under the integral sign if $t > 0$,

$$\check{\phi}^{(k)}(t) = \int_0^\infty \left(\frac{1}{2\tau}\frac{\partial}{\partial\tau}\right)^k \frac{\overline{\phi}(\tau, r)}{\tau}\bigg|_{\tau=(r^2+t)^{\frac{1}{2}}} r^{n-2}\, dr$$

$$= \int_0^\infty \left(c_k \frac{\overline{\phi}((r^2+t)^{\frac{1}{2}}, r)}{(r^2+t)^{k+\frac{1}{2}}} = \ldots\right) r^{n-2}\, dr, \qquad (6.1.13)$$

Here, c_k is a constant, and the denominators of the terms omitted are r^2+t raised to the powers $k+\tfrac{1}{2}-\tfrac{1}{2}l$, $1 \leqslant l \leqslant k$. If $\operatorname{supp}\phi \subset \mathbf{R}^n\backslash(0)$, then $\mathbf{\bar{R}}^+ \ni r \to \overline{\phi}(r, r)$ vanishes in a neighbourhood of $r = 0$, and so one can always make $t \downarrow 0$. In this case, $\check{\phi}(t) \in C^\infty(\mathbf{\bar{R}}^+)$, and so (6.1.11) can be continued analytically to the whole complex plane. This proves (i).

But if $(0) \in \operatorname{supp}\phi$, then $\check{\phi}^{(k)} \to \infty$ when $t \downarrow 0$ if $k \geqslant \tfrac{1}{2}(n-2)$, and (6.1.13) is only valid in $\mathbf{\bar{R}}^+$ if $k < \tfrac{1}{2}(n-2)$. Suppose that n is even, and put $n = 2m+2$, where $m \geqslant 1$. Then it follows from (6.1.13) that

$$\check{\phi}^{(m-1)}(t) = c_{m-1}\int_0^\infty \frac{\overline{\phi}((r^2+t)^{\frac{1}{2}}, r)}{(r^2+t)^{m-\frac{1}{2}}} r^{2m}\, dr + \psi(t),$$

where $\psi \in C^1(\mathbf{\bar{R}}^+)$. Hence

$$\check{\phi}^{(m-1)}(t) - \check{\phi}^{(m-1)}(0) = c_{m-1}\int_0^\infty \frac{\overline{\phi}((r^2+t)^{\frac{1}{2}}, r) - \overline{\phi}(r, r)}{(r^2+t)^{m-\frac{1}{2}}} r^{2m}\, dr$$

$$+ c_{m-1}\int_0^\infty \overline{\phi}(r, r)\left(\frac{r^{2m}}{(r^2+t)^{m-\frac{1}{2}}} - r\right) dr + \psi(t) - \psi(0).$$

But it is obvious that there is some constant a such that $\overline{\phi} = 0$ for $(t) > a$ and $r > a$; hence

$$|\check{\phi}^{(m-1)}(t) - \check{\phi}^{(m-1)}(0)| \leqslant K_1 \int_0^a \frac{(r^2+t)^{\frac{1}{2}} - r}{(r^2+t)^{m-\frac{1}{2}}} r^{2m}\, dr$$

$$+ K_2 \int_0^a \left(1 - \frac{r^{2m-1}}{(r^2+t)^{m-\frac{1}{2}}}\right) r\, dr + K_3 t,$$

where K_1, K_2 and K_3 are constants. Put, in each integral, $r = t^{\frac{1}{2}}s$; then

$$|\tilde{\phi}^{(m-1)}(t) - \tilde{\phi}^{(m-1)}(0)| \leqslant K_1 t^{\frac{3}{2}} \int_0^{at^{-\frac{1}{2}}} \frac{(s^2+1)^{\frac{1}{2}} - s}{(s^2+1)^{m-\frac{1}{2}}} s^{2m}\, ds$$

$$+ K_2 t \int_0^{at^{-\frac{1}{2}}} \left(1 - \frac{s^{2m-1}}{(s^2+1)^{m-\frac{1}{2}}}\right) s\, ds + K_3 t.$$

The first integrand remains bounded when $s \to \infty$, and the second one is $O(1/s)$. Hence

$$|\tilde{\phi}^{(m-1)}(t) - \tilde{\phi}^{(m-1)}(0)| \leqslant K_1' t + K_2' t\,|\log t| + K_3 t,$$

when $t \downarrow 0$, where K_1', K_2' are constants, and this implies that

$$-Kt\,|\log t| \leqslant \tilde{\phi}^{(m-1)}(t) - \tilde{\phi}^{(m-1)}(0) \leqslant Kt\,|\log t|,$$

with a new constant K. By repeated integration, one obtains

$$\tilde{\phi}(t) - \sum_{k=0}^{m-1} \tilde{\phi}^{(k)}(0)\frac{t^k}{k!} = O(t^m\,|\log t|).$$

But this implies, by (6.1.12), that $(t_+^\mu/\mu!, \tilde{\phi}(t))$ can be defined by analytic continuation for $\mathrm{Re}\,\mu > -m-1$, and as $\lambda - \frac{1}{2}n = \lambda - m - 1$, it follows that (6.1.11) can be continued analytically into $\mathrm{Re}\,\lambda > 0$.

The proof for odd n is similar. With $n = 2m+1$, $m \geqslant 1$, one obtains, arguing as in the even case, an estimate

$$|\tilde{\phi}^{(m-1)}(t) - \tilde{\phi}^{(m-1)}(0)| \leqslant K_1 t \int_0^{at^{-\frac{1}{2}}} \frac{(s^2+1)^{\frac{1}{2}} - s}{(s^2+1)^{m-\frac{1}{2}}} s^{2m-1}\, ds$$

$$+ K_2 t \int_0^{at^{-\frac{1}{2}}} \left(1 - \frac{s^{2m-1}}{(s^2+1)^{m-\frac{1}{2}}}\right) ds + K_3 t,$$

which implies that

$$\tilde{\phi}^{(m-1)}(t) - \tilde{\phi}^{(m-1)}(0) = O(t^{\frac{1}{2}}),$$

whence

$$\tilde{\phi}(t) - \sum_{k=0}^{m-1} \tilde{\phi}^{(k)}(0)\frac{t^k}{k!} = O(t^{m-\frac{1}{2}}),$$

when $t \downarrow 0$. So, by (6.1.12), $(t_+^\mu/\mu!, \tilde{\phi}(t))$ can be defined by analytic continuation in $\mathrm{Re}\,\mu > -m-\frac{1}{2}$, and as $\lambda - \frac{1}{2}n = \lambda - m - \frac{1}{2}$ one sees again that (6.1.11) can be extended to $\mathrm{Re}\,\lambda > 0$. The proof is therefore complete. $\square$

It is now easy to establish the principal result of this section.

Theorem 6.1.1. *The distribution*

$$G^+(x) = \lim_{\lambda \to 1} G_\lambda^+(x) \qquad (6.1.14)$$

is a fundamental solution of the wave equation in $\mathbf{R}^n$. *If* $n = 2m + 2, m \geqslant 1$,
then

$$G^+(x) = \frac{1}{2\pi^m}\,\delta_+^{(m-1)}(\gamma), \qquad (6.1.15)$$

and if $n = 2m + 1$, $m \geqslant 1$, *then*

$$G^+(x) = \frac{1}{2\pi^{m-\frac{1}{2}}(\frac{1}{2} - m)!}\,\gamma_+^{\frac{1}{2}-m}. \qquad (6.1.16)$$

PROOF. We first show that

$$G_0^+(x) = \lim_{\lambda \to 0} G_\lambda^+(x) = \delta(x). \qquad (6.1.17)$$

If $\operatorname{supp}\phi \subset \mathbf{R}^n\backslash(0)$, then (6.1.11) holds for all λ. The second factor on
the right-hand side is a well-defined distribution for all λ, and $C'_{n,\lambda} \to 0$
when $\lambda \to 0$. Hence $G_0^+(x) = 0$ in $\mathbf{R}^n\backslash(0)$. By Theorem 2.3.3, this
implies that G_0^+ is a finite linear combination of $\delta(x)$ and of its derivatives. (The existence of G_0^+ follows from Lemma 6.1.1.) Now G_λ^+ is,
clearly, a positively homogeneous function of x, of degree $2\lambda - n$, and
$\delta(x)$ is homogeneous of degree $-n$, $\partial^\alpha \delta(x)$ homogeneous of degree
$-n - |\alpha|$. It follows that

$$G_0^+(x) = A\,\delta(x),$$

where A is a constant. Also, by (6.1.4),

$$(G_\lambda^+(x), e^{-x^1}) = 1$$

for $\operatorname{Re}\lambda > \frac{1}{2}n - 1$. Although e^{-x^1} is not a test function, it is evident that
this identity has a meaning, and is preserved, when G_λ^+ is continued
analytically. In particular, it follows that $A = 1$, and so (6.1.17) is
proved.

Again, the identity (6.1.8) holds for all $\lambda \in \mathbf{C}$, by analytic continuation. As differentiation of distributions is a continuous operation, it
follows that

$$\Box\, G^+ = \lim_{\lambda \to 1} \Box\, G_\lambda^+ = G_0^+ = \delta(x).$$

Finally, Lemma 6.1.2 shows that G^+ can be computed by means of
(6.1.11). When n is odd, this gives (6.1.16). When n is even, say
$n = 2m + 2$, $m \geqslant 1$, then

$$\lim_{\lambda \to 1} \frac{t_+^{\lambda - m - 1}}{(\lambda - m - 1)!} = \delta^{(m-1)}(t)$$

and

$$C'_{n,1} = \frac{1}{2\pi^m},$$

and so one obtains (6.1.15). For $n = 4$, (6.1.15) becomes

$$G^+ = (1/2\pi)\,\delta_+(\gamma),$$

as in section 4.1. $\square$

Remark. By reversing the time-orientation, one obtains a second fundamental solution of the wave equation, $G^-(x)$, whose support is contained in $J^-(0)$.

One can assimilate (6.1.15) and (6.1.16), by writing the second member of (6.1.16) as a derivative of fractional order $m - \frac{3}{2} = \frac{1}{2}n - 2$ of $\delta_+(\gamma)$. But the fact remains that there are essential differences between the two cases; in particular, the support of G^+ is $C^+(0)$ when n is even, and $J^+(0)$ when n is odd.

6.2   Fundamental solutions

As in the four-dimensional case, each of the two basic fundamental solutions of a wave equation in an n-dimensional space–time M_n can be derived from a C^∞ parametrix. We shall only discuss scalar wave equations; the extension of the theory to tensor wave equations can be effected by a straightforward generalization of the arguments in section 5.5.

Let $\Omega_0 \subset M_n$ be a causal domain. This means, as before, that there is a geodesically convex domain $\Omega \subset M_n$ such that $\Omega_0 \subset \Omega$, and that, if p and q are points in Ω_0, then $J^+(q) \cap J^-(p)$ is a compact subset of Ω_0 (which may be empty). Let

$$Pu = \square u + \langle a, \nabla u\rangle + bu = f \tag{6.2.1}$$

be a scalar wave equation, defined on Ω_0, the vector field a and the scalar field b being in $C^\infty(\Omega_0)$. We shall construct a parametrix $\tilde{G}_q^+$ such that $P\tilde{G}_q^+ - \delta_q \in C^\infty(\Omega_0)$, and the support of $\tilde{G}_q^+$ is contained in $J^+(q)$. The method that will be used will utilize the properties of the Riesz integral which have just been derived, and will therefore be different from that employed in chapter 4.

The first step is to define a sequence of functions

$$U_\nu(p, q) \in C^\infty(\Omega_0 \times \Omega_0), \quad \nu = 0, 1, \ldots,$$

by means of transport equations which are similar to (4.2.12) and (4.3.11). The first of these equations is

$$2\langle \nabla\Gamma, \nabla U_0\rangle + (\square\,\Gamma + \langle a, \nabla\Gamma\rangle - 2n)\,U_0 = 0, \tag{6.2.2}$$

where $\Gamma = \Gamma(p, q)$, $a = a(p)$, and the differential operators ∇ and $\Box$ act at p. This can be integrated like (4.2.12), in a coordinate system that is normal at q, so that $\nabla^i\Gamma = 2x^i$; then

$$\Box\,\Gamma = |g|^{-\frac{1}{2}}\,\partial_i(2\,|g|^{\frac{1}{2}}\,x^i) = 2n + 2x^i\,\partial_i\log|g|^{\frac{1}{2}}.$$

The essential point is that $\Box\,\Gamma|_q = 2n$, as this ensures that (6.2.2) has a solution, unique up to a constant factor, that remains bounded when $p \to q$. In normal coordinates, the calculation is exactly as in the four-dimensional case, and one finds, with the subsidiary condition $U_0(q, q) = 1$, that

$$U_0(x) = \left|\frac{g(0)}{g(x)}\right|^{\frac{1}{4}} \exp\left(-\frac{1}{2}\int_0^1 a_i(xr)\,x^i\,dr\right),$$

since $x = 0$ at q. In a general local coordinate system, this is replaced by

$$\left.\begin{aligned} U_0(x, y) &= \kappa(x, y)\exp\left(-\frac{1}{2}\int_0^1 \left\langle a(z(r)), \frac{\partial z}{\partial r}\right\rangle dr\right),\\ \kappa(x, y) &= \left|\frac{g(y)}{g(x)}\right|^{\frac{1}{4}}\left|\frac{Dx}{D\theta}\right|^{-\frac{1}{2}}. \end{aligned}\right\} \tag{6.2.3}$$

Here, x and y are the local coordinates of p and q respectively; $[0, 1] \ni r \to z(r)$ is the equation of the geodesic $\widehat{qp}$, r being an affine parameter; and $\theta \to x = h(y, \theta)$ is a coordinate transformation, such that the θ^i are normal coordinates at q. One can also generalize (4.2.19), which becomes

$$\kappa(x, y) = \frac{1}{2^{\frac{1}{2}n}}\,\frac{|\det \partial^2\Gamma/\partial x^i\,\partial y^j|^{\frac{1}{2}}}{|g(x)\,g(y)|^{\frac{1}{4}}}. \tag{6.2.4}$$

For $\nu \geqslant 1$, the U_ν are determined by the transport equations

$$2\langle\nabla\Gamma, \nabla U_\nu\rangle + (\Box\,\Gamma + \langle a, \nabla\Gamma\rangle + 4\nu - 2n)\,U_\nu = -PU_{\nu-1}, \tag{6.2.5}$$

and the condition that each U_ν is to remain bounded when $p \to q$. For $n = 4$, these equations reduce to (4.3.11), if one puts $U_\nu = V_{\nu-1}$, and replaces ν by $\nu + 1$. Allowing for the change of notation, U_ν is again given in terms of $U_{\nu-1}$ by (4.3.13), which becomes

$$U_\nu(x, y) = -\tfrac{1}{4}U_0(x, y)\int_0^1 \left.\frac{PU_{\nu-1}}{U_0}\right|_{x=z(s)} s^{\nu-1}\,ds \quad (\nu = 1, 2, \ldots). \tag{6.2.6}$$

One can now introduce the distributions

$$Z_\lambda = C_{n,\,\lambda}\,\Gamma_+^{\lambda-\frac{1}{2}n} = \begin{cases} C_{n,\,\lambda}\,\Gamma^{\lambda-\frac{1}{2}n}, & p \in D^+(q),\\ 0 & p \notin D^+(q), \end{cases} \tag{6.2.7}$$

where the $C_{n,\lambda}$ are the constants (6.1.6) and, for $p \in D^+(q)$,

$$\Gamma^{\lambda-\frac{1}{2}n} = \exp((\lambda - \tfrac{1}{2}n)\log\Gamma),$$

$\log\Gamma$ being real. For fixed q and $p \in D^+(q)$, $\lambda \to Z_\lambda$ is an analytic function of λ; for fixed λ and $\operatorname{Re}\lambda > \tfrac{1}{2}n - 1$, $p \to Z_\lambda$ is a locally integrable function. As a distribution, acting on test functions $\phi(p) \in C_0^\infty(\Omega_0)$, Z_λ (which is also a function of q) is defined for all $\lambda \in \mathbf{C}$ by analytic continuation. For, in local coordinates that are normal and Minkowskian at q, and such that $(1, 0, ..., 0) \in TM_q$ is future-directed, one has

$$(Z_\lambda, \phi) = (C_{n,\lambda}(\gamma(x)_+)^{\lambda-\frac{1}{2}n}, \phi(x)\,|g(x)|^{\frac{1}{2}})$$

and so the results derived in the last section can be used; note that, as Ω_0 is by hypothesis contained in a geodesically convex domain, such a coordinate chart covers Ω_0. In particular, it follows from (6.1.17) that

$$Z_0 = \lim_{\lambda \to 0} Z_\lambda = \delta_q(p), \qquad (6.2.8)$$

and from (6.1.11) that, for $n = 2m + 2$, $m \geqslant 1$,

$$Z_\nu = \frac{1}{\pi^m 2^{2\nu-1}(\nu-1)!}\delta_+^{(m-\nu)}(\Gamma) \quad (\nu = 1, 2, ..., m). \qquad (6.2.9)$$

It also follows from (6.1.11) that, if $\operatorname{Re}\lambda > 0$ and $\lambda - \tfrac{1}{2}n$ is not a negative integer, then

$$(Z_\lambda, \phi) = C_{n,\lambda}(t_+^{\lambda-\frac{1}{2}n}, \tilde\phi(t)), \qquad (6.2.10)$$

where now

$$\tilde\phi(t) = \int_{\Sigma_t} \phi(p)\,\mu_{\Gamma(p,q)}(p), \qquad (6.2.11)$$

with

$$\Sigma_t = \{p;\, \Gamma(p,q) = t \geqslant 0,\, p \in J^+(q)\}.$$

To construct a parametrix, we first consider the series

$$\tilde G_{q,\lambda}^+(p) = \sum_{\nu=0}^\infty 2^{2\nu}\frac{(\lambda+\nu-1)!}{(\lambda-1)!}U_\nu Z_{\lambda+\nu}\sigma(k_\nu\Gamma)$$

$$= \frac{1}{\pi^{\frac{1}{2}n-1}2^{2\lambda-1}(\lambda-1)!}\sum_{\nu=0}^\infty U_\nu\frac{\Gamma_+^{\lambda+\nu-\frac{1}{2}n}}{(\lambda+\nu-\frac{1}{2}n)!}\sigma(k_\nu\Gamma), \qquad (6.2.12)$$

where $\sigma(t) \in C_0^\infty(\mathbf{R})$ satisfies the hypotheses of Lemma 4.3.2: $0 \leqslant \sigma \leqslant 1$, $\sigma = 1$ for $|t| \leqslant \tfrac{1}{2}$, and $\sigma = 0$ for $|t| \geqslant 1$. Let $A > \tfrac{1}{2}n + 2$ be a positive number, and let N be a positive integer such that $N > A + \tfrac{1}{2}n - 1$. Then the factors $1/(\lambda + \nu - \tfrac{1}{2}n)!$ are, for $\nu \geqslant N$, analytic functions of λ in $|\lambda| \leqslant A$. By an obvious extension of Lemma 4.3.2, there then exists a sequence $\{k_\nu\}_{0 \leqslant \nu < \infty}$, strictly increasing and tending to infinity, such that the series

$$\sum_{\nu=N}^{\infty} \partial^\alpha \left(\frac{U_\nu \Gamma^\nu}{(\lambda + \nu - \frac{1}{2}n)!} \sigma(k_\nu \Gamma) \right)$$

converge, for each multi-index $\alpha \geqslant 0$, uniformly on any compact subset of $\Omega_0 \times \Omega_0$ and $|\lambda| \leqslant A$. The sum will be a function in $C^\infty(\Omega_0 \times \Omega_0)$ that is an analytic function of λ for $|\lambda| \leqslant A$. With this choice of the k_ν, the second member of (6.1.12) is a distribution (acting on $\phi(p) \in C_0^\infty(\Omega_0)$) which is also an analytic function of λ on $|\lambda| \leqslant A$; if also $\operatorname{Re}\lambda > \frac{1}{2}n + 2$, then it is a C^2 function.

Lemma 6.2.1. *The distribution*

$$\tilde{G}_q^+(p) = \lim_{\lambda \to 1} \tilde{G}_{q,\lambda}^+(p) \tag{6.2.13}$$

is a C^∞ parametrix of the differential operator P. If $n = 2m+2$, $m \geqslant 1$, then

$$\tilde{G}_q^+(p) = \frac{1}{2\pi^m} \sum_{\nu=0}^{m-1} U_\nu \delta_+^{(m-\nu-1)}(\Gamma) + \frac{1}{2\pi^m} \sum_{\nu=m}^{\infty} U_\nu \frac{\Gamma_+^{\nu-m}}{(\nu-m)!} \sigma(k_\nu \Gamma), \tag{6.2.14}$$

and if $n = 2m+1$, $m \geqslant 1$, then

$$\tilde{G}_q^+(p) = \frac{1}{2\pi^{m-\frac{1}{2}}} \sum_{\nu=0}^{\infty} U_\nu \frac{\Gamma_+^{\nu+\frac{1}{2}-m}}{(\nu+\frac{1}{2}-m)!} \sigma(k_\nu \Gamma). \tag{6.2.15}$$

PROOF. We first show that, for $|\lambda| \leqslant A$,

$$P\tilde{G}_{q,\lambda}^+ - \tilde{G}_{q,\lambda-1}^+ \in C^\infty(\Omega_0 \times \Omega_0). \tag{6.2.16}$$

It is sufficient to prove this for $\{\lambda; \lambda \in \mathbf{C}, |\lambda| \leqslant A, \operatorname{Re}\lambda > \frac{1}{2}n + 2\}$, as it then holds in $|\lambda| \leqslant A$, by analytic continuation. When $\operatorname{Re}\lambda > \frac{1}{2}n + 2$, then all the terms of the series (6.1.12) are C^2 functions of p, and $P\tilde{G}_{q,\lambda}^+$ can be computed by term-by-term differentiation. Now

$$P\tilde{G}_{q,\lambda}^+ = \frac{1}{\pi^{\frac{1}{2}n-1} 2^{2\lambda-1}(\lambda-1)!} \sum_{\nu=0}^{\infty} \frac{P(U_\nu \Gamma_+^{\lambda+\nu-\frac{1}{2}n})}{(\lambda+\nu-\frac{1}{2}n)!} \sigma(k_\nu \Gamma) + \dots, \tag{6.2.17}$$

where each of the terms omitted contains either a factor $\sigma'(k_\nu \Gamma)$ or $\sigma''(k_\nu \Gamma)$, so that their sum is a function in $C^\infty(\Omega_0 \times \Omega_0)$ whose support is contained in $\{(p,q); p \in D^+(q)\}$. Again, as $\langle \nabla\Gamma, \nabla\Gamma \rangle = 4\Gamma$, and because the U_ν satisfy the transport equations (6.2.2) and (6.2.5), one finds, after a little manipulation, that

$$P(U_\nu \Gamma_+^{\lambda+\nu-\frac{1}{2}n}) = P(U_0) \Gamma_+^{\lambda+\nu-\frac{1}{2}n}$$
$$+ (\lambda+\nu-\tfrac{1}{2}n)(-P(U_{\nu-1}) + 4(\lambda-1) U_\nu) \Gamma_+^{\lambda+\nu-\frac{1}{2}n-1},$$

with the convention that $U_{-1} = 0$. When this is substituted in (6.2.17) one obtains in view of (6.2.12), after some rearrangement of the terms,

$$PG\tilde{}^+_{q,\lambda} = \tilde{G}^+_{q,\lambda-1}$$
$$+ \frac{1}{\pi^{\frac{1}{2}n-1}2^{2\lambda-1}(\lambda-1)!} \sum_0^\infty P(U_\nu) \frac{\Gamma^{\lambda+\nu-\frac{1}{2}n}_+}{(\lambda+\nu-\frac{1}{2}n)!} (\sigma(k_\nu\Gamma) - \sigma(k_{\nu+1}\Gamma)).$$

As $\sigma(t) = 1$ for $|t| \leqslant \frac{1}{2}$, this implies (6.2.16).

Now it follows from (6.2.8) and (6.2.12) that $\tilde{G}^+_{q,0} = \delta_q(p)$, and so $\tilde{G}^+_{q,1}$ is a C^∞ parametrix of P, whose support, for fixed q, is contained in $J^+(q)$. The detailed expressions (6.2.14) and (6.2.15) then follow from (6.2.9) and (6.2.10); the factors $\sigma(k_\nu\Gamma)$ can be omitted in the delta function terms in (6.2.14) because $\sigma(t) = 1$ for $|t| \leqslant \frac{1}{2}$. Thus the lemma is proved. $\square$

The construction of a fundamental solution $G^+_q(p)$ whose support is contained in $J^+(q)$ now proceeds exactly as in the four-dimensional case. In fact, Lemmas 4.4.1 and 4.4.2, and Theorem 4.4.2, obviously also hold in the n-dimensional case, with the necessary modifications. Now

$$P\tilde{G}_q(p) = \delta_q(p) + K(p,q),$$

say, where $K(p,q) \in C^\infty(\Omega_0 \times \Omega_0)$ and supp $K \subset \{(p,q); p \in J^+(q)\}$. Thus if $\phi \in C_0^\infty(\Omega_0)$, then

$$\phi(p) + \int K(p,q)\phi(p)\mu(p) = (\tilde{G}^+_q(p), {}^tP\phi(p)).$$

By Theorem 4.4.2 one can solve this integral equation for ϕ, to obtain

$$\phi(p) = (\tilde{G}^+_q(p), {}^tP\phi(p)) + \int L(p',q)(\tilde{G}^+_{p'}(p), {}^tP\phi(p))\mu(p'),$$

where L is the resolving kernel of K, which is also in $C^\infty(\Omega_0 \times \Omega_0)$ and supported in $\{p,q; p \in J^+(p)\}$, so that it vanishes to all orders when $\Gamma(p,q) \downarrow 0$, $p \in J^+(q)$. The salient point is now that the integral in the second member can be put into the form

$$\int W(p,q){}^tP\phi(p)\mu(p),$$

where $W(p,q) \in C^\infty(\Omega_0 \times \Omega_0)$ and supp $W \subset \{(p,q); p \in J^+(q)\}$. For this gives

$$\phi(p) = (G^+_q, {}^tP\phi), \quad G^+_q(p) = \tilde{G}^+_q(p) + W(p,q).$$

In section 4.5, this step was carried out by an inversion of orders of integration, followed by a detailed examination of the resulting integrals. This would be rather laborious in the general case. However, there is in fact a simpler way, as the required result follows from an obvious modification of Lemma 6.3.2, which will be proved in the next section. We omit the details, and only state the resulting theorem.

Theorem 6.2.1. *In a causal domain Ω_0, the differential operator P has a fundamental solution G_q^+ whose support is contained in $J^+(q)$. If $n = 2m+2$, $m \geqslant 1$, then*

$$G_q^+ = \frac{1}{2\pi^m} \left(\sum_{\nu=0}^{m-1} U_\nu \delta_+^{(m-\nu-1)}(\Gamma) + V^+ \right), \qquad (6.2.18)$$

where $V^+ = V^+(p,q) \in C^\infty(\{p,q; p \in J^+(q)\})$, and $V^+ = 0$ for $p \notin J^+(q)$. If $p \in J^+(q)$ and $\Gamma(p,q) \downarrow 0$, then

$$\Gamma^{-\mu} \left(V^+ - \sum_{\nu=m}^{m+\mu} U_\nu \frac{\Gamma^{\nu-m}}{(\nu-m)!} \right) \to 0 \qquad (\mu = 0, 1, \ldots). \qquad (6.2.19)$$

If $n = 2m+1$, $m \geqslant 1$, then

$$G_q^+ = \frac{1}{2\pi^{m-\frac{1}{2}}} W^+ \Gamma_+^{\frac{1}{2}-m}, \qquad (6.2.20)$$

where $W^+ = W^+(p,q) \in C^\infty(\{(p,q); p \in J^+(q)\})$. If $p \in J^+(q)$ and $\Gamma(p,q) \downarrow 0$, then

$$\Gamma^{-\mu} \left(W^+ - \sum_{\nu=0}^{\mu} \tilde{V}_\nu \frac{\Gamma^\nu}{(\nu+\frac{1}{2}-m)!} \right) \to 0 \qquad (\mu = 0, 1, \ldots). \qquad (6.2.21)$$

In the analytic case, there is a neighbourhood of the diagonal of $\Omega_0 \times \Omega_0$ in which G_q^+ is given by the second member of (6.2.14) or (6.2.15) respectively, with the factors $\sigma(k_\nu, \Gamma)$ omitted.

Remark. By reversing the time-orientation, one obtains a second fundamental solution, G_q^-, whose support is contained in $J^-(q)$, which has similar properties.

As in the four-dimensional case, the fundamental solution G_q^+ for even n splits into a singular part and a regular part. The regular part V^+, considered as a function of p, is again the solution of the characteristic initial value problem

$$PV^+ = 0, \; p \in D^+(q); \quad V^+ = U_0, \; p \in C^+(q). \qquad (6.2.22)$$

The two-dimensional case is also covered by Theorem 6.2.1, as (6.2.18) holds for $m = 0$, if the terms in $\delta_+(\Gamma)$ are omitted. Then G_q^+ reduces to the regular part, which is determined by (6.2.22). (The solution of this problem is unique.) But it is better to treat this case independently, as it is simpler than $n > 2$, and as there are also some essential differences.

In the first place, characteristics, null geodesics and null curves all coincide. In local coordinates, they are the integral curves of the differential equations

$$g_{11}(dx^1)^2 + 2g_{12}dx^1 dx^2 + g_{22}(dx^2)^2 = 0.$$

As the metric is, by hypothesis, indefinite and non-degenerate, every point of a two-dimensional space–time M_2 has a neighbourhood in which there are just two characteristic curves that go through the point. Suppose that $\Omega_0 \subset M_2$ is a connected open set in which these two families of characteristic curves can be introduced as coordinate curves. In such a coordinate system, one must have $g_{11} = g_{22} = 0$ and $g_{12} \neq 0$. The null cone $C(y)$ with vertex $y \in \Omega_0$ consists of the two straight lines $x^1 = y^1$ and $x^2 = y^2$, and $\Omega_0 \backslash C(y)$ consists of four components, each of which is a quadrant bounded by characteristics. In addition to reversing the time-orientation, one can now also interchange x^1 and x^2. So the four components of $\Omega_0 \backslash C(y)$ are on an equal footing; the distinction between the 'interior' and the 'exterior' of the null cone is artificial, and mathematically meaningless. A Cauchy problem with data on a curve S, is well posed if S is met at most once by a characteristic $x^1 = $ const. and by a characteristic $x^2 = $ const. One can say that the wave equation $Pu = f$ is hyperbolic with respect to S, but it does not make sense to speak of space-like and time-like curves.

The reduction of the line element to $2g_{12}dx^1dx^2$ also shows that a two-dimensional space–time is locally conformal to $\mathbf{R}^2$. The general wave equation $Pu = f$ becomes

$$Pu = (g_{12})^{-1}\partial_1\partial_2 u + a^1\partial_1 u + a^2\partial_2 u + bu = f, \qquad (6.2.23)$$

while the equation $\square\, u = f$ always reduces locally to the trivial case $\partial_1\partial_2 u = g_{12}f$. One can multiply (6.2.23) by g_{12}; this is equivalent to the passage from M_2 to a (locally) conformal flat space–time. It is evidently sufficient to consider this case only, which amounts to setting $g_{12} = 1$ in (6.2.23). Then the coordinates are normal, at all points of Ω_0, and

$$\Gamma = 2(x^1 - y^1)(x^2 - y^2).$$

The characteristic initial value problem (6.2.22), which now gives the fundamental solution outright, becomes

$$\left.\begin{array}{l} PV^+ = (\partial_1\partial_2 + a^1\partial_1 + a^2\partial_2 + b)\,V^+ = 0, \ x^1 > y^1, \ x^2 > y^2, \\[4pt] V^+ = U_0, \ x^1 = y^1, \ x^2 \geqslant y^2 \quad \text{and} \quad x^2 = y^2, \ x^1 \geqslant y^1, \end{array}\right\} \qquad (6.2.24)$$

if one assumes that the support of V^+ is in the quadrant

$$\{x; x^1 \geqslant y^1, x^2 \geqslant y^2\}.$$

Also, U_0 is determined by the transport equation of order zero,

$$4((x^1-y^1)\, \partial_1 U_0 + (x^2-y^2)\, \partial_2 U_0) + 2(a^1(x^1-y^1) + a^2(x^2-y^2))\, U_0 = 0,$$

and the condition $U_0(y,y) = 1$. Hence

$$
\left.
\begin{aligned}
\partial_2 U_0 + \tfrac{1}{2}a^2 U_0 = 0, \quad &\text{if} \quad x^1 = y^1, \\
\partial_1 U_0 + \tfrac{1}{2}a^1 U_0 = 0, \quad &\text{if} \quad x^2 = y^2, \\
U_0|_{x=y} = 1.
\end{aligned}
\right\}
\tag{6.2.25}
$$

But these are the boundary conditions for the Riemann function of the differential operator P. Let us denote this by $V = V(x,y)$. Then, by (6.2.18), the function that is equal to $\tfrac{1}{2}V$ in $\{x; x^1 \geqslant y', x^2 \geqslant y^2\}$ and zero elsewhere is a fundamental solution of P. There are three other fundamental solutions of this type, whose supports are the characteristic quadrants determined by y. We can list all four solutions:

$$
\left.
\begin{aligned}
\tfrac{1}{2}VH(x^1-y^1)\,H(x^2-y^2), \quad -\tfrac{1}{2}VH(y^1-x^1)\,H(x^2-y^2), \\
\tfrac{1}{2}VH(y^1-x^1)\,H(y^2-x^2), \quad -\tfrac{1}{2}VH(x^1-y^1)\,H(y^2-x^2).
\end{aligned}
\right\}
\tag{6.2.26}
$$

Note the change of sign in the second and fourth quadrants.

It is not proposed to carry the discussion of wave equations for two-dimensional space–times any further, as they can be treated, locally, by Riemann's method, which is well known. It must be emphasized, however, that Riemann's method (like the theory for $n > 2$ with which this book is concerned) is a local method, and that difficulties may arise in global problems. It should also be pointed out that, because of the special character of the two-dimensional case, two-dimensional models of space–time must be treated with caution, as they may lack some of the essential causal features of space–times of higher dimension.

6.3. Existence, uniqueness and representation theorems

We can now establish an existence theorem similar to Theorem 5.1.2. Although the argument is basically the same as before, it will be presented in a different form, and some preliminary definitions and lemmas are needed.

A linear map $\phi \to \tau_q(\phi)$ of $C_0^\infty(\Omega_0)$ into $C^\infty(\Omega_0)$ is called a continuous linear map $C_0^\infty(\Omega_0) \to C^\infty(\Omega_0)$ if the following condition is satisfied. Let

(Ω, π) be a coordinate chart, such that $\Omega_0 \subset \Omega$. (Such charts exist, as Ω_0 is supposed to be contained in a geodesically convex domain.) Then, for all pairs K, K' of compact subsets of Ω_0, there exist, for $N = 0, 1, \ldots$, constants $C_N > 0$ and non-negative integers k_N, such that

$$\sum_{|\alpha| \leqslant N} \sup_{\pi^{-1}y \in K'} |\partial_y^\alpha \tau_q(\phi) \circ \pi^{-1}(y)| \leqslant C_N \sum_{|\beta| \leqslant k_N} \sup |\partial_x^\beta \phi \circ \pi^{-1}(x)|, \quad (6.3.1)$$

for all $\phi \in C_0^\infty(\Omega_0)$ whose supports are contained in K.

It is evident that if (6.3.1) holds, and $(\tilde{\Omega}, \tilde{\pi})$ is another coordinate chart covering Ω_0, then a similar set of estimates holds in $(\tilde{\Omega}, \tilde{\pi})$, with different constants C_N, and the same k_N. The condition means that $C_0^\infty(\Omega_0)$, equipped with the usual inductive limit topology, is mapped continuously by τ_q into $C^\infty(\Omega_0)$, which becomes a Fréchet space when it is given the topology generated by the semi-norms on the right-hand side of (6.3.1). (The map τ_q is also called a vector valued distribution, with values in $C^\infty(\Omega_0)$.) For our purposes, the most important consequence of the continuity of τ_q is the following one.

Lemma 6.3.1. *Suppose that $\tau_q: C_0^\infty(\Omega_0) \to C^\infty(\Omega_0)$ is a continuous linear map, and that $f \in \mathscr{E}'(\Omega_0)$. Then $\phi \to (f(q), \tau_q(\phi))$ is a distribution.*

PROOF. It will be recalled that $\mathscr{E}'(\Omega_0)$ is the class of distributions with compact support, on Ω_0. By Definition 2.8.1 and Theorem 2.3.1, $\phi \to (f, \phi)$ is a continuous linear map $C^\infty(\Omega_0) \to \mathbf{C}$. Hence the composite map $f \circ \tau_q$: $\phi \to (f(q), \tau_q(\phi))$ is a continuous linear map $C_0^\infty(\Omega_0) \to \mathbf{C}$, which means that it is a distribution. □

Suppose now that $T(p, q) \in C^\infty(\Omega_0 \times \Omega_0)$, and put

$$\tau_\lambda(p, q) = T \frac{\Gamma_+^{\lambda - \frac{1}{2}n}}{(\lambda - \frac{1}{2}n)!} = T(p, q) \frac{(\Gamma(p, q))^{\lambda - \frac{1}{2}n}}{(\lambda - \frac{1}{2}n)!} \quad \text{when} \quad p \in D^+(q), \left.\begin{array}{c} \\ \\ \end{array}\right\} \quad (6.3.2)$$
$$= 0 \qquad\qquad\qquad \text{when} \quad p \notin D^+(q).$$

For $\mathrm{Re}\, \lambda > \frac{1}{2}n - 1$, one can either identify τ_λ with the distribution

$$(\tau_\lambda(p, q), \phi(p))$$
$$= \frac{1}{(\lambda - \frac{1}{2}n)!} \int_{D^+(q)} T(p, q) (\Gamma(p, q))^{\lambda - \frac{1}{2}n} \phi(p) \mu(p), \quad \phi \in C_0^\infty(\Omega_0), \quad (6.3.3)$$

which is a function of q, or with the distribution

$$(\tau_\lambda(p, q), \psi(p))$$
$$= \frac{1}{(\lambda - \frac{1}{2}n)!} \int_{D^-(q)} T(p, q) (\Gamma(p, q))^{\lambda - \frac{1}{2}n} \psi(q) \mu(q), \quad \psi \in C_0^\infty(\Omega_0), \quad (6.3.4)$$

which is a function of p. By introducing local coordinates that are normal and Minkowskian at q or at p respectively, and appealing to Lemma 6.1.1, one sees at once that both these distributions, which are analytic functions of λ for Re $\lambda > \frac{1}{2}n - 1$, can also be defined by analytic continuation for $\lambda \in \mathbf{C}$, $\lambda \neq 0, -1, \dots$. (These points are simple poles.) In the notation of the last section, (6.3.3) is

$$(\tau_\lambda(p,q), \phi(p)) = \frac{1}{(\lambda - \frac{1}{2}n)!}(T\Gamma_+^{\lambda - \frac{1}{2}n}, \phi)$$

and (6.3.4) is, with p and q interchanged,

$$(\tau_\lambda(q,p), \psi(p)) = \frac{1}{(\lambda - \frac{1}{2}n)!}(T^*\Gamma_+^{\lambda - \frac{1}{2}n}, \psi),$$

where $T^* = T(q,p)$. However, we shall now drop this notation, and use the first members of (6.3.3) and (6.3.4) instead, for

$$\lambda \in \mathbf{C}, \quad \lambda \neq 0, -1, \dots.$$

Lemma 6.3.2. *Both $\phi \to (\tau_\lambda(p,q), \phi(p))$ and $\psi \to (\tau_\lambda(p,q), \psi(q))$ are continuous linear maps $C_0^\infty(\Omega_0) \to C^\infty(\Omega_0)$. Furthermore, the identity*

$$\int(\tau_\lambda(p,q), \phi(p))\,\psi(q)\,\mu(q) = \int(\tau_\lambda(p,q), \psi(q))\,\phi(p)\,\mu(p) \quad (6.3.5)$$

holds for all $\phi \in C_0^\infty(\Omega_0)$, $\psi \in C_0^\infty(\Omega_0)$.

PROOF. Let (Ω, π) be a coordinate chart covering Ω_0; put

$$\pi p = x, \quad \pi q = y,$$

and write $T(x,y)$ for $T_0(\pi \times \pi)^{-1}(x,y)$, $\phi(x)$ for $\phi \circ \pi^{-1}(x)$. As Ω_0 is contained in a geodesically convex domain, there is a coordinate transformation $\tilde{x} \to x = h(y, \tilde{x})$ such that the $\tilde{x}^i$ are local coordinates admissible in Ω_0 which are normal and Minkowskian at q; one may also suppose that $(1, 0, \dots, 0) \in TM_q$ is future-directed. Note that h is a C^∞ function of y and $\tilde{x}$. Now

$$(\tau_\lambda(p,q), \phi(p)) \circ \pi^{-1}(y) = \left(\frac{\tilde{\gamma}_+^{\lambda - \frac{1}{2}n}}{(\lambda - \frac{1}{2}n)!}, T(h(y,\tilde{x}), y)\,k(y,\tilde{x})\,\phi(h(y,\tilde{x}))\right),$$

$$(6.3.6)$$

where $\tilde{\gamma} = \eta_{ij}\tilde{x}^i\tilde{x}^j$, and $k(y, \tilde{x})$ is just $|g|^{\frac{1}{2}}$, in the local coordinates $\tilde{x}^i$. By Lemma 6.1.1, $\tilde{\gamma}_+^{\lambda - \frac{1}{2}n}/(\lambda - \frac{1}{2}n)!$ is defined, as a distribution in $\mathscr{D}'(\mathbf{R}^n)$, for $\lambda \in \mathbf{C}$, $\lambda \neq 0, -1, \dots$. As in the proof of Lemma 2.4.1, one can therefore show that the second member of (6.3.6) is a C^∞ function of y, and that its derivatives with respect to y can be computed by differentiating with respect to y inside the duality brackets. But then the usual semi-norm estimates for the distribution $\tilde{\gamma}_+^{\lambda-\frac{1}{2}n}/(\lambda-\frac{1}{2}n)! \in \mathscr{D}'(\mathbf{R}^n)$ imply the requisite semi-norm estimates for the map $\phi \to (\tau_\lambda(p,q), \phi(p))$. The continuity of the map $\psi \to (\tau_\lambda(p,q), \psi(q))$ is proved by a similar argument.

Again, for $\mathrm{Re}\,\lambda > \frac{1}{2}n - 1$, it follows from (6.3.3), (6.2.4), and Fubini's theorem, that

$$\int (\tau_\lambda(p,q), \phi(p))\,\psi(q)\,\mu(q) = \int (\tau_\lambda(p,q), \psi(q))\,\phi(p)\,\mu(p)$$

$$= \frac{1}{(\lambda-\frac{1}{2}n)!}\iint_{p\in D^+(q)} T(p,q)\,(\Gamma(p,q))^{\lambda-\frac{1}{2}n}\,\phi(p)\,\psi(q)\,\mu(p)\,\mu(q).$$

$$(6.3.7)$$

As all the members of this identity are analytic functions of λ, it follows by analytic continuation that (6.3.5) holds for

$$\lambda \in \mathbf{C}, \quad \lambda \neq 0, -1, \dots .$$

This completes the proof of the lemma. $\square$

Remark. One can also consider the maps $\phi \to (\tau_\lambda(p,q), \phi(p))$ and $\psi \to (\tau_\lambda(p,q), \psi(q))$ as linear maps $C_0^\infty(\Omega_0) \to \mathscr{D}'(\Omega_0)$; both are continuous, if $\mathscr{D}'(\Omega_0)$ is given the usual weak topology. The identity (6.3.5) then means that these two maps are adjoints of each other. The common value of both sides of (6.3.5) is a bilinear form on $C_0^\infty(\Omega_0)$, continuous if either ϕ or ψ is kept fixed, or if the supports of ϕ and ψ are in fixed compact sets. This bilinear form has a unique extension to a continuous linear form (τ_λ, χ) on the space $C_0^\infty(\Omega_0 \times \Omega_0)$ of C^∞ biscalar fields on Ω_0, with compact support. This is called a *kernel distribution*. Obviously

$$(\tau_\lambda(p,q), \chi(p,q))$$

$$= \frac{1}{(\lambda-\frac{1}{2}n)!}\iint_{p\in D^+(q)} \tau(p,q)\,(\Gamma(p,q))^{\lambda-\frac{1}{2}n}\,\chi(p,q)\,\mu(p)\,\mu(q),$$

for $\chi \in C_0^\infty(\Omega_0 \times \Omega_0)$ and $\mathrm{Re}\,\lambda > \frac{1}{2}n - 1$, and one can again use analytic continuation to extend this kernel distribution to

$$\{\lambda;\, \lambda \in \mathbf{C}, \lambda \neq 0, -1, \dots\}.$$

One then has, for ϕ, ψ in $C_0^\infty(\Omega_0)$,

$$(\tau_\lambda(p,q), \phi(p)\,\psi(q)) = ((\tau_\lambda(p,q), \phi(p)), \psi(q)) = ((\tau_\lambda(p,q), \psi(q)), \phi(p)).$$

This is an example of a general scheme, which associates, with a given kernel distribution K, a bilinear form $(K(p,q), \phi(p)\,\psi(q))$. For fixed ϕ, this is a distribution, so that one obtains a continuous linear map $\phi \to (K(p,q), \phi(p))$ of $C_0^\infty(\Omega_0)$ into $\mathscr{D}'(\Omega_0)$; for fixed ψ, one obtains another continuous linear map $\psi \to (K(p,q), \psi(q))$, and these two maps are adjoints of each other. In the particular case of $K = \tau_\lambda$, both of these maps are also continuous maps $C_0^\infty(\Omega_0) \to C^\infty(\Omega_0)$. Such a kernel is called regular.

The fundamental solution $G_q^+(p)$ constructed in the last section is the sum of a finite number of terms of the type we have just considered. In conformity with the notation introduced in the present section, we shall, from now on, write $G^+(p,q)$ for $G_q^+(p)$; this can also be identified with a kernel distribution. For odd n, we therefore write (6.2.20) as

$$G^+(p,q) = \frac{1}{2\pi^{m-\frac{1}{2}}}\, W^+(p,q)\,(\Gamma_+(p,q))^{\frac{1}{2}-m}, \qquad (6.3.8)$$

where $m = \frac{1}{2}(n-1)$, and this is already of the form (6.3.3), provided that W^+ is understood to be a C^∞ extension of the corresponding function in (6.2.20) to $\Omega_0 \times \Omega_0$. (Such an extension is obtained, for example, by putting

$$W^+ = \sum_{\nu=0}^\infty U_\nu \frac{\Gamma^\nu}{(\nu+\frac{1}{2}-m)!}\, \sigma(k_\nu\,\Gamma), \quad p \notin D^+(q),$$

where $\sigma(t)$ and the k_ν are as in Lemma 6.2.1.) For even n, we write (6.2.18) as

$$G^+(p,q) = \left(\frac{1}{2\pi^m}\sum_{\nu=0}^{m-1} U_\nu \delta_+^{(m-1-\nu)}(\Gamma) + V^+ H_+(\Gamma)\right), \qquad (6.3.9)$$

where $m = \frac{1}{2}(n-2) \geqslant 1$ and V^+ is a C^∞ extension of the corresponding function in the second member of (6.2.18) to $\Omega_0 \times \Omega_0$. Each term is then a distribution of the form (6.3.3), with $\lambda = 1, 2, \ldots, m+1$. In the new notation, (G_q^+, ϕ) is denoted by $(G^+(p,q), \phi(p))$. One can now evidently also define $(G^+(p,q), \psi(q))$ for $\psi \in C_0^\infty(\Omega_0)$, and it follows from (6.3.5) that

$$\int (G^+(p,q), \phi(p))\,\psi(q)\,\mu(q) = \int (G^+(p,q), \psi(q))\,\phi(p)\,\mu(p) \quad (6.3.10)$$

for all ϕ and ψ in $C_0^\infty(\Omega_0)$.

If Γ_+ is replaced by Γ_- in (6.3.2), then one obtains, instead of (6.3.3) and (6.3.4), two distributions $(\tau_\lambda^-(p,q), \phi(p))$ and $(\tau_\lambda^-(p,q), \psi(q))$ which are, for $\operatorname{Re} \lambda > \frac{1}{2}n - 1$, equal to the second members of these equations, with $D^+(q)$ replaced by $D^-(q)$, and $D^-(q)$ replaced by $D^+(q)$ respectively.

For other $\lambda \in C$, $\lambda \neq 0, -1, \ldots$, they are defined by analytic continuation. As

$$\tau_{\overline{\lambda}}(p,q) = \tau_{\lambda}^*(q,p),$$

where T is replaced by $T^* = T(q,p)$, Lemma 6.3.2 also applies to $\tau_{\overline{\lambda}}(p,q)$. The second fundamental solution $G_q^-(p)$, which from now on will be denoted by $G^-(p,q)$, is a finite sum of distributions of this type. For $n = 2m+1$, $m \geqslant 1$, it is of the form

$$G^-(p,q) = \frac{1}{2\pi^{m-\frac{1}{2}}} W^-(p,q) (\Gamma_-(p,q))^{\frac{1}{2}-m}, \tag{6.3.11}$$

and for $n = 2m+2$, $m \geqslant 1$ it is of the form

$$G^-(p,q) = \frac{1}{2\pi^m} \left(\sum_{\nu=0}^{m-1} U_\nu \delta_-^{(m-\nu-1)}(\Gamma) + V^- H_-(\Gamma) \right). \tag{6.3.12}$$

One can then define $(G^-(p,q), \psi(q))$, and note that

$$\int (G^-(p,q), \phi(p)) \psi(p) \mu(q) = \int (G^-(p,q), \psi(q)) \phi(p) \mu(p), \tag{6.3.13}$$

for all ϕ and ψ in $C_0^\infty(\Omega_0)$. It is now easy to deal with the wave equation $Pu = f$, when f has compact support.

Lemma 6.3.3. *Suppose that $f \in \mathscr{E}'(\Omega_0)$. Then both*

$$(u^+, \phi) = (f(q), (G^+(p,q), \phi(q))), \quad (u^-, \phi) = (f(q), (G^-(p,q), \phi(p))) \tag{6.3.14}$$

are distributions that satisfy $Pu = f$. If f is a C^∞ function, so that $f \in C_0^\infty(\Omega_0)$, then

$$u^+(p) = (G^+(p,q), f(q)), \quad u^-(p) = (G^-(p,q), f(q)). \tag{6.3.15}$$

PROOF. It follows from Lemma 6.3.2 and Lemma 6.3.1 that u^+ and u^- are distributions, and as G^+ and G^- are fundamental solutions of P, both satisfy $Pu = f$. If $f \in C_0^\infty(\Omega_0)$, then

$$(u^+, \phi) = \int (G^+(p,q), \phi(p)) f(q) \mu(q) = \int (G^+(p,q), f(q)) \phi(p) \mu(p),$$

by (6.3.10), and as $p \to (G^+(p,q), f(q))$ is in $C^\infty(\Omega_0)$, this implies the first equation (6.3.15); the other one follows, similar, from (6.3.13). So the lemma is proved. □

The final step is the introduction of past-compact and future-compact sets in Ω_0, as in the four-dimensional case (Definition 5.1.1). It is evident that the properties of these sets, which were derived in section 5.1, and in particular Theorem 5.1.1 and Lemma 5.1.2, generalize to n dimensions. One can therefore repeat the proof of the basic existence theorem (Theorem 5.1.2). To state the result, we recall that $\mathscr{D}^+(\Omega_0)$ and $\mathscr{D}^-(\Omega_0)$ are the classes of C^∞ functions on Ω_0 with past-compact and future-compact support respectively; likewise, $\mathscr{D}'^+(\Omega_0)$ and $\mathscr{D}'^-(\Omega_0)$ are the spaces of distributions on Ω_0 with past-compact and future-compact support respectively.

Theorem 6.3.1. *If* $f \in \mathscr{D}'^+(\Omega_0)$, *then* $Pu = f$ *has one and only one solution* $u \in \mathscr{D}'^+(\Omega_0)$, *given by*

$$(u, \phi) = (f(q), (G^+(p,q), \phi(p))), \quad \phi \in C_0^\infty(\Omega_0); \qquad (6.3.16)$$

moreover, $\operatorname{supp} u \subset J^+(\operatorname{supp} f)$. *If* $f \in \mathscr{D}^+(\Omega_0)$, *then this solution is the* C^∞ *function*

$$u(p) = (G^+(p,q), f(q)), \qquad (6.3.17)$$

which is also in $\mathscr{D}^+(\Omega_0)$. *There are similar results when* $f \in \mathscr{D}'^-(\Omega_0)$ *or* $\mathscr{D}^-(\Omega_0)$, *with* G^+ *in* (6.3.16) *and* (6.3.17) *replaced by* $G^-(p,q)$ *respectively.*

PROOF. This is exactly like the proof of Theorem 5.1.2, and can therefore be omitted. Note, however, that (6.3.17) is now an immediate consequence of the second part of Lemma 6.3.3, which replaces the computations in part (ii) of the proof of Theorem 5.1.2.

By taking $f = \delta_q$, we also obtain:

Corollary 6.3.1. *The fundamental solutions* $G^+(p,q)$ *and* $G^-(p,q)$ *are the only fundamental solutions of* P *whose supports are contained in* $J^+(q)$ *and in* $J^-(q)$ *respectively.* □

The reciprocity theorem is also proved as in the four-dimensional case. Introducing the bilinear form $(G^+(p,q), \phi(p)\psi(q))$ which is the common value of the two members of (6.3.10), and the analogous bilinear forms $(G^-, \phi\psi)$, $({}^tG^+, \phi, \psi)$ and $({}^tG^-, \phi\psi)$, the last two of which refer to the adjoint differential operator ${}^tP$ one can state this theorem as:

Theorem 6.3.2. *The following relations hold between the fundamental solutions of* P *and those of its adjoint* ${}^tP$:

$$\left.\begin{aligned}({}^tG^+(p,q), \phi(p)\psi(q)) &= (G^-(p,q), \psi(p)\phi(q)), \\ ({}^tG^-(p,q), \phi(p)\psi(q)) &= (G^+(p,q), \psi(p)\phi(q)).\end{aligned}\right\} \qquad (6.3.18)$$

PROOF. This is exactly like the proof of Theorem 5.2.1. □

If one substitutes in (6.3.18) the detailed expressions (6.3.8) and (6.3.11), and the corresponding forms of $^tG^+$ and $^tG^-$, one can see that the reciprocity theorem implies that

$$
\left.
\begin{aligned}
{}^tW^+(p,q) &= W^-(q,p),\, {}^tW^-(p,q) = W^+(q,p),\, n \text{ odd,} \\
{}^tV^+(p,q) &= V^-(q,p),\, {}^tV^-(p,q) = V^+(q,p),\, n \text{ even,}
\end{aligned}
\right\} \quad (6.3.19)
$$

for $p \in J^+(q)$ and $p \in J^-(q)$ respectively. It will be shown in the next section that the coefficients U_ν and $^tU_\nu$ of the singular terms in the fundamental solutions in the even-dimensional case satisfy reciprocity relations valid throughout $\Omega_0 \times \Omega_0$.

It remains to derive explicit forms of the representations (6.3.17). As all the constituents of this are distributions of the type (6.3.3) with $\operatorname{Re} \lambda > 0$, it follows from (6.3.6) that they can be computed by means of Lemma 6.1.2. When n is odd, the result is, in general, the finite part of a divergent integral,

$$
u(p) = \frac{1}{2\pi^{m-\frac12}} Pf \int_{J^-(p)} W^+(p,q) f(q) \, (\Gamma(p,q))^{\frac12 - m} \mu(q). \quad (6.3.20)
$$

For $m = 1$ $(n = 3)$ this is actually an ordinary improper integral. When $m > 1$, then (6.3.20) means

$$
u(p) = \frac{1}{2\pi^{m-\frac12}} \lim_{\epsilon \downarrow 0} \left(\int_{\Sigma_\epsilon} W^+ f \Gamma^{\frac12 - m} \mu(q) - \sum_{\nu=1}^{m-1} A_\nu \epsilon^{\frac12 - m + \nu} \right), \quad (6.3.21)
$$

where $\Sigma_\epsilon = \{q;\, q \in D^-(p),\, \Gamma(p,q) = \epsilon > 0\}$, and the A_ν are constants chosen so that the limit exists; this rule determines both the A_ν and the limit uniquely.

When n is even, $n = 2m + 2$, $m \geqslant 1$, (6.3.17) is

$$
u(p) = \frac{1}{2\pi^m} \sum_{\nu=0}^{m-1} \left(-\frac{\partial}{\partial\epsilon} \right)^{m-\nu-1} \int_{\Sigma_\epsilon} U_\nu(p,q) f(q) \, \mu_{\Gamma(p,q)}(q) \big|_{\epsilon=0+}
$$

$$
+ \frac{1}{2\pi^m} \int_{J^-(p)} V^+(p,q) f(q) \mu(q), \quad (6.3.22)
$$

which reduces to (5.1.13) for $m = 1$.

Finally, we shall give a brief account of the representation theorems for the Cauchy problem. Suppose, as in section 5.3, that S is a past-compact space-like hypersurface, and that $\partial J^+(S) = S$. As the proof of Theorem 5.3.2 evidently carries over to the n-dimensional case, the existence of a unique C^∞ solution of the Cauchy problem relative to S, with C^∞ Cauchy data, is assured in $J^+(S)$.

When n is odd, one can argue as follows. Any $u \in C^\infty(\Omega_0)$ can be split up as
$$u = u_1 + u_2,$$

$$u_1 \in C^\infty(\Omega_0), \operatorname{supp} u \subset J^+(S) \backslash S; \quad u_2 \in C^\infty(\Omega_0), p \notin \operatorname{supp} u_2. \quad (6.3.23)$$

Note that this implies that $u_1(p) = u(p)$, and that $u_2(q) = u(q)$ in a neighbourhood of S. Clearly, $u_1 \in \mathscr{D}^+(\Omega_0)$, and so it follows from (6.3.17), with $f = Pu$, that

$$u(p) = u_1(p) = Pf \int G^+(p,q) \, Pu_1(q) \, \mu(q). \quad (6.3.24)$$

Again, $q \to G^+(p,q)$ is C^∞ when restricted to $D^-(p)$, and, by the reciprocity theorem, $q \to G^+(p,q)$ satisfies

$$^tPG^+ = 0,$$

when $q \in D^-(p)$. For $q \in D^-(p)$ one therefore has the identity

$$G^+(p,q) \, Pu_2(q)$$
$$= \nabla_i(G^+(p,q) \, \nabla^i u_2(q) - u_2(q) \, \nabla^i G^+(p,q) + a^i(q) \, u_2(q) \, G^+(p,q));$$

note that $u_2 = 0$ in a neighbourhood of p. Let Σ_ϵ again denote the pseudo-sphere $\{q; \, q \in D^-(p), \, \Gamma(p,q) = \epsilon > 0\}$, and let

$$D_\epsilon = \{q; \, q \in D^-(p), q \in J^+(S), \, \Gamma(p,q) \geqslant \epsilon\}$$

be the closure of the domain bounded by S and Σ_ϵ. One can then integrate the above identity over D_ϵ, and deduce, by the divergence theorem, that

$$\int_{D_\epsilon} G^+(p,q) \, Pu_2(q) \, \mu(q) = -\int_{S \cap D_\epsilon} *v - \int_{\Sigma_\epsilon \cap D_\epsilon} *v,$$

where
$$v = G^+(p,q) \, \nabla u_2(q) - u_2(q) \, \nabla G^+(p,q) + a(q) \, \mu_2(q) \, G^+(p,q); \quad (6.3.25)$$

∇ acts at q, and both S and Σ_ϵ have the orientations induced by the normal drawn into D_ϵ. The integral over Σ_ϵ is a fractional infinity. Hence

$$Pf \int G^+(p,q) \, Pu_2(q) \, \mu(q) + Pf \int_{S \cap D_0} v = \sum_{\nu=0}^{m-1} B_\nu \epsilon^{\frac{1}{2}-m+\nu} + O(\epsilon^{\frac{1}{2}}),$$

where the B_ν are constants. Multiplying this by $\epsilon^{m-\frac{1}{2}}$ and making $\epsilon \downarrow 0$, one concludes that $B_0 = 0$; repeating this argument with $\epsilon^{m-\frac{3}{2}}, ..., \epsilon^{\frac{1}{2}}$, in turn, one finds that all the B_ν vanish. Hence one can let $\epsilon \downarrow 0$ to conclude that

$$0 = Pf \int_{D_0} G^+(p,q) \, Pu_2(q) \, \mu(q) + Pf \int_{S \cap D^-(p)} *v.$$

One can now add this to (6.3.24). Taking (6.2.23) and (6.2.25) into account, and noting that $u_2 = u$ in a neighbourhood of S, one obtains the classical representation due to Hadamard,

$$u(p) = Pf\int_{J^+(S)\cap D^-(p)} G^+(p,q)\,Pu(q)\,\mu(q)$$

$$+ Pf\int_{S\cap D^-(p)} *\,(G^+(p,q)\,\nabla u(q) - u(q)\,\nabla G^+(p,q) + a(q)\,u(q)\,G^+(p,q)),$$
$$(6.3.26)$$

where S has the orientation induced by the normal drawn into $J^+(S)$, which is future-directed.

When n is even, $n = 2m+2$, $m \geqslant 1$, one can repeat the manipulations that gave, in the four-dimensional case, the representation (5.3.8). We omit the details, which are similar to the proof of Theorem 5.3.3. One finds that

$$u(p) = u^{(1)}(p) + u^{(2)}(p) + u^{(3)}(p). \qquad (6.3.27)$$

Here, $u^{(1)}(p)$ is just (6.3.22), with f replaced by Pu. The term $u^{(2)}$ is an integral over $S_p = S \cap D^-(p)$,

$$u^{(2)}(p) = \frac{1}{2\pi^m}\int_{S_p} *\,(V^+\nabla u - u\nabla V^+ + aV^+u), \qquad (6.3.28)$$

where $V^+ = V^+(p,q)$. The third term is an integral over the intersection σ_p of $C^-(p)$ and S, and will be stated in a form corresponding to (5.3.16). Let $t(q)$ be a null field, such that $t(q) > 0$ in $J^+(S)\backslash S$, and that $T_p = \{q;\, t(q) = 0\}$ meets S in σ_p; let $\sigma_{p,\epsilon}$ denote the intersection of the pseudo-sphere Σ_ϵ and of T_p (ϵ need not now be positive). Then

$$u^{(3)}(p) = \frac{1}{2\pi^m}\sum_{\nu=0}^{m-2}\left(-\frac{\partial}{\partial\epsilon}\right)^{m-\nu-1}$$

$$\times \int_{\sigma_{p,\epsilon}} U_\nu(2\langle\nabla t,\nabla u\rangle + (\Box t + \langle q,\nabla t\rangle)\,u)\,\mu_{t,\Gamma}|_{\epsilon=0}$$

$$+ \frac{1}{2\pi^m}\int_{\sigma_p} (U_{m-1}(2\langle\nabla t,\nabla u\rangle + \Box t + \langle a,\nabla t\rangle u)$$
$$- \langle\nabla t,\nabla\Gamma\rangle\,V^+u)\,\mu_{t,\Gamma}. \qquad (6.3.29)$$

6.4 The method of descent

Consider the following initial value problem for the ordinary wave equation in $\mathbf{R}^n$,

$$(\partial_1^2 - \partial_2^2 - \ldots - \partial_n^2)\,u = f(x^1,\ldots,x^k), \quad x^1 > 0,$$

$$u = u_0(x^2,\ldots,x^k), \quad \partial_1 u = u_1(x^2,\ldots,x^k), \quad x^1 = 0,$$

where $k < n$, and f, u_0 and u_1 are, say, C^∞ functions. The problem has a unique solution, and it is an immediate consequence of the uniqueness theorem that this solution is a function of $(x^1, ..., x^k)$ only. So, if one can solve the problem in $\mathbf{R}^n$, then one can also solve it in $\mathbf{R}^k$, where $k < n$. It was pointed out by Hadamard that this trivial observation, which can of course be extended to any partial differential equation on $\mathbf{R}^n$ whose coefficients depend only on $(x^1, ..., x^k)$, leads to many interesting results; he called it the method (or principle) of *descent*. We shall now discuss two applications of this method, to a differential operator

$$Pu = \Box u + \langle a, \nabla u \rangle + bu \qquad (6.4.1)$$

defined on an n-dimensional space–time M_n. In each case, the basic idea is to associate, with P, another differential operator P', on a space–time of higher dimension, from which P can be derived by descent.

We suppose that M_n is given, and construct an $(n+m)$-dimensional space–time M'_{n+m}, as follows. (In fact, only $m = 1$ and $m = 2$ will be used.) The underlying topological space is $M_n \times \mathbf{R}^m$, with the product topology. The points of M'_{n+m} are thus ordered pairs $p' = (p, z)$, where p is a point of M_n, and $z \in \mathbf{R}^m$; the projection $p' \to p$ will be denoted by Π. If (Ω, π) is a coordinate chart in M_n, then a coordinate chart (Ω', π') can be defined in M'_{n+m}, with $\Omega' = \Pi^{-1}\Omega$, by setting

$$\pi' p' = (x, x^{n+1}, ..., x^{n+m}),$$

where $x = \pi p = (x^1, ..., x^n)$. Such a chart will be called a standard chart. If (Ω, π) and $(\tilde{\Omega}, \tilde{\pi})$ are two charts in M_n, such that $\Omega \cap \tilde{\Omega} \neq \varnothing$, then the coordinate transformation between the corresponding standard charts (Ω', π') and $(\tilde{\Omega}', \tilde{\pi}')$ is specified by

$$\pi'(\Omega' \cap \tilde{\Omega}') \ni x'$$
$$= (x, x^{n+1}, ..., x^{n+m}) \to (\tilde{\pi} \circ \pi^{-1}x, x^{n+1}, ..., x^{n+m}) = \tilde{x}' \in \tilde{\pi}'(\Omega' \cap \tilde{\Omega}'),$$

with the extra coordinates x^{n+A}, $A = 1, ..., m$, unchanged. The collection of standard charts covers M'_{n+m}, and can be enlarged to a complete C^∞ structure. However, for the method of descent one must work with standard charts.

The metric on M'_{n+m} is defined by setting

$$\langle dx', dx' \rangle = \langle dx, dx \rangle - \sum_{A=1}^{m} (dx^{n+A})^2. \qquad (6.4.2)$$

The canonical form of the differential equations of the geodesics is therefore

$$\frac{dx^i}{dr} = g^{ij}\xi_j, \quad \frac{d\xi_i}{dr} = -\tfrac{1}{2}\xi_j\xi_k\partial_i g^{jk} \quad (i,j,k = 1,...,n),$$

$$\frac{dx^{n+A}}{dr} = -\xi_{n+A}, \quad \frac{d\xi_{n+A}}{dr} = 0 \quad (A = 1,...,m).$$

As also $\xi^{n+A} = -\xi_{n+A}, A = 1,...,m$, it follows that the exponential map is, in local coordinates,

$$x^i = h^i(y,\eta) \quad (i = 1,...,n), \qquad x^{n+A} = y^{n+A} + r\eta^{n+A} \quad (A = 1,...,m),$$

where $\eta \to h(y,\eta)$ is the exponential map in M_n. Hence if Ω is a geodesically convex domain in M_n then $\Omega' = \Pi^{-1}\Omega$ is a geodesically convex domain in M'_{n+m}. It also follows from the equations of the geodesics, and (6.4.2), that the squares of the geodesic distances in Ω and Ω' are related by

$$\Gamma'(x',y') = \Gamma(x,y) - \sum_{A=1}^{m} (x^{n+A} - y^{n+A})^2, \tag{6.4.3}$$

in local coordinates. One can easily deduce from this that, if Ω_0 is a causal domain in M_n, then $\Omega'_0 = \Pi^{-1}\Omega_0$ is a causal domain in M'_{n+m}; likewise, if Φ is past-compact (or future-compact) in Ω_0 then $\Pi^{-1}\Phi$ is past-compact (or, respectively, future-compact) in Ω'_0. Finally, we note that it follows from (6.4.2) that

$$\square' = \square - \partial_{n+1}^2 - \cdots - \partial_{n+m}^2. \tag{6.4.4}$$

One can now write down the differential operator P' which will be associated with P. Given a vector field a in M_n, one can define a vector field a' in M'_{n+M} by setting $a' = (a^1,...,a^n, 0,...,0)$ in each standard chart; the scalar field b can be considered as a scalar field (still denoted by b) in M'_{n+m}. We set

$$P'u' = \square'u' + \langle a', \nabla'u'\rangle + bu'$$

$$= (\square - \partial_{n+1}^2 - \cdots - \partial_{n+m}^2)\, u' + \sum_{i=1}^{n} a^i\partial_i u' + bu'. \tag{6.4.5}$$

It will be recalled that the biscalars U_ν ($\nu = 0, 1,...$), which figure in the construction of the fundamental solutions of P, are defined by the equations (6.2.2) and (6.2.5); these can be written, jointly, as

$$2\langle\nabla\Gamma, \nabla U_\nu\rangle + (\square\Gamma + \langle a, \nabla\Gamma\rangle + 4\nu - 2n)\, U_\nu = -PU_{\nu-1} \quad (\nu = 0, 1,...), \tag{6.4.6}$$

with the convention $U_{-1} = 0$. Furthermore, the U_ν are determined uniquely in Ω_0 by these equations, and by the conditions

$$U_0(y,y) = 1, \quad U_\nu(x,y) = O(1) \quad \text{when} \quad x \to y \quad (\nu = 1, 2, \ldots). \quad (6.4.7)$$

Lemma 6.4.1. *Let $U_\nu'(x',y'), \nu = 0, 1, \ldots$ denote the biscalars that are defined by the equations and boundary conditions similar to (6.4.6) and (6.4.7), which are obtained when Ω_0 is replaced by $\Omega_0' = \Pi^{-1}\Omega_0$ and P by P'. Then $U_\nu'(x',y') = U_\nu(x,y)$, for $\nu = 0, 1, \ldots$.*

PROOF. The new transport equations are

$$2\langle \nabla'\Gamma', \nabla'U_\nu'\rangle + (\square'\Gamma' + \langle a', \nabla'\Gamma'\rangle + 4\nu - 2(n+m))\, U_\nu'$$
$$= -P'U_{\nu-1}' \quad (\nu = 0, 1, \ldots).$$

Now, by (6.4.3),

$$\nabla'^{n+A}\Gamma' = 2(x^{n+A} - y^{n+A}) \quad (A = 1, \ldots, m),$$

and so it follows from (6.4.4) that

$$\square'\,\Gamma' = \square\,\Gamma + 2m.$$

Also, $\langle a', \nabla'\Gamma'\rangle = \langle a, \nabla\Gamma\rangle$ as $a' = (a, 0, \ldots, 0)$. Hence, and in view of (6.4.5), we have

$$2\langle \nabla\Gamma, \nabla U_\nu'\rangle + 2\sum_{A=1}^{m} (x^{n+A} - y^{n+A})\,\partial_{n+A}U_\nu'$$

$$+ (\square\,\Gamma + \langle a, \nabla\Gamma\rangle + 4\nu - 2n)\, U_\nu' = -(P - \partial_{n+1}^2 - \ldots - \partial_{n+m}^2)\, U_{\nu-1}',$$

for all ν. These equations can obviously be satisfied by taking $U_\nu'(x',y') = U_\nu(x,y), \nu = 0, 1, \ldots$. But then the conditions at y', $U_0'(y',y') = 1$ and $U_\nu(x',y') = O(1)$ when $x' \to y'$, $\nu \geqslant 1$, also hold. As these conditions, and the transport equations, determine the U_ν' uniquely, this implies the lemma. $\square$

As a first application of this lemma, we shall prove reciprocity relations for the U_ν.

Theorem 6.4.1. *Let ${}^tU_\nu(p,q), \nu = 0, 1, \ldots$, denote the biscalars corresponding to the U_ν, for the adjoint ${}^tP$ of P. Then $U_\nu(p,q) = {}^tU_\nu(q,p)$ for all ν, provided that $\Gamma(p,q) \geqslant 0$.*

PROOF. Suppose first that n is even, say $n = 2m$, $m \geqslant 2$. Then $n+1$ is odd. We take $m = 1$ in Lemma 6.4.1, and note that the forward fundamental solution of P' then has the form (6.2.20),

$$\frac{1}{2\pi^{m-\frac{1}{2}}}\,\Gamma_+'^{m+\frac{1}{2}}F(p',q').$$

Here $F(p', q')$ is a C^∞ function on $\{(p', q'); p' \in J^+(q')\}$ (which is a closed subset of $\Omega'_0 \times \Omega'_0$), and has the asymptotic expansion

$$F \sim \sum_{\nu=0}^{\infty} U'_\nu \frac{\Gamma'^\nu}{(\nu - m - \frac{1}{2})!},$$

when $\Gamma' \downarrow 0$. By (6.4.3) and Lemma 6.4.1 this is

$$F \sim \sum_{\nu=0}^{\infty} U_\nu(p, q) \frac{(\Gamma(p, q) - z^2)^\nu}{(\nu - m - \frac{1}{2})!},$$

where $z = x^{n+1} - y^{n+1}$. Similarly, the backward fundamental solution of the adjoint of P', which is ${}^t P - \partial_{n+1}^2$, is of the form

$$\frac{1}{2\pi^{m-\frac{1}{2}}} \Gamma'^{\frac{1}{2}-m}_{-} {}^t F(p', q'),$$

and

$$ {}^t F \sim \sum_{\nu=0}^{\infty} {}^t U_\nu(p, q) \frac{(\Gamma(p, q) - z^2)^\nu}{(\nu - m - \frac{1}{2})!},$$

when $\Gamma' \downarrow 0$ in $\{(p', q'); p' \in J^-(q')\}$. But it follows from the reciprocity theorem that $F(p', q') = {}^t F(q', p')$ (compare (6.3.19)), and so

$$(\Gamma - z^2)^{-\mu} \sum_{\nu=0}^{\mu} (U_\nu(p, q) - {}^t U(q, p)) \frac{(\Gamma(p, q) - z^2)^\nu}{(\nu - m - \frac{1}{2})!} \to 0, \qquad (6.4.8)$$

when $\Gamma' = \Gamma - z^2 \downarrow 0$ in $\{(p', q'); p' \in J^+(q')\}$. So one can conclude, by letting $z \uparrow \Gamma^{\frac{1}{2}}$ for fixed (p, q), and taking $\mu = 0, 1, \ldots$, in turn, that $U_\nu(p, q) = {}^t U_\nu(q, p)$ holds for all ν, if $p \in J^+(q)$. The same argument, applied to the other pair of reciprocally related fundamental solutions of P' and its adjoint, gives the assertion for $p \in J^-(q)$. So the theorem is proved for even n.

The proof for odd n is similar, except that one must take $m = 2$. One then obtains (6.4.8) with

$$z^2 = (x^{n+1} - y^{n+1})^2 + (x^{n+2} - y^{n+2})^2,$$

and can again deduce that $U_\nu(p, q) = {}^t U_\nu(q, p)$ for all ν and $\Gamma(p, q) \geqslant 0$. The proof is complete. $\square$

Remark. The $U_\nu(p, q)$ are not needed for $\Gamma(p, q) < 0$. In fact, it is easy to show that $U_\nu(p, q) = {}^t U_\nu(q, p)$ holds for all $(p, q) \in \Omega_0 \times \Omega_0$. For in the analytic case, this follows from Theorem 6.4.1, by analytic continuation; the C^∞ case can then be settled by approximating by analytic equations, in a fixed coordinate chart. For $\nu = 0$, the reciprocity relation can also be deduced from (6.2.3) and (6.2.4), as in the four-dimensional case.

For the second application of Lemma 6.4.1, we take $m = 1$ and write z for x^{n+1}, so that $P' = P - \partial_z^2$. The method of descent provides a link between the fundamental solutions of P' and P. But one can obtain more, by means of a Fourier–Laplace transform in z; this gives the fundamental solutions of $P + \zeta^2$, where ζ is a complex constant, in terms of the fundamental solutions of P'.

As Fourier transforms are not used elsewhere in this book, an *ad hoc* definition will be given. Let $q' = (q, 0)$, where $q \in \Omega_0$, be a point in $\Omega_0' = \Pi^{-1}\Omega_0$. Let $H^+(p, z; q)$ denote the fundamental solution of P' with pole q' and support contained in $J^+(q')$ (the forward fundamental solution), so that

$$P'H^+ = \delta_q(p) \otimes \partial(z), \left.\begin{array}{c} \\ \\ \end{array}\right\}$$
$$\operatorname{supp} H^+ \subset J^+(q') = \{(p, z); \; |z| \leqslant |\Gamma(p, q)|^{\frac{1}{2}}, \; p \in J^+(q)\}. \quad (6.4.9)$$

Formally, the Fourier–Laplace transform of H^+ is

$$\hat{H}^+(p, q; \zeta) = \int H^+ e^{i\zeta z}\, dz = \int_{-\Gamma^{\frac{1}{2}}}^{\Gamma^{\frac{1}{2}}} H^+ e^{i\zeta z}\, dz,$$

where ζ can be allowed to be complex, as the interval of integration is finite. To define $\hat{H}^+$ as a distribution, one can proceed as follows. For any $\phi(p) \in C_0^\infty(\Omega_0)$, one can define a function $\phi'(p') \in C^\infty(\Omega_0')$ by setting $\phi'(p') = \phi(\Pi p)$. As $|z| < |\Gamma(p, q)|^{\frac{1}{2}}$ in $J^+(q')$, the intersection of $\operatorname{supp} \phi'$ and $J^+(q')$ is a compact subset of Ω_0'. Let $\sigma(p') \in C_0^\infty(\Omega_0')$ be a cut-off function, such that $\sigma = 1$ on a neighbourhood of $\operatorname{supp} \phi' \cap J^+(q')$. Then, for fixed $\zeta \in \mathbf{C}$, $\phi'(p')\, \sigma(p') \exp(i\zeta z) \in C_0^\infty(\Omega_0')$, and so one can set

$$(\hat{H}^+(p, q; \zeta), \phi(p)) = (H^+(p, z; q)\, \phi'(p')\, \sigma(p')\, \sigma(p')\, e^{i\zeta z}). \quad (6.4.10)$$

The second member is a linear form on $C_0^\infty(\Omega_0)$, and tends to zero when $\phi \to 0$ in $C_0^\infty(\Omega_0)$; hence $\phi \to (\hat{H}^+, \phi) \in \mathscr{D}'(\Omega_0)$. It is obviously independent of the choice of σ.

Lemma 6.4.2. *The distribution $\hat{H}^+$ is the forward fundamental solution of $P + \zeta^2$ with source point q.*

PROOF. It follows from (6.4.9) and (6.4.10) that $\operatorname{supp} \hat{H}^+ \subset J^+(q)$. To compute $(P + \zeta^2)\hat{H}^+$, we note that the adjoint of P' is ${}^tP - \partial_z^2$, where ${}^tP$ is the adjoint of P. Now

$$({}^tP - \partial_z^2)\, \phi'(p')\, \sigma(p')\, e^{i\zeta z} = \sigma(p')\, e^{i\zeta z}({}^tP + \zeta^2)\, \phi(p) + \ldots,$$

where the terms omitted vanish on a neighbourhood of

$$\operatorname{supp} \phi' \cap J^+(q').$$

Hence
$$((P + \zeta^2)\hat{H}^+, \phi) = (\hat{H}^+, ({}^tP + \zeta^2)\phi) = (H^+, {}^tP'\phi'\sigma\,e^{i\zeta z}).$$

But this implies, by (6.4.9),

$$((P + \zeta^2)\hat{H}^+, \phi) = (P'H^+, \phi'\sigma\,e^{i\zeta z})$$
$$= (\delta_q(p) \otimes \delta(z), \phi'\sigma\,e^{i\zeta z}) = \phi(q),$$

since $\sigma(q') = 1$ and $\phi'(q') = \phi(q)$. So the lemma is proved. $\square$

Suppose now that M_n is an analytic space–time, and that P is an analytic differential operator on Ω_0. This means that the vector field a and the scalar field b in (6.4.1) are analytic functions in Ω_0. Then every point $q \in \Omega_0$ has a neighbourhood $\omega_q \subset \Omega_0$ in which the parametrix of Lemma 6.2.1, with the factors $\sigma(k_\nu \Gamma)$ omitted, converges uniformly to $G^+(p, q)$. Thus

$$\left.\begin{aligned}
G^+(p,q) &= \frac{1}{2\pi^m}\left(\sum_{\nu=0}^{m-1} U_\nu \delta_+^{(m-\nu-1)}(\Gamma) + \sum_{\nu=m}^{\infty} U_\nu \frac{\Gamma_+^{\nu-m}}{(\nu-m)!}\right),\ n = 2m+2, \\
G^+(p,q) &= \frac{1}{2\pi^{m-\frac{1}{2}}}\sum_{\nu=0}^{\infty} U_\nu \frac{\Gamma_+^{\frac{1}{2}-m+\nu}}{(\frac{1}{2}-m+\nu)!},\ n = 2m+1,
\end{aligned}\right\}$$

$$(6.4.11)$$

where (G^+, ϕ) is restricted to ω_q. There is a similar result for $H^+(p, z; q)$, with n replaced by $n+1$, Γ by $\Gamma' = \Gamma - z^2$, and the U_ν replaced by the U_ν'. Now $U_\nu' = U_\nu$ for all ν, by Lemma 6.4.1, and $0 \leqslant \Gamma' \leqslant \Gamma$ in $J^+(q')$. It is therefore easily shown that the series expansions for H^+ converge in $\omega_{q'}' = \Pi^{-1}\omega_q = \{(p, z);\ p \in \omega_q\}$. They are, with n still denoting the dimension of the space–time M_n on which P is defined,

$$\left.\begin{aligned}
H^+(p,z;q) &= \frac{1}{2\pi^{m+\frac{1}{2}}}\sum_{\nu=0}^{\infty} U_\nu \frac{(\Gamma-z^2)_+^{-\frac{1}{2}-m+\nu}}{(-\frac{1}{2}-m+\nu)!},\ n = 2m+2, \\
H^+(p,z;q) &= \frac{1}{2\pi^m}\left(\sum_{\nu=0}^{m-1} U_\nu \delta_+^{(m-\nu-1)}(\Gamma-z^2)\right. \\
&\qquad\left. + \sum_{\nu=m}^{\infty} U_\nu \frac{(\Gamma-z^2)_+^{\nu-m}}{(\nu-m)!}\right),\ n = 2m+1.
\end{aligned}\right\}\ (6.4.12)$$

As the series converge uniformly, the Fourier–Laplace transform of H^+ can be evaluated term-by-term. So one has to compute the Fourier–Laplace transform of the distributions by which the U_ν are multiplied in (6.4.12), as the U_ν are independent of z. All these are of the form

$$Z_{n+1,\lambda}(\Gamma') = \frac{1}{2\pi^{\frac{1}{2}(n-1)}}\frac{(\Gamma-z^2)_+^{\lambda-\frac{1}{2}(n+1)}}{(\lambda-\frac{1}{2}(n+1))!},\qquad (6.4.13)$$

with $\mathrm{Re}\,\lambda > 0$, and the usual interpretation when $\lambda - \tfrac{1}{2}(n+1)$ is a negative integer.

Lemma 6.4.3. *The Fourier–Laplace transform of* $Z_{n+1,\,\lambda}(\Gamma')$ *is the distribution in* $\mathscr{D}'(\Omega_0)$ *which is equal to the function*

$$\frac{1}{2\pi^{\frac{1}{2}n-1}}\left(\frac{2\Gamma_+^{\frac{1}{2}}}{\zeta}\right)^{\lambda-\frac{1}{2}n} J_{\lambda-\frac{1}{2}n}(\zeta\Gamma^{\frac{1}{2}}) = \frac{1}{2\pi^{\frac{1}{2}n-1}}\sum_{\mu=0}^{\infty}(-1)^{\mu}\frac{(\tfrac{1}{2}\zeta)^{\mu}\,\Gamma_+^{\lambda+\mu-\frac{1}{2}n}}{\mu!\,(\lambda+\mu-\tfrac{1}{2}n)!}$$

(6.4.14)

when $\mathrm{Re}\,\lambda > \tfrac{1}{2}(n-1)$, *and which is thence obtained by analytic continuation to* $\{\lambda;\ \lambda \in \mathbf{C},\ \lambda \neq 0, -1, \ldots\}$.

PROOF. We have

$$(\hat{Z}_{n+1,\,\lambda}, \phi) = (Z_{n+1,\,\lambda},\ \phi'(p')\,\sigma(p')\,e^{i\zeta z}).$$

In a coordinate system that is normal and Minkowskian at q', one can deduce from Lemma 6.1.1, with n replaced by $n+1$, that $(\hat{Z}_{n+1,\,\lambda}, \phi)$ is analytic on $\{\lambda;\ \lambda \in \mathbf{C},\ \lambda \neq 0, -1, \ldots\}$. When $\mathrm{Re}\,\lambda > \tfrac{1}{2}(n-1)$, then $p' = (p,z) \to Z_{n+1,\,\lambda}$ is locally integrable, and so it follows from Fubini's theorem that $\hat{Z}_{n+1,\,\lambda}$ is just the usual integral transform

$$\hat{Z}_{n+1,\,\lambda}(p,q) = \frac{1}{2\pi^{\frac{1}{2}(n-1)}(\lambda-\frac{1}{2}(n+1))!}\int_{-\Gamma^{\frac{1}{2}}}^{\Gamma^{\frac{1}{2}}}(\Gamma-z^2)^{\lambda-\frac{1}{2}(n+1)}e^{i\zeta z}\,dz$$

if $p \in D^+(q)$, and of course zero if $p \notin D^+(q)$. If one puts $z = \Gamma^{\frac{1}{2}}t$, then this becomes

$$\hat{Z}_{n+1,\,\lambda}(p,q) = \frac{\Gamma_+^{\lambda-\frac{1}{2}n}}{2\pi^{\frac{1}{2}(n-1)}(\lambda-\frac{1}{2}(n+1))!}\int_{-1}^{1}(1-t^2)^{\lambda-\frac{1}{2}(n+1)}e^{i\zeta\Gamma^{\frac{1}{2}}t}\,dt,$$

and (6.4.14) follows from the Hankel–Poisson representation of the Bessel function $J_{\lambda-\frac{1}{2}n}(\zeta\Gamma^{\frac{1}{2}})$. One can also verify this by expanding $\exp(i\zeta\Gamma^{\frac{1}{2}}t)$ as a power series and integrating term-by-term, noting that the convergence is uniform on $|t| \leqslant 1$. $\square$

It is now a simple matter to compute the Fourier–Laplace transform of H from (6.4.12) and (6.4.14), to obtain the following result.

Theorem 6.4.2. *Suppose that the space–time and the differential operator* P *are analytic and let* ζ *be a complex number. Then every point* $q \in \Omega_0$ *has a neighbourhood* $\omega_q \subset \Omega_0$ *in which the forward fundamental* *solution $G^+(p, q; \zeta)$ of $P + \zeta^2$ can be expanded as a uniformly convergent series, as follows. If $n = 2m+1$, $m \geqslant 1$, then*

$$G^+(p, q; \zeta) = \frac{1}{2\pi^{m-\frac{1}{2}}} \sum_{\nu=0}^{\infty} U_\nu \left(\frac{2\Gamma_+^{\frac{1}{2}}}{\zeta}\right)^{\frac{1}{2}-m+\nu} J_{\frac{1}{2}-m+\nu}(\zeta\Gamma_+^{\frac{1}{2}}). \quad (6.4.15)$$

If $n = 2m+2$, $m \geqslant 1$, then

$$G^+(p, q; \zeta) = \frac{1}{2\pi^m} \sum_{\nu=0}^{m-1} U_\nu \left(\sum_{\mu=0}^{m-\nu-1} (-1)^\mu (\tfrac{1}{2}\zeta)^\mu \frac{\delta_+^{(m-\nu-1)}(\Gamma)}{\mu!}\right.$$

$$\left. + (-1)^{m-\nu} \left(\frac{\zeta}{2\Gamma_+^{\frac{1}{2}}}\right)^{m-\nu} J_{m-\nu}(\zeta\Gamma_+^{\frac{1}{2}})\right) + \frac{1}{2\pi^m} \sum_{\nu=m}^{\infty} U_\nu \left(\frac{2\Gamma_+^{\frac{1}{2}}}{\zeta}\right)^{\nu-m} J_{\nu-m}(\zeta\Gamma_+^{\frac{1}{2}}).$$

$$(6.4.16)$$

In each case, the U_ν are the biscalars defined by (6.2.3) and (6.2.6).

Remark. For $\zeta = 0$, (6.4.12) and (6.4.16) reduce to the expansions (6.4.11).

PROOF. It has already been pointed out that G^+ can be obtained by calculating the Fourier–Laplace transforms of the series (6.4.12) term-by-term, when $p \in \omega_q$. When n is odd, $n = 2m+1$, the second expansion in (6.4.12) must be used, and it is immediate from Lemma 6.4.3 that (6.4.15) results. For even $n = 2m+2$, $m \geqslant 1$, one has to use the first series, and this means that one must put $\lambda = \nu+1$ and $n = 2m+2$ in (6.4.14), where $\nu = 0, 1, \dots$. For $\nu \geqslant m$, (6.4.14) can be used as it stands. For $\nu < m$, the terms in question must be replaced by the appropriate delta function terms; so one obtains

$$\frac{1}{2\pi^m} \sum_{\mu=0}^{m-\nu-1} (-1)^\mu (\tfrac{1}{2}\zeta)^{2\mu} \frac{\delta_+^{(m-\mu-\nu-1)}(\Gamma)}{\mu!}$$

$$+ \frac{1}{2\pi^m} \sum_{\mu=m-\nu}^{\infty} (-1)^\mu (\tfrac{1}{2}\zeta)^{2\mu} \frac{\Gamma_+^{\mu+\nu-m}}{\mu!(\mu+\nu-m)!}.$$

If one replaces μ by $\mu+\nu-m$ in the second series, then this series becomes

$$\frac{1}{2\pi^m} \sum_{\mu=0}^{\infty} (-1)^{\nu-m+\mu} (\tfrac{1}{2}\zeta)^{2(\nu+\mu-m)} \frac{\Gamma_+^\mu}{\mu!(\mu+\nu-m)}$$

$$= (-1)^{m-\nu} \left(\frac{\zeta}{2\Gamma_+}\right)^{m-\nu} J_{m-\nu}(\zeta\Gamma_+^{\frac{1}{2}}),$$

and so (6.4.16) follows. □

In point of fact, (6.4.16) also holds when $m = 0$, so that $n = 2$. As the cases $n = 2$, 3 and 4 are of particular interest in applications, we list them separately. If $n = 2$, then

$$G^+(p, q; \zeta) = \frac{1}{2} \sum_{\nu=0}^{\infty} U_\nu \left(\frac{2\Gamma_+^{\frac{1}{2}}}{\zeta} \right)^\nu J_\nu(\zeta \Gamma_+^{\frac{1}{2}}). \qquad (6.4.17)$$

If $n = 3$, then the Bessel function of half-integral order can be expressed in terms of circular functions, and one finds that

$$G^+(p, q; \zeta) = \frac{1}{2\pi} \left(U_0 \frac{\cos \zeta \Gamma_+^{\frac{1}{2}}}{\Gamma_+^{\frac{1}{2}}} + \sum_{\nu=0}^{\infty} (-1)^\nu 2^\nu U_{\nu+1} \Gamma_+^{\nu+\frac{1}{2}} \left(\frac{d}{z\,dz} \right)^\nu \frac{\sin z}{z} \right) \bigg|_{z = \zeta \Gamma_+^{\frac{1}{2}}}. \qquad (6.4.18)$$

For $n = 4$, we revert to the notation used in chapters 4 and 5,

$$U_0 = U, \quad U_\nu = V_{\nu-1} \quad (\nu = 1, 2, \ldots).$$

It then follows from (6.4.16) that

$$G^+(p, q; \zeta) = \frac{1}{2\pi} U \left(\delta_+(\Gamma) - \frac{\zeta}{2\Gamma_+^{\frac{1}{2}}} J_1(\zeta \Gamma^{\frac{1}{2}}) \right)$$

$$+ \frac{1}{2\pi} \sum_{\nu=0}^{\infty} V_\nu \left(\frac{2\Gamma_+^{\frac{1}{2}}}{\zeta} \right)^\nu J_\nu(\zeta \Gamma^{\frac{1}{2}}). \quad (6.4.19)$$

In the Minkowskian case, where $U_0 = U = 1$ and $U_\nu \neq V_{\nu-1}, = 0$, for $\nu = 1, 2, \ldots$, (6.4.17)–(6.4.19) reduce to well-known results.

We conclude the section with a remark about the C^∞ case, taking $n = 4$. One then has, by Theorem 6.2.1,

$$H^+(p, z; q) = \frac{U}{2\pi^{\frac{3}{2}}} (\Gamma - z^2)_+^{-\frac{3}{2}} + W_1(p, z; q)(\Gamma - z^2)_+^{-\frac{1}{2}},$$

where $U \equiv U_0$ and W_1 is C^∞ in $\{p, q; p = J^+(q)\}$. It therefore follows from Lemmas 6.4.2 and 6.4.3 that

$$G^+(p, q: \zeta) = \frac{1}{2\pi} U \left(\delta_+(\Gamma) - \frac{\zeta}{2\Gamma_+^{\frac{1}{2}}} J_1(\zeta \Gamma^{\frac{1}{2}}) \right)$$

$$+ H_+(f) \int_{-\Gamma^{\frac{1}{2}}}^{\Gamma^{\frac{1}{2}}} W_1(p, z; q)(\Gamma - z^2)^{-\frac{1}{2}} e^{i\zeta z} dz. \quad (6.4.20)$$

Suppose that ζ is real and positive. Then one can show that the integral is $O(\zeta^{-\frac{1}{2}})$ when $\zeta \to \infty$, and so one obtains an asymptotic formula for G^+,

$$G^+(p, q; \zeta) = \frac{1}{2\pi} U \left(\delta_+(\Gamma) - \left(\frac{\zeta}{2\pi} \right)^{\frac{1}{2}} \Gamma_+^{-\frac{3}{4}} \cos \left(\zeta \Gamma^{\frac{1}{2}} - \frac{3\pi}{4} \right) \right) + H_+(\Gamma) O(\zeta^{-\frac{1}{2}}).$$

$$(6.4.21)$$

This result, which could not have been obtained directly from the Hadamard series, shows that the regular part (the 'tail') of the fundamental solution of $G^+(p, q; \zeta)$ gives rise to dispersion.

NOTES

The method of analytic continuation is due to M. Riesz (1949); it is developed in terms of distribution theory in Gelfand and Shilov (1964) see also Duff (1956). In Combet (1965), it is also used to derive fundamental solutions for the d'Alembertian on a space–time of arbitrary (indefinite) signature, following Fourès-Bruhat (1956). For kernel distributions, see Schwartz (1966).

The importance of the method of descent was emphasized by Hadamard (1923, 1932). Theorem 6.4.2 was proved, by a different method, by Zauderer (1971).

Appendix

This appendix is a brief account of some of the definitions and results in topology which are used in the book.

A *topological space* is a pair (X, T) consisting of a set X and a family T of subsets of X that has the following properties:

(i) the empty set and X itself are members of T;

(ii) the intersection of every finite sub-family of members of T is a member of T;

(iii) the union of every sub-family of members of T is a member of T.

The family T is called a *topology* for the set X, and its members are called *open sets*. If S and T are different topologies for the same set X, and every member of S is also a member of T, then S is called *smaller* (or *coarser*) than T, and T is called *larger* (or *finer*) than S.

Let (X, T) be a topological space. A *neighbourhood* of a point x in X is a set $U \subset X$ which contains an open set G such that $x \in G$. (Likewise, a neighbourhood of a set $V \subset X$ is a set $U \subset X$ that contains an open set G with $V \subset G \subset U$.) It is easy to see that a set $U \subset X$ is open if and and only if every point of U has a neighbourhood that is contained in U. The collection of the neighbourhoods of a point x is the *neighbourhood system* of x; note that a topology for a set X can be specified by giving the neighbourhood system of each point of X.

A family B of subsets of X is a *base* for the topology T if every member of T is the union of members of B; an equivalent condition is that, for each point $x \in X$, and each neighbourhood U of x, there is a member V of B such that $x \in V \subset U$. Different bases may give the same topology. For instance, the usual topology of $\mathbf{R}^n$ can be built up from the base consisting of open balls, or from that consisting of open n-cubes. A necessary and sufficient condition for a family B of subsets of X to be the base of a topology is that, for every pair of members U and V of B, and each point $x \in U \cap V$, there is a $W \in B$ such that $x \in W \subset U \cap V$. So a family of subsets of X need not be the base of a topology. But, given such a (non-void) family S, one can show that the family of all intersections of finite sub-families of S is a base for a topology. This is the topology *generated* by S, and S is called a *sub-base*. Bases and sub-bases of the neighbourhood system of a point are defined in the same way.

Another way of generating topologies is by relativization. Let (X, T) be a topological space, and let Y be a subset of X. The family S consisting of the intersections of the members of T with Y is a topology for Y, which is called the *relative* or *induced* topology.

Again, let (X, T) and (Y, S) be two topological spaces. It is easy to show that the sets of the form $U \times V$, where U is open in X and V is open in Y, are the base for a topology of the product set $X \times Y$. This is called the *product topology*. The construction obviously extends to the product of a finite number of topological spaces; as a matter of fact, one can equally define a product topology for an arbitrary family of topological spaces.

Let us again consider a fixed topological space X, so that 'set' will mean 'subset of X'. A set V is called *closed* if its complement $X \backslash V$ is open. A point x is called an *accumulation point* of a set V if every neighbourhood of x contains points of V other than x. The union of a set V and of the set of its accumulation points is a closed set, which is called the *closure* of V and is denoted by $\bar{V}$. It is also the intersection of the family of all closed sets that contain V. So a set V is closed if and only if $V = \bar{V}$.

A point x of a set V is called an *interior point* if it has a neighbourhood that is contained in V; the set of all interior points of V is the *interior* of V. An open set consists entirely of interior points. The *boundary* of V is the set of all points $x \in X$ with the property that every neighbourhood of x meets both V and $X \backslash V$. Thus a closed set contains its boundary, while an open set is disjoint from its boundary.

A sequence of points $\{x_j\}_{1 \leqslant j < \infty}$ *converges* to point x if every neighbourhood of x contains all but a finite number of the x_j. In a general topological space, a sequence may converge to two (or more) points. But the limit of a convergent sequence is unique if the space in question is a *Hausdorff* space; this means that, for every pair of distinct points x and y, there are disjoint neighbourhoods of x and y respectively. If x is the limit of a convergent sequence of points belonging to a set V, then it is either in V or is an accumulation point of V. But in general, an accumulation point of a set need not be the limit of a sequence of points of the set. It is for this reason that one cannot specify the topology of a space by giving the class of convergent sequences. (One has to use nets, or filters; see for instance Kelley

(1955) Chapter 2.) But this difficulty does not arise in a space that satisfies the *first countability axiom*: the neighbourhood of each point has a countable base.

An important class of Hausdorff topological spaces that satisfy this axiom are the *metric spaces*. A *metric* for a set X is a real-valued function d on the product $X \times X$ with the following properties:

(i) $d(x,y) \geqslant 0$;
(ii) $d(x,y) = d(y,x)$;
(iii) $d(x,z) \leqslant d(x,y) + d(y,z)$, (triangle inequality);
(iv) $d(x,y) = 0$ if $x = y$;
(v) if $d(x,y) = 0$ then $x = y$.

One naturally calls $d(x,y)$ the *distance* from x to y. Given a metric d, one can define a topology for X by taking as a base of the neighbourhood system of each point x the *open balls* $\{y; d(x,y) < r, r > 0\}$. It is first countable because it is in fact enough to let r range over the positive rational numbers. The paradigm of metric spaces is of course $\mathbf{R}^n$, with the usual Euclidean distance.

A topological space (X, T) is called *metrizable* if its topology can be generated by a metric. Now a space is called a T_1-space if the sets consisting of single points are closed; every Hausdorff space is a T_1-space. It is called regular if the family of closed neighbourhoods of each point is a neighbourhood base of the point. *Urysohn's metrization theorem* asserts that a regular T_1-space is metrizable if its topology has a countable base. (A space whose topology has a countable base is said to satisfy the *second countability axiom*.)

In a metric space, a sequence $\{x_j\}_{1 \leqslant j < \infty}$ converges to a point x if and only if $d(x, x_j) \to 0$ as $j \to \infty$. It is called a *Cauchy sequence* if $d(x_j, x_k) \to 0$ as $j, k \to \infty$. A convergent sequence is a Cauchy sequence. If, conversely, every Cauchy sequence converges, then the space is called *complete*. So $\mathbf{R}^n$ is complete, but the set consisting of points with rational coordinates, with the induced topology, is not complete. A metric space can be imbedded isometrically in a complete space, which is called its *completion*.

An *open cover* of a set A in a topological space X is a family of open sets of X whose union contains A; a *sub-cover* is a sub-family of this cover whose union also contains A. A set $K \subset X$ is said to be *compact* if every open cover of K contains a finite sub-cover. In a Hausdorff space, compact sets are closed. In $\mathbf{R}^n$, a set is compact if and only if it is closed and bounded; this is, in effect, the classic *Heine–Borel–Lebesgue lemma.* A set K is called *sequentially compact* if every sequence of points of K contains a convergent subsequence whose limit is a point of K. In a metric space, compactness and sequential compactness are equivalent. Finally, a set is called *relatively compact* if its closure is compact.

We must next consider *functions* (or *maps*: the terms are interchangeable). Let X and Y be sets; a function $f: X \to Y$ is a collection of ordered pairs (x,y), where $x \in X$ and $y \in Y$ (i.e., a subset of $X \times Y$), such that, whenever (x,y) and (x,y') are members of f, then $y = y'$. The set of all x that are first coordinates of f is the *domain* of f; if this is all of X, one says that f is a function *on* X. The set of all y that are second coordinates of f is the *range* or *image* of f. We say that f maps or sends x to $y = f(x)$ and write $x \to f(x)$. The *inverse image* $f^{-1}(A)$ of a set $A \subset Y$ is the set $\{x; x \in X, f(x) \in A\}$. (The set A need not be contained in the range of f.)

If X' is a subset of X, then the *restriction* of f to X', here denoted by $f|_{X'}$, is just the function $f|_{X'}: X' \to Y$ whose value at every $x \in X'$ is $f(x)$. Conversely, if one is given a function $f': X' \to Y$, and $f: X \to Y$ is such that $f|_{X'} = f'$, then f is called an *extension* of f'.

The *composition* of functions is defined in the obvious way. Let $f: X \to Y$ and $g: Y \to Z$ be two functions. Then the composite function $g \circ f: X \to Z$ can be defined, simply be setting $g \circ f(x) = g(f(x))$, provided that the domain of g contains the range of f. Composition is associative.

Let $f: X \to Y$ be a function on X. Then f is *injective* if $f(x') = f(x'')$ implies that $x' = x''$ (or, equivalently, if $x' \neq x''$ implies that

$$f(x') \neq f(x'')).$$

An important example is the *canonical injection* or *inclusion map* $j: X' \to X$, where X' is a subset of X and $j(x) = x$ for all $x \in X'$. The function is *surjective* (*onto*) if its range is Y. A function that is both injective and surjective is called *bijective* (or one-to-one and onto). A bijective function has a unique *inverse* $f^{-1}: Y \to X$, which is characterized by $f^{-1} \circ f = j_X$ and $f \circ f^{-1} = j_Y$, where $j_X: X \to X$ and $j_Y: Y \to Y$ are the respective identity maps. Note also the identity $(f \circ g)^{-1} = g^{-1} \circ f^{-1}$, valid for two bijections f and g.

Suppose now that X and Y are topological spaces, and that $f: X \to Y$ is a function on X. Then f is *continuous* if the inverse image of every open set in Y is open in X. It is of course sufficient for this that the inverse image of every member of an (open) basis for the topology of Y is an open set in X. This, in turn, can be replaced by the condition that, for each $x \in X$ and each neighbourhood U of $f(x)$, there is a neighbourhood V of x such that $f(V) \subset U$. (Here,

$$f(V) = \{y; y = f(x), x \in V\}.)$$

If this condition holds at a particular point, then f is said to be continuous at that point. Finally, continuous functions map convergent sequences to convergent sequences. In a space in which the first countability axiom is satisfied, this property can also serve to characterize continuous functions. In general, a function that has this property is called *sequentially continuous*.

It is easy to see that the composition of two continuous functions is again continuous.

It is obvious that the inverse images of closed sets under a continuous map are closed. But the direct images of open or closed sets need not be open or closed respectively. Simple counter examples can be constructed by means of projections in $\mathbf{R}^2$. On the other hand, the direct image of a compact set is compact, under a continuous map, while the inverse image of a compact set need not be compact.

A map (function) $f: X \to Y$ is called a *homeomorphism* if it is bijective, and both f and its inverse f^{-1} are continuous. Its inverse is then also a homeomorphism. Obviously, a homeomorphism maps open sets to open sets, closed sets to closed sets, and compact sets to compact sets. So if M is an n-dimensional manifold, and (Ω, π) is a coordinate chart in M, then one can, for instance, use properties of compact sets in $\mathbf{R}^n$ when one is dealing with compact subsets of Ω.

The proper background for the theory of distributions is the theory of topological vector spaces. A knowledge of this subject is not needed in chapter 2, or for the applications of distribution theory that are made in this book. However, it is hoped that the following summary of some of the elements – which is far from complete – will help to explain how concepts such as convergence in C_0^∞ or in $\mathscr{D}'$ arise. For a detailed account, the reader is referred to the literature, for instance to Yosida (1971).

Let X be a vector space over the field $\mathbf{C}$ of complex numbers, and suppose that the underlying set consisting of the members of X has been equipped with a topology T. The pair (X, T) is called a *topological vector space* if vector addition is a continuous map $X \times X \to X$ and multiplication by a complex number is a continuous map $\mathbf{C} \times X \to X$. (The topologies of $X \times X$ and $\mathbf{C} \times X$ are the product topologies, with the usual topology for **C**.) One can, of course, also consider topological vector spaces over the real numbers. To specify such a topology, it is sufficient to give a base of (open) neighbourhoods of the zero element of X. For this determines the neighbourhood system of zero, and the neighbourhood system of any other point of X can then be defined by translation; thus a set V is a neighbourhood of a point x if and only if $V - x = \{y; y = z - x, z \in V\}$ is a neighbourhood of zero.

A *semi-norm* is a function $p\colon X \to \mathbf{R}$ which satisfies the conditions

(i) $p(\alpha x) = |\alpha|\, p(x)$ $(\alpha \in \mathbf{C}, x \in X)$,

(ii) $p(x+y) \leqslant p(x) + p(y)$ $(x, y \text{ in } X)$.

It follows from (i) that $p(0) = 0$, and from (i) and (ii) that

$$|p(x) - p(y)| \leqslant p(x - y),$$

so that $p(x) \geqslant 0$. A set $U = \{x; p(x) < \delta, \delta > 0\}$ has three characteristic properties. It is convex: if $x \in U$, $y \in U$ and $0 < \alpha < 1$, then

$$\alpha x + (1 - \alpha) y \in U.$$

It is balanced: if $|\alpha| \leqslant 1$ and $x \in U$, then $\alpha x \in U$. It is also absorbing: for every $x \in X$, there is an $\alpha > 0$ such that $\alpha^{-1} x \in U$. Conversely, if U is a convex, balanced and absorbing set, then $p_U(x) = \inf\{\rho > 0; x \in \rho U\}$, which is called the *Minkowski functional* of U, is a semi-norm, for which $U = \{x; p_U(x) < 1\}$. A topological vector space is called *locally convex* if it has a base of neighbourhoods of zero that consists of convex, balanced and absorbing sets. Such a space can be constructed as follows.

Let P be a family of semi-norms that satisfies the *separation axiom*: for every $x \in X$, $x \neq 0$, there is a $p \in P$ such that $p(x) \neq 0$. Take the sets $\{x; p(x) < \delta, p \in P, \delta > 0\}$ as a sub-base for the neighbourhoods of zero. The corresponding base consists of finite intersections of such sets, that is to say of sets of the form

$$V = \{x; p_1(x) < \delta_1, \ldots, p_k(x) < \delta_k\},$$

where the δ_j are positive real numbers and the p_j are members of P. The topology for X obtained in this way is compatible with the vector space structure. It is obviously locally convex, and also Hausdorff, because of the separation axiom. It is called the *topology generated by P*, and the topological space in question can be denoted by (X, P). The members of P are often labelled by an index set J, so that P consists of semi-norms p_j with $j \in J$; the index set need not be countable. Note that two different families of semi-norms may generate the same topology.

It is easy to show that a semi-norm p' on a locally convex topological vector space (X, P) is continuous if and only if there is a constant $C > 0$ and a $p \in P$ such that

$$p'(x) \leqslant Cp(x).$$

In the first place, it follows from $|p'(x) - p'(y)| \leqslant p'(x - y)$ that p' is continuous if and only if it is continuous at $x = 0$. So the condition is obviously sufficient. To prove that it is also necessary, we note that, as $p(0) = 0$, the set $\{x; p'(x) < \epsilon\}$ must be an open neighbourhood of zero, and must hence contain a set $\{x; p(x) < \epsilon\}$ where $p \in P$ and $\epsilon > 0$. Thus $p(x) < \epsilon$ implies that $p'(x) < 1$. Now, given $x \in X$, one can find a $\lambda > 0$ such that $p(\lambda x) < \epsilon$. For if $p(x) \neq 0$, one can take $\lambda = \epsilon/2p(x)$, and if $p(x) \neq 0$, one can take λ arbitrary. But then $p'(\lambda x) = \lambda p'(x) < 1$, and so $p'(x) = 0$ if $p(x) = 0$ (make $\lambda \to \infty$) and $p'(x) < (2/\epsilon)p(x)$ if $p(x) \neq 0$. So the condition follows, with $C = 2/\epsilon$. It will be seen that the condition implies that the topology generated by the family of all continuous semi-norms of (X, P) is identical with the topology generated by P.

Let (X, P) and (Y, Q) be two locally convex topological vector spaces, and let $f: X \to Y$ be a linear map on X; the value of such a map at x may be denoted by (f, x). Linearity means that f is a homomorphism of the underlying vector spaces,

$$(f, x_1 + x_2) = (f, x_1) + (f, x_2), \quad (f, \alpha x) = \alpha(f, x).$$

If q' is a semi-norm on Y, then its pull-back $q' \circ f(x) = q'((f, x))$ is a semi-norm on X. If f is a continuous linear map, then the pull-back of a continuous semi-norm on Y must be a continuous semi-norm on X. So it follows, in particular, that if $f: X \to Y$ is a continuous linear map, then for every $q \in Q$ there is a $p \in P$ and a constant $C > 0$ such that

$$q \circ f(x) \leqslant Cp(x).$$

It is evident that this condition is also sufficient for the continuity of f.

An important special case is obtained by taking $Y = \mathbf{C}$, with its usual topology. A linear map $f: X \to \mathbf{C}$ is called a *linear form* (or a linear functional) on X. The criterion just derived shows that a linear form f is continuous if and only if there is a $p \in P$ and a $C > 0$ such that

$$|(f, x)| \leqslant Cp(x).$$

The continuous linear forms on X can be made into a vector space over $\mathbf{C}$ by setting

$$(f_1+f_2, x) = (f_1, x) + (f_2, x), (\alpha f_1, x) = \alpha(f_1, x).$$

This space is called the *dual* of X, and usually denoted by X'. The functions $f \to |(f, x)|$ are obviously semi-norms on X'. The topology which they generate is called the *weak topology* for X'; a base of the neighbourhoods of zero consists of the sets

$$\{f; f \in X', |(f, x_1)| < \delta_1, \dots, |(f, x_k)| < \delta_k\},$$

where the δ_j are positive real numbers, and the $\{x_j\}_{1 \leqslant j \leqslant k}$ range over all finite subsets of X.

It follows from the general definition that a sequence $\{x_j\}_{1 \leqslant j < \infty}$ in a locally convex topological vector space converges to a point $x \in X$ if and only if, for all $p \in P$, $p(x_j - x) \to 0$ as $j \to \infty$. It will be a Cauchy sequence if and only if, for every $p \in P$, $p(x_j - x_k) \to 0$ as $j, k \to \infty$. If every Cauchy sequence has a limit, then (X, P) is called *sequentially complete*.

We shall now consider some examples. First, suppose that P consists of a single semi-norm p. By the separation axiom, one must then have $x = 0$ when $p(x) = 0$. A semi-norm with this property is called a *norm*, and is usually written as $\|x\|$ or as $\|x\|_X$. (Semi-norms are also often written as $\|x\|_j$, where j runs through some index set.) The space $(X, (p))$ is called a *normed space*. A normed space can also be considered as a metric space, as one can define a metric by setting $d(x, y) = \|x - y\|$. A complete normed space is called a *Banach space*, or a *B-space*. For instance, let $K \subset R^n$ be a compact set, and let $C^m(K)$ denote the vector space of m times continuously differentiable complex valued functions on K. (This means that each $\phi \in C^m(K)$ is the restriction to K of a C^m function defined on an open neighbourhood of K.) It can be made into a normed space by putting

$$\|\phi\| = \sum_{|\alpha| \leqslant m} \sup_K |D^\alpha \phi|;$$

one could equally well use $\max_{|\alpha| \leqslant m} \sup_K |D^\alpha \phi|$ as a norm. Convergence with respect to this norm is just uniform convergence of ϕ and of its derivatives of all orders not exceeding m, and so $C^m(K)$ is in fact a Banach space. If K has a non-void interior (and this is really the only case of interest), then the space $C_0^m(K)$ which consists of all $\phi \in C^m(K)$

with supports in K, is a closed (vector) sub-space of $C^m(K)$, and is also a Banach space.

A linear form $u: X \to \mathbf{C}$ on a normed space is continuous if and only if there is a constant $C > 0$ such that

$$|(u, x)| \leqslant C \|x\|.$$

The dual X' can therefore be made into a normed space by setting, for all $u \in X'$,

$$\|u\| = \sup_{x \in X} \frac{|(u, x)|}{\|x\|}.$$

The *strong topology* defined by this norm is larger than the weak topology of X', If X is a Banach space, then its strong dual is also a Banach space.

Next, suppose that the family P of semi-norms that generates the topology is countable. Then (X, P) is called a *Fréchet space* (or *F-space*) if it is complete; otherwise, it is a *pre-Fréchet* (or *pre-F*) *space*. Such a space satisfies the first countability axiom, so that one can often work with convergent sequences only. It is also metrizable. For, if the members of P are ordered as $p_1, p_2, \ldots$, then

$$d(x, y) = \sum_{j=1}^{\infty} \frac{1}{2^j} \frac{p_j(x - y)}{1 + p_i(x - y)}$$

defines a distance function on (X, P). But it is usually easier to use the semi-norms.

An example of a Fréchet space is the space where K is again a compact set in $\mathbf{R}^n$ with non-void interior. It is topologized by the semi-norms $\phi \to \sup |D^\alpha \phi|$. Alternatively, one can use the equivalent semi-norms

$$\|\phi\|_m = \sum_{|\alpha| \leqslant m} \sup |D^\alpha \phi| \quad (m = 0, 1, 2, \ldots),$$

(which are actually norms); then the sets $\{\phi;\ \|\phi\|_m < \delta,\ \delta > 0\}$, where $m = 0, 1, 2, \ldots$ are a base for the neighbourhoods of zero. Roughly speaking, this space is the intersection of the $C_0^m(K)$; technically, it is the projective limit of these spaces.

Another example is $C^\infty(\Omega)$, where $\Omega \subset \mathbf{R}^n$ is an open set. Choose an expanding sequence of compact sets $K_j, j = 1, 2, \ldots$, whose union is equal to Ω; for instance, if $\Omega = \mathbf{R}^n$, one can take $K_j = \{x; |x| \leqslant j\}$. Then a system of semi-norms generating the topology of $C^\infty(\Omega)$ is

$$\phi \to \sum_{|\alpha| \leqslant j} \sup_{K_j} |D^\alpha \phi|.$$

This topology is also generated by the semi-norms $\phi \rightarrow \sum\limits_{|\alpha|\leqslant m} \sup\limits_{K} |D^\alpha \phi|$, where K ranges over the compact subsets of Ω and $m = 0, 1, 2, \ldots$. So a linear form u on $C^\infty(\Omega)$ is continuous if and only if there is a compact set $K \subset \Omega$, an integer $N \geqslant 0$ and a constant $C > 0$, such that

$$|(u, \phi)| \leqslant C \sum\limits_{|\alpha|\leqslant m} \sup\limits_{K} |D^\alpha \phi|.$$

So the dual of $C^\infty(\Omega)$ is $\mathscr{E}'(\Omega)$ (see Definition 2.3.1). As this example shows, the (weak) dual of a Fréchet space need not be a Fréchet space.

Finally, we shall consider $C_0^\infty(\Omega)$, where $\Omega \subset \mathbf{R}^n$ is an open set. For every compact set $K \subset \Omega$, $C_0^\infty(K)$ is a sub-space of $C_0^\infty(\Omega)$ (as a vector space). The topology for $C_0^\infty(\Omega)$ used in distribution theory is the largest locally convex topology for which the inclusion maps

$$C_0^\infty(K) \rightarrow C_0^\infty(\Omega)$$

are continuous; it is called the topology of the *inductive limit*. So a semi-norm p on $C_0^\infty(\Omega)$ is continuous if and only if, for every compact set $K \subset \Omega$ there is an integer $N \geqslant 0$ and a constant $C > 0$ such that

$$p(\phi) \leqslant C \sum\limits_{|\alpha|\leqslant N} \sup\limits_{K} |D^\alpha \phi|$$

for all $\phi \in C_0^\infty(K)$. An example of such a semi-norm is

$$p(\phi) = \Sigma \sup |\rho_\alpha D^\alpha \phi|,$$

where α runs over all multi-indices, and the ρ_α are continuous functions whose supports are a locally finite covering of Ω. As a matter of fact, the topology of $C_0^\infty(\Omega)$ is generated by these semi-norms.

It is now easy to see the genesis of the definition of a distribution given in chapter 2 (Definition 2.1.1). Let $u\colon C_0^\infty(\Omega) \rightarrow \mathbf{C}$ be a linear form. Then u is continuous if and only if its restriction to every $C_0^\infty(K)$ is continuous, that is to say if, for every compact set $K \subset \Omega$, there is an integer $N \geqslant 0$ and a constant $C > 0$ such that

$$|(u, \phi)| \leqslant C \sum\limits_{|\alpha|\leqslant N} \sup |D^\alpha \phi|$$

for all $\phi \in C_0^\infty(K)$.

Generally, a *vector valued distribution* is a continuous linear map $C_0^\infty(\Omega) \rightarrow X$, where X stands for a locally convex topological vector space (X, P). It follows from the characterization of continuous linear maps derived above that such a linear map u is a vector-valued distribution if and only if, for every $p \in P$ and every compact set $K \subset \Omega$, there is an $N \geqslant 0$ and a $C > 0$ such that

$$p \circ u(\phi) \leqslant C \sum_{|\alpha| \leqslant N} \sup |D^{\alpha}\phi|$$

for all $\phi \in C_0^{\infty}(K)$. In particular, if $X = C^{\infty}(\Omega')$, where $\Omega' \subset \mathbf{R}^n$ is an open set, then one can take $p = \sum_{|\beta| \leqslant k} \sup_{K'} |D^{\beta}u(\phi)|$, where K' ranges over the compact subsets of Ω', and $k = 0, 1, \ldots.$ Generalized in the obvious way to manifolds, this gives the set of conditions (6.3.1).

Bibliography

ATIYAH, M. F., BOTT, R. and GÅRDING, L. (1970). Lacunas for hyperbolic differential equations with constant coefficients I. *Acta Math.* **124**, 109–189.

BRUHAT, Y. (1962). The Cauchy problem, in *Gravitation; an introduction to current research*, ed. L. Witten (Wiley, New York).

BRUHAT, Y. (1964). Sur la théorie des propagateurs. *Ann. di Mat. Pura Appl.* **64**, 191–228.

CARATHÉODORY, C. (1935). *Variationsrechnung und partielle Differentialgleichungen erster Ordnung* (Teubner, Berlin).

COMBET, E. (1965). *Solutions élémentaires des dalembertiens généralisées.* Mém. Sc. Math. Fasc. CLX (Gauthiers-Villars, Paris).

COURANT, R. and HILBERT, D. (1962). *Methods of mathematical physics*, Vol. II (Interscience, New York).

DIEUDONNÉ, J. (1960). *Foundations of modern analysis*, (Academic Press, London).

DOUGLIS, A. (1954). The problem of Cauchy for linear hyperbolic equations of the second order. *Comm. Pure Appl. Math.* **7**, 271–95.

DUISTERMAAT, J. J. and HÖRMANDER, L. (1972). Fourier integral operators. II. *Acta Math.* **128**, 183–269.

DUFF, G. F. D. (1956). *Partial differential equations* (University of Toronto Press, Toronto).

FOURÈS-BRUHAT, Y. (1952). Théorèmes d'existence pour certains systèmes d'équations aux dérivées partielles non linéaires. *Acta Math.* **88**, 141–225.

FOURÈS-BRUHAT, Y. (1956). Solutions élémentaires d'équations du second ordre de type quelconque, in *Coll. int, sur la théorie des équations aux dérivées partielles*, Nancy (C.N.R.S., Paris).

FRIEDLANDER, F. G. (1946). Simple progressive wave solutions of the wave equation. *Proc. Camb. Phil. Soc.* **43**, 360–73.

FRIEDLANDER, F. G. (1958). *Sound pulses* (Cambridge University Press, London).

GÅRDING, L. (1956). Solution directe du problème de Cauchy pour les équations hyperboliques, in *Coll. int. sur la théorie des équations aux dérivées partielles*, Nancy (C.N.R.S., Paris).

GÅRDING, L., KOTAKE, T. and LERAY, J. (1964). Uniformisation et développement asymptotique de la solution du problème de Cauchy linéaire, à données holomorphes (Problème de Cauchy I bis et VI). *Bull. Soc. Math. France* **92**, 263–361.

GELFAND, I. M. and SHILOV, G. E. (1964). Generalized functions, Vol. I (Academic Press, London).

GOLUBITSKY, M. and GUILLEMIN, V. (1973). *Stable mappings and their singularities*, (Springer, New York).

GUILLEMIN, V. and SCHAEFFER, D. (1973). Remarks on a paper by D. Ludwig. *Bull. Amer. Math. Soc.* **79**, 381–5.

GÜNTHER, P. (1952). Zur Gültigkeit des Huygensschen Princips bei partiellen Differentialgleichungen vom normalen hyperbolischen Typus. Sitz.-Ber. *Sächs. Akad. Wiss. Leipzig, Math. Naturw. Kl.* **100**, 1–43.

GÜNTHER, P. (1965). Ein Beispiel einer nichttrivialen Huygensschen Differentialgleichung mit vier unabhängigen Variablen. *Arch. Rat. Mech. Anal.* **18**, 103–6.

HADAMARD, J. (1923). *Lectures on Cauchy's problem in linear partial differential equations* (Yale University Press).

HADAMARD, J. (1932). Le problème de Cauchy et les équations aux dérivées partielles linéaires hyperboliques, (Hermann, Paris).

HADAMARD, J. (1954). The psychology of invention in the mathematical field (Dover, New York).

HAWKING, S. W. and ELLIS, G. F. R. (1973). *The large scale structure of space–time* (Cambridge University Press, London).

HÖRMANDER, L. (1963). *Linear partial differential operators* (Springer, Berlin).

HÖRMANDER, L. (1971a) On the existence and regularity of solutions of linear pseudo-differential equations. *L'Enseignement Mathématique*, IIe Sér. XVII, 99–163.

HÖRMANDER, L. (1971b). Fourier integral operators I. *Acta Math.* **127**, 79–183.

KELLEY, J. L. (1955). *General topology* (van Nostrand Reinhold, New York).

KLINE, M. and KAY, I. W. (1965). *Electromagnetic theory and geometrical optics* (Interscience, New York).

LANG, S. (1972). *Differential manifolds* (Addison Wesley, Reading, Mass).

LAX, P. D. (1955). On Cauchy's problem for hyperbolic equations and the differentiability of solutions of elliptic equations. *Comm. Pure Appl. Math.* **8**, 615–33.

LAX, P. D. (1957). Asymptotic solutions of oscillatory initial value problems. *Duke Math. J.* **24**, 627–46.

LERAY, J. (1952). Hyperbolic differential equations (Institute of Advanced Studies, Princeton).

LICHNÉROWICZ, A. (1961). Propagateurs et commutateurs en rélativité générale (Publ. Math. I.H.E.S. No 10, Paris).

LICHNÉROWICZ, A. and WALKER, A. G. (1945). Sur les espaces riemanniens harmoniques de type hyperbolique normal. *C.R. Acad. Sci.* **221**, 394–6.

LUDWIG, D. (1960). Exact and asymptotic solutions of the Cauchy problem. *Comm. Pure Appl. Math.* **13**, 473–508.

LUDWIG, D. (1966). Uniform asymptotic expansion at a caustic. *Comm. Pure Appl. Math.* **19**, 215–50.

MATTISSON, M. (1939). Le problème de M. Hadamard relatif à la diffusion des ondes. *Acta Math.* **71**, 249–82.

MCLENAGHAN, R. G. (1969). An explicit determination of the empty space–times on which the wave equation satisfies Huygens' principle. *Proc. Camb. Phil. Soc.* **65**, 139–55.

METHÉE, P. D. (1954). Sur les distributions invariantes dans le groupe des rotations de Lorentz. *Comm. Math. Helv.* **28**, 255–69.

MÉTHÉE, P. D. (1957). L'équation des ondes avec second membre invariante. *Comm. Math. Helv.* **32**, 153–69.

PETROVSKY, I. G. (1945). On the diffusion of waves and the lacunas for hyperbolic equations. *Mat. Sb.* **17** (**59**), 289–370.

DE RHAM, G. (1960). *Variétes différentiables* (Hermann, Paris).

RIESZ, M. (1949). L'intégrale de Riemann–Liouville et le problème de Cauchy. *Acta Math.* **81**, 1–223.

RIESZ, M. (1957). A special characteristic surface. Report No. 25, Dept. of Mathematics, Univ. of Maryland.

RIESZ, M. (1960). A geometric solution of the wave equation in space–time of even dimension. *Comm. Pure Appl. Math.* **13**, 329–51.

RUSE, H. S., WALKER, A. G. and WILLMORE, T. (1961). Harmonic spaces. Cons. Naz. d. Ric., Monogr. matem. 8, Rome.

SCHWARTZ, L. (1966). Théorie des distributions (Hermann, Paris).

SOBOLEV, S. L. (1936). Méthode nouvelle à resoudre le problème de Cauchy pour les équations linéaires hyperboliques normales. *Mat. Sb.* (*N.S.*) **1**, 39–71.

STELLMACHER, K. L. (1953). Ein Beispiel einer Huygensschen Differentialgleichung. *Nachr. Akad. Wiss. Gottingen Math.-Phys. Kl. II.* **10**, 133–8.

STERNBERG, S. (1964). Lectures on differential geometry (Prentice Hall, Englewood Cliffs, N. J.).

WHITEHEAD, J. H. C. (1932). Convex regions in the geometry of paths. *Quart. J. of Math.* (Oxford) **3**, 33–42.

WHITNEY, H. (1957). *Geometric integration theory* (Princeton University Press, Princeton, N.J.).

DE WITT, B. S. and BREHME, R. W. (1960). Radiation damping in a gravitational field. *Ann. Phys.* **9**, 220–59.

WÜNSCH, V. (1970). Über eine selbstadjungierte Huygenssche Differentialgleichung mit vier unabhängigen Variablen. *Math. Nachr.* **47**, 131–54.

YOSIDA, K. (1971). *Functional analysis*, 3rd ed. (Springer, Berlin).

ZAUDERER, E. (1971). On a modification of Hadamard's method for obtaining fundamental solutions for hyperbolic and parabolic equations. *J. Inst. Math. and its Appl.* **8**, 8–15.

Notation

General

$x \in A$ ($x \notin A$): x is a member (is not a member) of the set A

$\{x; \mathscr{P}(x)\}$: all x such that $\mathscr{P}(x)$ holds

$\varnothing$: empty set. $B \subset A$ (or $A \supset B$): B is a subset of A

$A \cup B$: union; $A \cap B$: intersection; $A \times B$: topological product;
  $A \backslash B = \{x; x \in A, x \notin B\}$: difference, of two sets A and B

$\bar{A}$: closure, ∂A: boundary, of the set A

$f: x \to y = f(x): f$ maps x to $y = f(x)$

$f: A \to B: f$ maps the set A into the set B

$f: A \ni x \to y \in B: f$ maps $x \in A$ to $y \in B$

$f|_A$: restriction of the function (map) f to the set A

$g \circ f$: composition of the functions f and g

inf: least upper bound; sup: greatest lower bound

$\mathbf{R}$: real line; $\mathbf{R}^n$: n-dimensional Euclidean space;

$\mathbf{R}^+ = \{x \in \mathbf{R}; x > 0\}$: half line; $\mathbf{R}_+^n = \{x \in \mathbf{R}^n; x^n > 0\}$: half space;
  $\mathbf{C}$: complex plane; $\mathbf{S}^n$: n-dimensional Euclidean sphere

$\downarrow 0$ ($\uparrow 0$): tends to zero from above (below)

Spaces of functions and distributions

$C^k, C^\infty, C_0^k, C_0^\infty$, 29

$C^k(\Omega), C^\infty(\Omega), C_0^k(\Omega), C_0^\infty(\Omega)$, $\Omega \subset \mathbf{R}^n$ an open set, 59

$C^k(M), C^\infty(M)$, M a C^∞ manifold, 3

$\mathscr{D}'$, 32

$\mathscr{D}'(\Omega)$, 59

$\mathscr{D}'(M)$, 61

$\mathscr{D}^{(s,r)}(M)$, 63

$\mathscr{D}'^{(r, s)}(M)$, 63

$\mathscr{D}'^+(R)$, 48

$\mathscr{D}^\pm(\Omega_0)$, 172

$\mathscr{D}'^\pm(\Omega_0)$, 172

$\mathscr{E}'$, 39

$\mathscr{E}'(\Omega)$, 59

L_1, L_1^{loc}, 30

Other symbols used frequentley

$C(p)$: null cone with vertex p, 129

$C^{\pm}(p)$: null semi-cones with vertex p, 129

d: exterior differential, 21

$\operatorname{div} v$: divergence of a vector field v, 11

$D^{\pm}(p)$: dependence domains of p, 129

$\exp_q v$: exponential map, 14

$g_{jk}(g^{jk})$: covariant (contravariant) components of the metric tensor, 9

g: determinant of the g_{jk}, 11

$\operatorname{grad} u \ (\equiv \nabla u)$: gradient of a scalar field u, 7

$G_q^{\pm}(p)$: fundamental solutions, 155, 207

${}^t G_q^{\pm}(p)$: fundamental solutions of the adjoint equation, 178

$G^+(p, q)$: fundamental kernels, 246

H: Heaviside function, 121

$H_{\pm}(\Gamma)$, 138

$J^{\pm}(p)$: emissions of p, 138

P: differential operator $\square u + \langle a, \nabla u \rangle + bu$, 65

${}^t P$: adjoint of the differential operator P, 65

pq: geodesic joining p and q, 17

supp: support, 4

$T^{i_1 \cdots i_r}{}_{j_1 \cdots j_s}$, components of tensor T of type (r, s), 9

TM_p: tangent space at $p \in M$, 6

TM: tangent bundle of M, 6

T^*M_p: cotangent space at $p \in M$, 7

T^*M: cotangent bundle of M, 8

δ, δ_q: Dirac delta function, 32, 65

$\delta_{\pm}(\Gamma)$, 131

$\Gamma(p, q)$: square of geodesic distance of q from p, 17

$\mu, \mu(p)$: invariant volume element, 25

$\mu_S, \mu_S(p)$: Leray form such that $dS \wedge \mu_S = \mu$, 66

η_{jk}: components of Minkowskian metric tensor, 72

$\partial_1, \ldots, \partial_n$: partial derivatives, 29

$\partial^{\alpha} = \partial_1^{\alpha_1} \cdots \partial_n^{\alpha_n}$: partial derivative of order $|\alpha|$, 29

$\nabla = (\nabla_1, \ldots, \nabla_n)$: covariant derivative, 11

$\nabla^i = g^{ij} \nabla_j, \ i = 1, \ldots, n$, 11

$\langle , \rangle$: inner (scalar) product, 7, 10

$\wedge$: exterior product, 20

$\otimes$: tensor product, 8, 40

$(\,,\,)$: duality of distributions and test functions, 32
$u * v$: convolution of u and v, 43
$* v$: de Rham operator, 26
$\square$: d'Alembertian, 12
$\triangle$: Laplacian, 12

Index

adjoint
  of a differential operator, 49, 65
  of a tensor differential operator, 203
advanced potential, 117
affine connection, 10
affine parameter, 12
analytic continuation, 53
analytic space–time, 141

bicharacteristic, 79
bijective, 265
bitensor, 204
boundary
  of a manifold, 5
  of a set, 263

C^∞ structure, 3
Cauchy problem, 50
  for a scalar wave equation, 181, 249
  for a tensor wave equation, 211
causal curve, 147
causal domain, 146
caustic, 88
characteristic, 51, 77
characteristic initial value problem, 76, 193
chart, 3
closed set, 263
closure, 263
codimension, 5
compact set, 264
continuous function (map), 265
convolution, 43
coordinate neighbourhood, 3
coordinate system, 3
coordinate transformation, 3
cotangent bundle, 8
cotangent space, 7
covector, 5
cross-section, 6
curvature, 11
curve, 5

d'Alembertian, 12
de Rham operator, 27
delta function, see Dirac delta function,
dependence domain, 129
descent, method of, 251

diffeomorphism, 4
differential operator, 49
Dirac delta function, 32
  on a manifold, 65
discontinuity of order k, 92
distribution
  on $\mathbf{R}^n$, 32
  of finite order, 34
  on a manifold, 60, 63
  with compact support, 38
  topology, see weak topology,
divergence, 11
divergence theorem, 26
dual, of a topological vector space, 269

emission, 138
exponential map, 14
exterior
  differential, 21
  form, 20
  product, 20

finite part, 54
flux of a vector field, 26
focal point, 88
Fourier–Laplace transform, 256
function, 265
fundamental solution, 52, 65
  of a scalar wave equation, 154, 240
  of a tensor wave equation, 207
future-directed, 73
future-compact, 169

geodesic, 12
geodesic distance, 17
geodesically convex domain, 15
gradient, 7

Hadamard series, 146
Hausdorff space, 263
Heaviside function, 121
homeomorphism, 266
Huygens' principle, 221
hypersurface, 5

imbedding, 4
improperly posed, 123

induced orientation, 26
initial value problem, 50
    for characteristics and null fields 81,
injective, 265
inner product, 9
integral of an exterior form, 23
invariant volume element, 25

kernel distribution, 246
Kirchhoff's formula, 124

Laplacian, 12
Leray form, 68
Liénard–Wiechert potential, 219
line source, 212
linear form, 268
local coordinates, 3
locally finite, 4
locally integrable, 23
Lorentzian metric, 9

m-surface, 5
manifold, 2
manifold-with-boundary, 5
metric, 9
Minkowskian chart, 72
multi-index, 29

neighbourhood, 262
normal coordinates, 16
normal neighbourhood, 15
null
    cone, 129
    field, 76
    geodesic, 80
    hypersurface, 75
    semi-cone, 129
    vector, 73

open set, 262
ordinary wave equation, 72, 229
orientable, 23
orientation, 22
orientation-preserving chart, 23
orthogonal vectors, 74

paracompact, 4
parametrix, 144
partition of unity, 4
past-compact, 169
past-directed, 73
progressive wave, 101, 104
propagator, see transport bitensor
proper map, 48
properly posed, 123
pseudo-Riemannian metric, 9

regularization, 47

resolving kernel, 151
restriction, 265
retarded potential, 117
Riemann function, 242
Riemannian metric, 9

scalar product, 7
semi-norm, 267
semi-norm estimate, 32
sequentially continuous, 266
signature, 9
space-like field, 76
space-like hypersurface, 75
space-like vector, 73
space–time, 72
spatial boundary, 175
sub-manifold, 4
support
    of a distribution, 33
    of an exterior form, 20
    of a function, 4
surjective, 265

tail term, 158
tangent bundle, 6
tangent space, 6
tangent vector, 6
tensor, 8
tensor distribution, 63
tensor product
    of tensors, 8
    of distributions, 40
tensor wave equation, 109, 203
test function, 30
time-like field, 76
time-like hypersurface, 75
time-like vector, 73
time orientation, 73
topology, 262
    induced, 263
    product, 263
topological space, 262
    vector space, 266
transport bitensor, 205
transport equation
    scalar, 96, 104
    tensor, 109
trivial transformation, 159

vector, see tangent vector
vector field, 6
volume element, see invariant volume
    element

wave equation, 72
wave front, 177
weak solution, 49
weak topology, 34